NATO Science for Peace and Security Series A: Chemistry and Biology

NATO Science for Peace and Security Series

Sub-Series A: Chemistry and Biology

Volumes in the Chemistry and Biology Sub-Series deal with scientific solutions to security threats arising from nuclear, biological, chemical and conventional agents and weapons, covering topics such as CBRN agent detection, protection, destruction or decontamination. Covered also are topics dealing with medical countermeasures, infectious diseases or vaccines, as well as pharmacology and toxicology, biotechnology and genetics. Food security is another topic covered in this sub-series.

Special offer: For all clients with a print standing order we offer free access to the electronic volumes of the Series published in the current year.

Christopher L. Cummings · Benjamin D. Trump ·
Valentina Prado · Beth Ellinport · Igor Linkov
Editors

Biotechnology and AI: Technological Convergence and Information Hazards

Springer

Editors
Christopher L. Cummings
US Army Engineer Research
and Development Center
Vicksburg, USA

Benjamin D. Trump
Tor Intelligence
LLC
Apex, NC, USA

Valentina Prado
Universidad de los Andes
Bogotá, Colombia

Beth Ellinport
US Army Engineer Research
and Development Center
Concord, USA

Igor Linkov
US Army Engineer Research
and Development Center
Concord, USA

ISSN 1874-6489 ISSN 1874-6527 (electronic)
NATO Science for Peace and Security Series A: Chemistry and Biology
ISBN 978-3-032-05245-2 ISBN 978-3-032-05246-9 (eBook)
https://doi.org/10.1007/978-3-032-05246-9

This Springer imprint is published by the registered company Springer Nature Switzerland AG
The registered company address is: Gewerbestrasse 11, 6330 Cham, Switzerland

Foreword

Life itself may be the richest, least unexplored dataset that exists in the world. In its complexity, from the smallest not-quite-living microvirus to the 200-ton great blue whale, it exudes data, most of which is undetectable. Even when humans have found rudimentary means to detect these signals, the data is interpretable only at its most basic level. We know we can detect our genes in some order of A, T, C, and G, yet who looks in the mirror and sees DNA? Nevertheless, we know that DNA is connected, somehow, to life. Life and its multitude of data are so extraordinary that humans, and maybe other animals, have tried to explain it through myths and religion, medicine, psychology, philosophy, art, and now models. And those models are improving with artificial intelligence (AI) and machine learning (ML). Or are they?

Biological data, at least the type that we can detect and quantify right now, underpins the emergent properties we observe that constitute life. Somehow, from proteins expressed: a beating heart (which we can measure by electrocardiogram); from our neurotransmitters and electrochemical gradients: thinking and sleeping (which we can measure by electroencephalogram); and from a quasi-species of virus: infection (which we can measure by host white blood cell count and fever). Scientists today can manipulate genes, proteins, neurotransmitters, electrochemical gradients, a quasi-species, all with exquisite precision. But scientists can't yet correlate the quasi-species that leads to a certain outcome in an influenza infection, the change in protein expression required to prevent the collapse in a pet boxer with cardiomyopathy, or the firing of neurons that led to my decision to write this Foreword.

As the contributors to this volume so masterfully relate, we also have tremendous capability to target and manipulate the genes that underpin these biological phenomena, if we can only make the link between phenomenon and gene. Science has made extensive leaps in manipulating DNA over the last 50 years. For many decades, targeting these genes occurred through breeding in the lab, reminiscent of breeding in agriculture. But with the advent of recombinant DNA technology, targeting genes became common by the 1990s, and today gene-editing is routine. However, changing genes is only useful if the scientists know which genes to target. That is, we can change the emergent property we see in an organism if we can make the link between the emergent property (modeled by data) and the underlying gene. Before the arrival of AI, biologists had created a handful of models, which they then correlated to genes, and have manipulated the genes to alter those emergent properties of life.

Until the very recent past, critical leaps in understanding have required the spark of human genius. In his seminal 1952 paper, Alan Turing applied his masterful mathematical mind to biological phenomena,[1] deducing the morphogenetic mechanism by which a leopard gets its spots from a homogenous bag of molecules of the developing embryo. A few decades later, Edward Lewis, Christiane Nüsslein-Volhard, and Eric Wieschaus identified a few genes that led to patterning of an organism, a profound scientific advancement for which they shared the Nobel Prize in Physiology and Medicine in 1995. "Your discoveries of genes that govern embryonic development have allowed us to comprehend how a single cell develops into a complex multicellular organism,"[2] said the Nobel Committee, recognizing their pioneering work linking genes to an emergent property.

The next rung was to target the genes and change the emergent property. In 2012, a team from the Instituto de Biomedicina y Biotecnología de Cantabria in Spain described how they manipulated these genes characterized in the 1980s, based on a model from the 1960s, to create extra fingers and toes in animals.[3] Today, this cycle of model, gene discovery, and gene manipulation, has sped up, and we do not need to wait for a Turing to "see" such patterns; we know AI can find them for us. Scientists can train machines to deduce the connections between emergent behaviors and the data they feed the model. These models can then predict the genes or other biological molecules from which the phenomenon emerges. The technology exists to manipulate the gene with ease. We can change life itself—an immense power that comes with immense responsibility.

As we prepare for AI to transform modeling, we should pause to reconsider one question: are we explaining life better now through these models? That depends on governance, which includes training. Turing the human developed his model and could and did reflect on the process of training data, noting that previous scientists had dismissed subtle but important features of the developing embryo, and as a result missed the underlying mechanism of patterning. Observable irregularities in the homogeneity of the developing embryo exist, but scientists before Turing had ignored them, assuming that minor fluctuations could not possibly lead to large patterning, such as zebra stripes. In his paper, Turing is explicit that we need to account for specific data, because some deviations do matter.[4]

---

[1] Turing, AM, "The Chemical Basis of Morphogenesis," *Philosophical Transactions of the Royal Society of London. Series B, Biological Sciences*, Vol. 237, No. 641. (August 14, 1952), pp. 37–72.

[2] Vennström, Björn, "Award Ceremony Speech for the Nobel Prize in Physiology and Medicine of 1995," *Nobel Lectures, Physiology or Medicine 1991–1995*, Editor Nils Ringertz, World Scientific Publishing Co., Singapore, 1997.

[3] Sheth R, Marcon L, Bastida MF, Junco M, Quintana L, Dahn R, Kmita M, Sharpe J, Ros MA. "Hox genes regulate digit patterning by controlling the wavelength of a Turing-type mechanism." *Science*. 2012 December 14; 338(6113):1476–1480.

[4] Turing, 1962.

Whether a Turing is doing the thinking, or we are training a model, we will certainly include some data and not others, not to bias the model but because we think it irrelevant. If we choose incorrectly, the model is erroneous. We choose the data we include and withhold because compute power and data storage are finite. Good models have good governance, and good governance starts with responsible training.

Responsible training, rooted in classical reinforcement learning,[5] is similar to that used to train animals. Norbert Wiener clearly made this connection when he developed his cybernetic theory[6] a decade before Turing published his paper on biological patterning. A responsible owner trains a puppy to participate in dog and human society, and there is extensive science surrounding best practices in training. Learning to sit alone requires repeated instruction and reinforcement of sitting and ignoring of non-sitting. We know that some dogs are easier to train than others, and we also know that sitting may be easier for the dog in certain environments, so we also practice sitting in a variety of contexts.

In addition to responsible training, governance entails confirming that models function as they are supposed to and for the problem sets for which they are designed. For example, it is the policy of the US Department of Defense (DoD) that when it uses models to support its work and decisions, such models are verified, validated, and accredited—the model needs to do what it says it's going to (verification), it must represent the real world within the context of its intended use (validation), and its use must be limited to the context for which it is defined (accredited).[7]

One of the first areas where the US DoD is looking to apply AI/ML models to biological data is in biomanufacturing. Biomanufacturing is the use of biological systems to produce goods at scale, and the DoD is employing biomanufacturing to create a host of defense-relevant products through several efforts that program simple organisms (e.g., bacteria, yeast) or cell-free systems to create industrial chemicals (e.g., heat resistant materials, specialty fuels, adhesives, propellants) critical to military systems.[8] As we refine the programming of microbes to manufacture needed chemicals and materials, we can use this data to create predictive models that assist in speeding up not only the front end synthetic biology, but also the fermentation conditions, downstream processing, etc. Good governance throughout this process is mostly about efficiency; improperly training a model to guide decision-making in manufacturing could lead to wasted time and resources.

[5] For a nice review of earlier applications to training AI and its roots in psychology of reinforcement learning, see Andrew G. Barto, Richard S. Sutton, "Reinforcement Learning in Artificial Intelligence," Editor(s): John W. Donahoe, Vivian Packard Dorsel, *Advances in Psychology*, North-Holland, Volume 121, 1997, pp. 358–386.

[6] Wiener created the term "cybernetics" from the Greek word meaning "governor." Norbert Wiener, *Cybernetics: Or Control and Communication in the Animal and Machine*, 1948.

[7] United States Department of Defense, DoD Instruction 5000.61, "DoD Modeling and Simulation Verification, Validation, and Accreditation," September 17, 2024.

[8] US Department of Defense. (2023). Department of Defense Biomanufacturing Strategy. United States Government. https://media.defense.gov/2023/Mar/22/2003184301/-1/-1/1/BIOMANUFACTURING-STRATEGY.PDF.

However, as we delve deeper into emergent properties of life—including cognition, infection, and disease—responsible training and verification, validation, and accreditation become more important. We will invariably make decisions about what data we exclude because of limited compute power, storage, or scarcity of good quality data itself. Those exclusions may be honest oversights, because we deem them unimportant or irrelevant (as Turing noted), but they bias the model and may make it erroneous.

Discovery resulting from this convergence of AI, biotechnology, and synthetic biology promises incredible human growth and benefit. But our own perceptions as humans trying to understand ourselves could influence the inclusion and exclusion of training data. Progress is never certain.

Washington, D.C., USA Kate Sixt

Acknowledgments

The editors would like to acknowledge the generous funding for June 2025 Advanced Research Workshop in Malta from NATO's Science for Peace and Security Programme, which inspired this book. The editors also thank the many participants who attended this meeting and contributed their ideas to improve the international assessment, coordination, and governance of biotechnologies. Additional gratitude is due to the US Army Engineer Research and Development Center, which supported Drs. Cummings, Trump, Linkov, and Ms. Ellinport in their work related to resilience and biotechnology modernization. Dr. Prado acknowledges Sphera's sponsorship of her work and engagement in this effort. We would like to acknowledge our internal production editors. Madison Horgan and George Siharulidze made considerable effort to prepare all chapters for submission.

The editors are grateful for the guidance and generous support of Dr. Eyup Turmus, whose support was critical to ARW success and the resulting scientific collaboration. The editors are also thankful to the Universidade de los Andes and Bart Van Hoof for their warm welcome, assistance in meeting support, and friendship. Significant thanks are due to the US Office of Naval Research Global (ONR), who provided support that enabled many leading global scientists to travel to Cartagena for his critical meeting. Likewise, all editors are extensively grateful for the DOD Office of Research and Engineering, including Principal Director for Biotechnology Dr. Kate Sixt and Deputy Dr. Peter Emanuel, who enabled this idea to come to fruition. Lastly, the editors and authors acknowledge the generous friendship and guidance of CAPT Thomas Janisko, USACE Command Surgeon, who provided critical leadership guidance.

Acknowledgments

Contents

About the Editors

Christopher L. Cummings Ph.D., is a Research Social Scientist and Lead for the Center for Health Engineering at the US Army Engineer Research and Development Center (ERDC). His research advances engagement with science and technological governance, risk communication theory, and health decision making across the life span. An experienced behavioral theorist and risk communication scholar, Dr. Cummings employs quantitative, qualitative, and mixed methods, with work featured in leading journals, global media, and over 70 applied science and health venues. He was recently appointed to the US National Academies of Sciences, Engineering, and Medicine (NASEM) Committee for the External Review of Environmental, Biosafety, and Biosecurity Considerations for Synthetic Cell Research and Development. His contributions have been recognized with multiple honors, including ERDC's 2024 International R&D Award, Technology Transition Award, and Lab Innovation of the Year, as well as the 2021 Excellence in Research and Practice Award for international coordination of COVID-19 risk communication from the Society for Risk Analysis, the 2019 Distinguished Alumni Award from North Carolina State University, and the 2017 Nanyang Education Award from Nanyang Technological University, Singapore.

Dr. Benjamin D. Trump Ph.D. is an entrepreneur pioneering social science and human factors research pertaining to organizational and societal resilience, the governance of extreme and systemic risks, and decision making for defense, homeland security, and emergency response. For his scientific contributions to many applications of human factors engineering, President Biden awarded Dr. Trump with a Presidential Early Career Award for Scientists and Engineers (PECASE) in January 2025—the highest honor bestowed by the United States federal government on outstanding scientists and engineers in the early stages of their independent research careers. Dr. Trump has published nine books and over 100 peer-reviewed papers and chapters, and has received multiple awards for his scientific contributions, including the Society for Risk Analysis' Chauncey Starr Award for outstanding achievement in science or public policy relating to risk analysis and exceptional promise for continued contributions to risk analysis. Dr. Trump was commended for his two years of service as an emergency responder for the COVID pandemic, including the receipt of the US Army Civilian Medal for Humanitarian Service, and the Army Superior Civilian Service Award for hands-on work that saved many lives in his area of responsibility. Likewise, he was a senior leader and contributor to a scientific team that earned USACE-ERDC an Army Superior Unit Award (AUSA) in January 2025—the first Army unit award in the history of ERDC. In December 2024, he was elected President of the Society for Risk Analysis, with a term from December 2025—December 2027. He is a longtime scholar-contributor to many multinational organizations, including NATO Science for Peace and Security, the United Nations Office for Disaster Risk Reduction, the OECD, the International Risk Governance Council, and many others.

Valentina Prado is Principal Consultant at Sphera's North America based in Vancouver, BC and has spent the last few years as an LCA consultant in North America. Besides her consulting role, Valentina serves as adjunct faculty in the School of Management in Los Andes University in Colombia. Prior to her consulting roles, Valentina was an assistant professor at Leiden University in the Netherlands. Valentina has a bachelor's in civil engineering from Jackson State University and a M.S. and Ph.D. focused in LCA and decision analysis from Arizona State University. Valentina's experience includes methodological innovations in interpretation of comparative and her work is published in the *International Journal of Life Cycle Assessment*, *Journal of Industrial Ecology* and the *Journal of Cleaner Production.*

Beth Ellinport is a Research Systems Engineer at the US Army Engineer Research and Development Center's (ERDC) Center for Health Engineering (CHE). Driving her career at the intersection of social science and engineering, she focuses in navigating complex systems to enhance resilience, efficiency, and workforce well-being. Her portfolio combines technical rigor with human-centered approaches in fields such as Safety and Occupational Health, Mental Resilience Training, and Emergency Management. She holds a B.S. in Systems Engineering with a concentration in Public Health and an M.S. in Systems Engineering with a concentration in Crisis emergency, and Risk Management, both from The George Washington University, respectively. Her research has been featured in peer-reviewed journals, including IEEE, where she first published probabilistic methods for Urban Search and Rescue in 2022. She continues to produce technical papers that translate intricate challenges into accessible frameworks, fostering collaboration across disciplines. An advocate for upstream thinking she excels in bridging engineering methodologies with the complexities of human systems. Her work reflects a dedication to building stronger and more adaptive solutions.

Dr. Igor Linkov is the Senior Scientific and Technical Manager with the US Army Engineer Research and Development Center, and Adjunct Professor with Carnegie Mellon University. Dr. Linkov has managed multiple risk and resilience assessments and management projects in many application domains, including cybersecurity, transportation, supply chain, homeland security and defense, and critical infrastructure. He was part of several Interagency Committees and Working Groups tasked with developing resilience metrics and resilience management approaches, including the US Army Corps of Engineers Resilience Roadmap. Dr. Linkov has organized more than 30 national and international conferences and continuing education workshops, including NATO workshops on Cyber Resilience in Estonia (2018) and Finland (2019), as well as Chaired Program Committee for 2015 and 2019 World Congresses on Risk in Singapore and Cape Town. He has published widely on environmental policy, environmental modeling, and risk analysis, including twenty five books and over 350 peer-reviewed papers and book chapters in top journals, like Nature, Nature Nanotechnology, Nature Climate Change, among others. He has served on many review and advisory panels for DOD, DHS, FDA, EPA, NSF, EU and other US and international agencies. Dr. Linkov is Society for Risk Analysis Fellow and recipient of 2005 Chauncey Starr Award for exceptional contribution to Risk Analysis, 2014 Outstanding Practitioner Award as well as 2019 Distinguished Educator Award. He is Elected Fellow with the American Association for the Advancement of Science (AAAS). Dr. Linkov has a B.S. and M.Sc. in Physics and Mathematics (Polytechnic Institute) and a Ph.D. in Environmental, Occupational and Radiation Health (University of Pittsburgh). He completed his postdoctoral training in Risk Assessment at Harvard University.

Chapter 1
Introduction to Biotechnology: Technological Convergence and Information Hazards

Christopher L. Cummings, Benjamin D. Trump, Valentina Prado, Beth Ellinport, and Igor Linkov

1.1 Introduction

Scientific and technological progress no longer follows a steady, linear trajectory. Instead, it unfolds at an exponential rate, driven by the convergence of artificial intelligence, biotechnology, and synthetic biology. These once-separate fields are now deeply interconnected, fueling a transformation that is reshaping medicine, agriculture, national security, and the very way humans interact with biological systems. What was once the realm of speculation—AI-powered drug discovery, machine learning-guided gene editing, and automated protein design—has moved from theoretical possibility to scientific reality.

Historically, technological progress developed in isolated domains, with researchers in different fields working independently of one another. Today, however, advances in computational power, data analysis, and machine learning have erased many of these boundaries. AI can now predict protein structures with remarkable accuracy, accelerate drug discovery by identifying promising molecular compounds in a fraction of the time, and optimize synthetic biology designs that allow microbes to produce biofuels, pharmaceuticals, and biodegradable materials. Each breakthrough feeds into the next, forming a self-reinforcing cycle of innovation. The speed at

C. L. Cummings (✉)
US Army Engineer Research and Development Center, Vicksburg, MS, USA
e-mail: Christopher.l.cummings@usace.army.mil

B. D. Trump
Tor Intelligence, LLC, Apex, NC, USA

V. Prado
Sphera, Vancouver, Canada

B. Ellinport · I. Linkov
US Army Engineer Research and Development Center, Concord, MA, USA

C. L. Cummings et al. (eds.), *Biotechnology and AI: Technological Convergence and Information Hazards*, NATO Science for Peace and Security Series A: Chemistry and Biology, https://doi.org/10.1007/978-3-032-05246-9_1

which discoveries occur is no longer dictated by the limits of human intuition or traditional trial-and-error experimentation; instead, machine-driven analysis and automated research pipelines are unlocking scientific frontiers that were previously inaccessible.

This rapid acceleration of discovery is not just an academic concern. The fusion of AI and biotechnology is reshaping industries, economies, and societies in ways that demand urgent attention. AI-driven precision medicine is tailoring treatments to individual genetic profiles, promising a future where disease prevention and therapy are customized at an unprecedented scale. Meanwhile, bioengineering is producing new agricultural solutions to combat food insecurity, developing crops that are more resilient to climate change and pests. Even environmental challenges, long thought to be beyond human control, are now being addressed through AI-optimized biological interventions, from carbon-capturing organisms to engineered bacteria capable of breaking down pollutants. Yet this acceleration is not without profound risks. The very speed at which AI and biotech are converging creates an increasingly urgent governance challenge: the ability of societies to regulate and manage these technologies is falling behind. Known as the pacing problem, this widening gap between innovation and oversight presents one of the most difficult dilemmas of modern technological governance.

Regulatory frameworks are often designed for stable, predictable domains, yet convergent technologies evolve in ways that defy traditional policymaking structures. The interdisciplinary nature of AI-driven biotechnology complicates efforts to categorize and regulate these developments. AI-designed genetic modifications raise questions that do not fit neatly within existing biomedical ethics or AI governance frameworks. At the same time, the rapid democratization of scientific tools—including open-source AI models and widely available gene-editing kits—means that powerful technologies are no longer confined to government-funded labs or major corporations. The growing presence of independent researchers, biotech startups, and even biohacking communities introduces both innovation potential and security concerns, as decentralized access to cutting-edge tools increases the possibility of unintended or malicious applications. Compounding these challenges is the reality that technological competition is global. No single country can regulate AI-biotech convergence in isolation, yet international governance mechanisms remain fragmented. As different nations race to harness the economic and strategic benefits of AI-driven biological engineering, inconsistencies in oversight create gaps that can be exploited. Without coordinated international action, regulatory disparities could lead to ethical lapses, security risks, and an uneven playing field where innovation flourishes in some regions while being hindered in others.

Perhaps the most difficult aspect of governing convergent technologies is their inherent unpredictability. Unlike traditional scientific advancements, AI-driven research is characterized by emergent behavior—outcomes that even developers and researchers may not fully anticipate. An AI system tasked with optimizing a biological process may arrive at solutions that are novel and effective but poorly understood. This unpredictability complicates risk assessment, making it difficult for policymakers to establish guidelines that account for unforeseen consequences. The accelerating pace of AI and biotech convergence leaves societies with two dangerous extremes: an environment of under-regulation, where rapid technological advancements outstrip oversight and create ethical and security vulnerabilities, or an over-regulated landscape, where fear of the unknown leads to excessive restrictions that stifle scientific progress. Neither approach is sustainable.

A new model of governance is urgently needed—one that is flexible, adaptive, and capable of evolving alongside technological change. Rigid, static policies will not suffice in a world where breakthroughs emerge at an unprecedented pace [1]. Instead, governance must be guided by dynamic regulatory principles that can adjust to new developments in real time [5]. Cross-disciplinary collaboration will be essential, as scientists, ethicists, policymakers, and industry leaders must work together to anticipate risks before they materialize. This moment presents both an extraordinary opportunity and a formidable challenge. The accelerating convergence of AI and biotechnology is not an unstoppable force with a predetermined outcome; it is a phenomenon that societies must actively shape. If managed wisely, these technologies have the potential to unlock extraordinary benefits, from life-saving medical treatments to sustainable solutions for environmental crises. But if left unchecked or governed poorly, they could introduce risks that extend far beyond the capabilities of current oversight mechanisms. The fundamental question is not whether technological convergence will continue—it will. The question is whether societies will develop the tools, institutions, and ethical frameworks necessary to guide its trajectory. The pace of innovation cannot be slowed, but the way it is governed can—and must—be thoughtfully designed. The choices made in this era will determine whether convergence becomes a force for global good or an unmanageable challenge that spirals beyond human control.

1.2 A Call for Adaptive Governance

The rapid convergence of artificial intelligence, biotechnology, and synthetic biology is not simply a scientific revolution; it is a transformation that demands urgent and deliberate governance. Unlike past technological breakthroughs that unfolded within well-established regulatory paradigms, today's advances defy traditional oversight mechanisms. The speed, complexity, and interdisciplinary nature of convergent technologies have created an urgent governance dilemma:

How can societies balance innovation and security, openness and control, progress and precaution in a world where technological change outpaces our ability to anticipate its consequences?

At the heart of this challenge lies the dual-use dilemma, the reality that powerful technologies can be harnessed for both beneficial and harmful purposes. AI-powered drug discovery has already accelerated the development of life-saving medicines, yet the same computational models could, in theory, be repurposed to design highly potent biological weapons. The genetic engineering tools that allow scientists to cure hereditary diseases could also be used to create dangerous modifications with unintended ecological or security risks. Synthetic biology promises revolutionary applications in agriculture, sustainability, and bio-based materials, but it also raises concerns about ecosystem disruptions, bioterrorism, and monopolization of biological resources. This convergence of possibility and peril creates an urgent need for governance structures that can adapt in real time—structures that neither suppress innovation nor leave humanity vulnerable to catastrophic misuses.

Compounding the governance challenge is the unprecedented accessibility of emerging biotechnologies. In previous technological revolutions—such as nuclear energy or space exploration—advancements were largely confined to government institutions and a small number of corporate entities. Today, however, breakthroughs in AI and biotech are occurring in academic research labs, private startups, and even community-based biohacker spaces. Cloud-based AI platforms allow independent researchers to analyze complex biological datasets with computational power that was once the domain of national research institutions. Gene-editing kits are now commercially available, enabling high school students to experiment with CRISPR techniques that were once limited to elite scientific institutions. While this democratization of technology fosters innovation and accelerates progress, it also raises profound ethical and security concerns. Without oversight, the possibility of unintended consequences, including biosafety risks, data privacy violations, and biosecurity threats—grows exponentially. A further complication arises from the global nature of AI-biotech convergence. Unlike traditional scientific disciplines that evolve within national boundaries and regulatory jurisdictions, today's technological advances are inherently international. AI models trained in one country can be instantly deployed across the globe. Genetic data shared through open-source platforms can be accessed and manipulated by researchers in different regulatory environments. Biotechnology supply chains span multiple continents, with research, development, and production often occurring in separate regions under vastly different oversight regimes. These factors make governance particularly challenging. A regulatory framework that is effective in one nation may be irrelevant in another, and a security risk in one jurisdiction could have cascading effects worldwide. Without global coordination, regulatory inconsistencies could create loopholes that allow for the exploitation of weak oversight structures, increasing the likelihood of technological misuses and governance failures.

Equally concerning is the issue of information hazards, the risks associated with making powerful scientific knowledge widely available without sufficient safeguards. In an era of open science and data transparency, many of the world's leading scientific institutions prioritize the free exchange of information. While this ethos has driven remarkable innovation, it also raises difficult questions about responsible disclosure. Should AI-designed genetic modification techniques be published in full, knowing that they could be misused? How can scientists balance the benefits of openness with the need for security in an era where high-risk biological research can be conducted with publicly available tools? These ethical dilemmas require governance frameworks that are not only reactive—responding to crises as they emerge—but also anticipatory, proactively identifying and mitigating risks before they materialize.

Traditional governance models are ill-equipped to handle the challenges posed by convergent technologies. Regulatory agencies operate on timelines measured in years, while AI and synthetic biology breakthroughs occur on timelines measured in weeks or months. Policy development is frequently hindered by bureaucratic inertia, political gridlock, and an inability to keep pace with the speed of innovation. Existing regulatory structures are often rigid, designed to govern stable and predictable industries rather than fast-moving fields characterized by continuous disruption. This misalignment between technological acceleration and policy stagnation creates vulnerabilities that must be urgently addressed. Governance in the age of AI-biotech convergence cannot be a matter of simple rule enforcement; it must be an adaptive, dynamic process that evolves alongside the technologies it seeks to regulate. Policymakers must embrace iterative, principle-based governance models rather than attempting to impose static rules on an inherently fluid landscape. Regulatory agencies must develop real-time monitoring systems capable of tracking emerging risks and adjusting policies accordingly. Ethical frameworks must be embedded within research and development processes, ensuring that considerations of security, equity, and long-term societal impact are not afterthoughts but foundational elements of scientific innovation.

To achieve this, governments, industry leaders, and scientific communities must prioritize cross-sector collaboration. AI researchers must work alongside bioethicists to develop guidelines that mitigate the risks of machine-learning-driven biological discovery. Legal scholars must engage with synthetic biologists to craft policies that balance safety with scientific freedom. Policymakers must consult with technologists to ensure that regulations remain technically informed and practically enforceable. Without this level of interdisciplinary engagement, governance efforts will remain fragmented, reactive, and ultimately ineffective in addressing the complexity of technological convergence. The urgency of this moment cannot be overstated. Societies are standing at a threshold where the choices made today will shape the future trajectory of AI and biotechnology for decades to come. The failure to implement adaptive governance mechanisms risks leaving humanity vulnerable to unpredictable and potentially irreversible consequences. A world in which biotechnological advancements proceed without sufficient oversight could lead to public health crises, environmental disasters, or geopolitical instability. Conversely, a world in which regulation is too restrictive and inflexible could stifle innovation, slowing progress in life-saving medical treatments, sustainable bioengineering, and other transformative applications.

The central task ahead is not merely to regulate, but to govern wisely—to create governance models that are as innovative, interdisciplinary, and forward-thinking as the technologies they seek to guide. The challenge is immense, but the stakes demand nothing less than deliberate, informed, and coordinated action. The future of AI and biotechnology is not predetermined; it is being shaped by decisions made today. Whether this future is one of equitable, responsible progress or one of chaotic and unmanageable risk will depend on the governance choices societies make in the present moment.

1.3 Beyond Regulation: Towards Resilient Governance

Traditional regulatory models were designed for technological landscapes that evolved gradually, with clear distinctions between industries and well-established mechanisms for oversight. Today, however, the accelerating convergence of artificial intelligence, biotechnology, and synthetic biology demands a fundamental shift in how governance is conceived and implemented. The challenge is no longer simply about regulating new technologies; it is about building resilience into governance itself—ensuring that policy frameworks remain flexible, adaptive, and responsive to the unpredictable nature of scientific progress. Rigid, prescriptive regulations are inherently limited in their ability to govern rapidly evolving fields. By the time policymakers draft and implement legal frameworks, new advancements often render them obsolete or insufficient. This misalignment creates governance gaps, where outdated policies either fail to address emerging risks or impose unnecessary constraints on innovation. In biotechnology and AI, where scientific breakthroughs can occur faster than legislative cycles, a more dynamic approach is required—one that prioritizes continuous monitoring, iterative policymaking, and cross-sector collaboration.

1.3.1 The Need for Adaptive, Anticipatory Governance

To address the risks of technological convergence without stifling its benefits, governance systems must move beyond static rule enforcement and embrace adaptive governance—a model that allows for real-time policy adjustments, scenario planning, and interdisciplinary coordination. In this framework, governance is not treated as a fixed set of rules but as an ongoing process of learning, adapting, and responding to technological changes as they unfold. A key feature of adaptive governance is its ability to incorporate anticipatory foresight—the proactive identification of potential risks and opportunities before they fully emerge. This approach recognizes that waiting until crises materialize before implementing regulations is a reactive and often ineffective strategy. Instead, policymakers, researchers, and industry leaders must engage in scenario planning, risk assessment exercises, and continuous engagement with technological trends to develop forward-looking policies. Regulatory sandboxes, for example, provide an environment where new biotechnologies can be tested under controlled conditions, allowing policymakers to observe potential risks before they become widespread. Ethical foresight committees, composed of scientists, ethicists, and legal experts, can provide ongoing guidance on the implications of emerging AI-biotech integrations. Another critical component of adaptive governance is real-time monitoring. Traditional regulatory models rely on periodic assessments, often years apart, to evaluate compliance and assess risks. However, in fast-moving technological domains, a lag in oversight can lead to significant harm before policymakers even become aware of emerging issues. To address this, governance structures should integrate AI-driven regulatory systems that leverage machine learning to detect anomalies, predict unintended consequences, and flag potential security risks as they arise. These systems can provide continuous oversight, ensuring that governance remains proactive rather than reactive.

1.3.2 Balancing Innovation and Security

One of the greatest challenges of governing convergent technologies is striking the right balance between enabling scientific progress and ensuring security. Overly restrictive regulations can slow down life-saving innovations, while insufficient oversight can create vulnerabilities that lead to biosecurity threats, ethical violations, or environmental harm. Resilient governance must recognize that regulation and innovation are not inherently opposed but must be harmonized through flexible policies that encourage responsible research and development. A critical aspect of this balance is embedding ethical considerations into the technological development process itself. Ethical governance should not be an afterthought—something addressed only after a technology has already been deployed—but rather a guiding principle throughout its lifecycle. This means integrating ethical and security assessments directly into the research and development pipeline, ensuring that risk mitigation is built into new innovations from the outset. By fostering a culture of responsible innovation, governments, industry leaders, and research institutions can ensure that technologies are developed with an awareness of their broader societal implications. At the same time, security and risk mitigation must extend beyond governmental intervention and into industry self-regulation. Private sector entities play a crucial role in AI-biotech convergence, yet self-regulation alone is insufficient. To ensure that companies act responsibly, resilient governance models should incorporate co-regulation strategies, where industry standards are developed in partnership with public agencies, academic institutions, and international organizations. This model allows for greater flexibility than traditional regulatory approaches while maintaining accountability and transparency.

1.3.3 The Role of Global Cooperation in Harnessing Resilience

Technological convergence is not confined by national borders. AI and biotechnology are being developed, deployed, and refined in research labs, private companies, and academic institutions around the world. However, regulatory and governance mechanisms remain largely nation-based, creating inconsistencies in oversight that can be exploited. A fragmented global approach to governance risks creating regulatory arbitrage, where companies and researchers seek out jurisdictions with the least restrictive regulations, potentially increasing global security risks. Resilient governance requires strong international coordination [4]. Without harmonized standards, cross-border information sharing, and cooperative enforcement mechanisms, regulatory gaps will persist. Countries must move beyond isolated national policies and instead engage in multilateral governance initiatives that reflect the borderless nature of technological convergence. International institutions such as the World Health Organization (WHO), the United Nations (UN), and intergovernmental biotech coalitions must take a proactive role in setting global standards for AI-driven biotechnology. Additionally, new public–private partnerships should be established to ensure that scientific innovation remains both ethical and secure on a global scale. One promising model for global cooperation is the use of international regulatory agreements, where nations commit to shared principles of responsible innovation. These agreements can prevent a regulatory arms race, ensuring that safety and ethical considerations remain a priority, even as nations compete for leadership in biotech and AI. At the same time, knowledge-sharing networks—which facilitate real-time data exchange between scientists, regulators, and policymakers across different countries—can help detect risks early and coordinate responses before problems escalate.

1.3.4 A Governance Framework for the Future

The future of AI-biotech governance must be built on resilience, adaptability, and collaboration. The goal is not to halt innovation, nor to impose rigid controls that stifle progress, but to ensure that governance is as dynamic and forward-thinking as the technologies themselves. This requires a multi-layered approach, one that combines anticipatory foresight, real-time monitoring, ethical responsibility, and global cooperation. A resilient governance system should include:

- Agile regulatory mechanisms that allow for continuous policy updates based on new developments.
- Integrated risk assessment frameworks that proactively identify and mitigate potential harm before they materialize.
- Public-private partnerships that engage scientists, policymakers, and industry leaders in shaping ethical and security guidelines.
- International collaboration that harmonizes global regulations, preventing regulatory loopholes and security vulnerabilities.
- AI-assisted governance tools that enhance oversight through automated risk detection and real-time analysis.

The challenge of governing AI-biotech convergence is immense, but it is not insurmountable. The key to success lies in recognizing that governance must evolve alongside technological progress [3]. The decisions made today will set the trajectory for the future—determining whether technological convergence serves as a force for societal good or a destabilizing force that introduces unmanageable risks. The time to act is now. A failure to build resilient governance structures will not slow the pace of technological change; it will only increase the likelihood of governance failures that could have far-reaching consequences. By designing governance frameworks that are adaptive, ethical, and internationally coordinated, societies can ensure that AI and biotechnology remain powerful tools for progress, security, and global well-being.

1.4 A Crucial Juncture in Technological History

We stand on the precipice of a technological revolution that may well define the trajectory of future generations. The convergence of artificial intelligence, biotechnology, and synthetic biology is not a distant possibility—it is happening now, reshaping industries, economies, and the very nature of human existence. The decisions made today regarding the governance of these transformative technologies will determine whether the future is one of prosperity, security, and ethical progress, or one of instability, inequity, and unintended consequences [2]. At the heart of this challenge is a fundamental tension: how to harness the unprecedented power of convergent technologies while ensuring they remain safe, ethical, and beneficial for all. If governed wisely, these innovations have the potential to revolutionize medicine, offering personalized treatments tailored to individual genetic profiles, curing diseases once thought untreatable, and extending human longevity. AI-driven biological engineering could provide sustainable solutions to climate change, developing carbon-sequestering organisms, bioengineered crops resistant to extreme weather, and synthetic materials that reduce our dependence on fossil fuels. The fusion of AI and biotech could even mitigate food insecurity by enabling precision agriculture and lab-grown proteins that reduce environmental strain. Yet these same technologies, if left unchecked or mismanaged, pose unprecedented risks. AI-powered genetic modification could be misused for unethical human enhancement, widening societal inequalities and raising profound bioethical dilemmas. The automation of biological research could enable the design of dangerous pathogens, whether through state-sponsored bioweapons programs or rogue actors exploiting gaps in oversight. If economic incentives drive the monopolization of biotechnology, access to life-saving treatments and bioengineered resources could be dictated by corporate interests rather than humanitarian needs, exacerbating global disparities.

The stakes of inaction are too great to ignore. If governance structures fail to keep pace with scientific advancement, societies may find themselves in a reactive, crisis-driven cycle—scrambling to regulate only after irreversible consequences have emerged. Past technological revolutions, from nuclear power to the digital age, have demonstrated the dangers of delayed regulatory responses, where failures to anticipate risks led to environmental catastrophes, ethical breaches, and geopolitical instability. The difference today is that AI and biotechnology are exponentially accelerating, and the window for proactive governance is closing rapidly. However, excessive or poorly designed regulation could also stifle the very innovations that hold the key to solving many of humanity's most pressing challenges. Overregulation driven by fear rather than foresight could slow down medical breakthroughs, delay the development of sustainable bioengineering solutions, and push scientific research underground, where it is even harder to monitor. The challenge is not simply to regulate, but to govern wisely—to strike a balance that protects against existential risks while fostering responsible, ethical progress.

This volume is not merely an academic exploration of convergent technologies; it is a call to action. The governance of AI and biotechnology is not just a technical issue for policymakers and scientists—it is a defining societal question about how we choose to shape our collective future. Will we build governance structures that are adaptable, inclusive, and forward-thinking, or will we allow technological change to dictate our trajectory without sufficient safeguards? Will nations cooperate globally to establish ethical frameworks, or will competitive fragmentation lead to governance loopholes and security vulnerabilities? The answers to these questions cannot be postponed. The time for decisive, informed action is now. The world cannot afford to wait until crises force reactive policies. Instead, policymakers, scientists, industry leaders, and civil society must engage proactively and collaboratively to build governance frameworks that are as dynamic and transformative as the technologies they seek to guide. The governance of AI-biotech convergence will determine whether humanity moves toward a future of equity, sustainability, and responsible innovation, or whether we find ourselves struggling to contain the unintended consequences of technologies that evolved beyond our ability to control them. The choices made in this moment will reverberate for generations to come. Whether the era of AI-biotech convergence is remembered as a time of unprecedented human advancement or one of avoidable crises depends entirely on the actions taken today. This book offers not only an examination of the challenges at hand but also a roadmap for responsible governance in an era of uncertainty. The future is not predetermined, it is being shaped now. The question is whether societies will rise to the occasion and meet this challenge with the urgency, wisdom, and foresight it demands.

References

Greer SL, Trump B (2019) Regulation and regime: the comparative politics of adaptive regulation in synthetic biology. Policy Sci 52(4):505–524

Groff-Vindman CS, Trump BD, Cummings CL, Smith M, Titus AJ, Oye K, Linkov I (2025) The convergence of AI and synthetic biology: the looming deluge. NPJ Biomed Innov 2(1):20

Trump BD, Cummings CL, Kuzma J, Linkov I (Eds.) (2020) Synthetic biology 2020: frontiers in risk analysis and governance. Cham: Springer

Trump BD, Cummings CL, Loschin N, Keisler JM, Wells EM, Linkov I (2023) The worsening divergence of biotechnology: the importance of risk culture. Front Bioeng Biotechnol 11:1250298

Trump BD, Trump J, Antunes D, Palma-Oliveira J, Keisler J (2025) Fusing technology readiness levels (TRLs) and safe-by-design (SbB) for guiding investment in emerging technologies. Environ Syst Decis 45(3):1–13

Chapter 2
A Brief Introduction to the Convergence of Artificial Intelligence and Biotechnology

Megan C. Marcellin, Gianluca Pescaroli, Markus Schmidt, Benjamin D. Trump, Maj Travis Tubbs, Maria Francisca Villegas-Torres, James H. Lambert, Christian Haggenmiller, Yaroslav Blume, Alexander Kamrud, and Igor Linkov

Abstract This chapter examines the accelerating convergence of artificial intelligence (AI) and biotechnology, a transformative development reshaping scientific discovery, technological innovation, and global governance. As AI methods, particularly machine learning (ML), deep learning (DL), and natural language processing (NLP), are increasingly applied to biological research and engineering, they enable

M. C. Marcellin (✉) · J. H. Lambert
University of Virginia, Charlottesville, VA, USA
e-mail: mcm5ft@virginia.edu

G. Pescaroli
University College London, London, UK

M. Schmidt
Biofaction, Vienna, Austria

B. D. Trump · I. Linkov
Tor Intelligence, LLC, Apex, NC, USA

M. T. Tubbs
Air Force Office of Scientific Research, Santiago, Chile

M. F. Villegas-Torres
Universidad de los Andes, Bogotá, Colombia

C. Haggenmiller
European Military Staff/European External Action Service, Brussels, Belgium

Y. Blume
Institute of Food Biotechnology and Genomics, National Academy of Sciences of Ukraine, Kyiv, Ukraine

A. Kamrud
United States Air Force Academy, Colorado Springs, CO, USA

C. L. Cummings et al. (eds.), *Biotechnology and AI: Technological Convergence and Information Hazards*, NATO Science for Peace and Security Series A: Chemistry and Biology, https://doi.org/10.1007/978-3-032-05246-9_2

new capabilities ranging from protein structure prediction and gene editing to AI-driven design of biological systems. These developments mark not only the integration of tools and techniques but the emergence of new paradigms at the intersection of computation and biology. We explore how this convergence unfolds across multiple domains, detailing both its profound opportunities and systemic risks. The chapter identifies four core integration paradigms: AI-augmented biological design, biological computing substrates, distributed biological intelligence, and synthetic systems at the AI-biology interface, highlighting the potential for cascading dynamics and identifying measures for assessing and mitigating risk. In parallel, we consider how diverse stakeholder perspectives shape interpretations of uncertainty, with implications for public trust, innovation policy, and regulatory design. A central contribution of the chapter is its ten governance principles grounded in an adaptive, pluralistic, and capability-based approach. Drawing on theories of complex systems, risk, and uncertainty, the chapter outlines mechanisms for anticipating emerging challenges, managing dual-use concerns, and fostering inclusive, participatory approaches to decision-making. The future of this interface relies not only on technological advances but on our ability to ensure that innovation proceeds in ways that are safe, equitable, and aligned with societal priorities and values. By integrating empirical case studies with conceptual frameworks, this chapter offers guidance for scholars, policymakers, and practitioners seeking to responsibly navigate the convergence of AI and biotechnology.

Keywords Convergence · Artificial intelligence · Biotechnology

2.1 Introduction

The convergence of artificial intelligence (AI) and biotechnology marks a profound transformation in modern science, opening new frontiers with far-reaching implications for engineering, medicine, and society. At the heart of this convergence lies a multifaceted challenge: understanding how these technologies interact to produce novel capabilities, uncertainties, and governance considerations that neither field presents independently. When AI techniques such as machine learning (ML) and natural language processing (NLP) merge with biological systems engineering, they generate capabilities that transcend traditional boundaries.

A defining characteristic of AI and biotechnology convergence is the bidirectional exchange of concepts and design principles between the two disciplines. Neural networks, drawing inspiration from biological systems, exemplify this change. Properties of biological systems (the human brain, in the case of neural networks) influence the design of algorithms that navigate complex problems within stochastic environments, while AI simultaneously enhances our ability to engineer biological systems with unprecedented precision. This remarkable integration holds significant potential for addressing global challenges in food security, drug discovery, agriculture, and precision medicine. Yet it also raises important questions about responsible development, deployment, and governance.

As AI capabilities expand and biotechnology becomes more accessible globally, the significance of addressing these questions grows. The AI-biotechnology interface accelerates innovation cycles, compressing timelines for discovery and development across multiple domains. This acceleration creates opportunities for rapid scientific advancement while simultaneously introducing uncertainties regarding long-term consequences and appropriate governance frameworks. Through increasing democratization of biotechnology tools, powered by AI-driven design and automation, we are witnessing a fundamental reshaping of who participates in biotechnology development and how these technologies disseminate globally. These developments warrant thoughtful examination as society balances innovation with responsible stewardship of technologies that interact with fundamental biological processes.

Formally established as a discipline in 1956 at the Dartmouth College conference, AI began with ambitious goals of creating machines capable of language use, conceptual reasoning, and self-improvement [1]. Despite this early enthusiasm, the field progressed intermittently for decades with limited practical applications [2]. The 2010s brought substantial revitalization, catalyzed by advancements in data infrastructure and the rise of deep learning (DL). These developments enabled voice assistant technologies such as Apple's Siri and Amazon's Alexa to become household fixtures [3]. Today's AI discourse centers predominantly on generative AI technologies that produce text, images, or audio with characteristics resembling human creative output [4, 5].

Life sciences have their own rich history of computational adoption. Early computational work in the 1950s and 1960s focused on protein structures, eventually leading to the creation of the first biological database of protein sequences [6–8]. By the late 1970s, attention shifted to DNA sequencing, spurring the development of new statistical methods for phylogenetic inference [9, 10]. The computational advances of the 1990s launched the "genomics era," highlighted by the Human Genome Project's completion in 2001 [11]. With the emergence of the internet, genomic databases became more accessible and useful, accelerating scientific discovery through unprecedented data sharing and collaborative research across the globe.

Breakthroughs at the AI-biotechnology interface now demonstrate transformative potential across multiple sectors. Consider DL models like AlphaFold and RoseTTAFold, which have revolutionized protein structure prediction by solving the "protein folding problem" in minutes—a task that previously required months or years of intensive laboratory investigation [12, 13]. The latest iteration, AlphaFold 3, extends these predictive capabilities to analyze diverse biological molecules with enhanced accuracy [12], enabling exploration of previously uncharacterized biological functions through computational approaches that dramatically compress experimental timelines.

In healthcare, AI-designed gene editing tools have introduced remarkable opportunities for precision medicine. ML models enhance CRISPR gene editing efficiency by predicting off-target effects with greater accuracy [13], while innovations like OpenCRISPR-1 demonstrate AI's ability to generate novel functional gene editors for use in human cells [14]. Within industrial applications, AI-powered laboratory automation enhances experimental efficiency through comprehensive data processing, accelerating the design-build-test-learn cycle fundamental to synthetic biology development [15, 16].

This chapter examines the current state of AI-biotechnology convergence by progressively moving from general concepts to specific implications, providing stakeholders with a nuanced understanding of this rapidly evolving landscape. We begin by defining artificial intelligence and convergence, distinguishing between risk and uncertainty in this context. From there, we analyze specific opportunities and threats across different sectors—healthcare, agriculture, industry, and biosecurity—examining how various stakeholders perceive these developments through different interpretative frameworks. The discussion then explores potential tipping points where incremental technological changes might trigger significant systemic shifts, along with cascading dynamics that could amplify both benefits and challenges across interconnected systems. The chapter concludes with avenues for addressing risk and resilience of emerging and future systems.

2.1.1 What Is Artificial Intelligence?

AI is often portrayed in media as a monolithic entity—a singular, all-encompassing technology with unified intelligence capable of solving any problem. This characterization, while convenient for storytelling, fundamentally misrepresents the diverse ecosystem of methodologies and approaches that constitute the field. In reality, artificial intelligence encompasses a broad spectrum of techniques, tools, and paradigms with decades of discussion and learning that continue to evolve at an accelerating pace [17]. This diversity becomes particularly significant when examining AI's transformative impact on biotechnology, where different AI approaches address distinct challenges in biological research and engineering.

Rather than representing a stable, fixed technology, AI constitutes a dynamic field of computational approaches unified by the goal of creating systems that can perform tasks typically requiring human intelligence. These tasks span perception, reasoning, learning, problem-solving, language understanding, and adaptation to new situations. The United States National Artificial Intelligence Act of 2020 defines AI as a "machine-based system that can, for a given set of human-defined objectives, make predictions, recommendations, or decisions influencing real or virtual environments" [18]. This definition acknowledges the outcome-oriented nature of AI while remaining deliberately technology-agnostic—a necessary approach given how rapidly underlying methods evolve.

The landscape of AI encompasses numerous methodological branches that have developed along different trajectories. Figure 2.1 provides a visual representation of the layered relationship between various artificial intelligence methodologies and their applications in biotechnology, illustrating both the hierarchical organization of the field and the continuous evolution occurring within each subdomain. This visualization highlights how different AI approaches serve complementary roles in biological research and engineering, from data processing and pattern recognition to complex system design and optimization. We first introduce the concepts of, and relationship between, machine learning (ML) and deep learning (DL), before diving deeper into two classes of DL that showcase especially transformative potential in biotechnology: large language models (LLMs) and biological design tools (BDTs).

ML represents a substantial subset of AI where computational algorithms extract patterns from data to improve performance over time without explicit programming [3, 19, 20]. ML itself branches into multiple paradigms: supervised learning (training on labeled examples), unsupervised learning (finding patterns in unlabeled data), and reinforcement learning (learning through interaction with environments). Each of these approaches offers distinct advantages for different types of

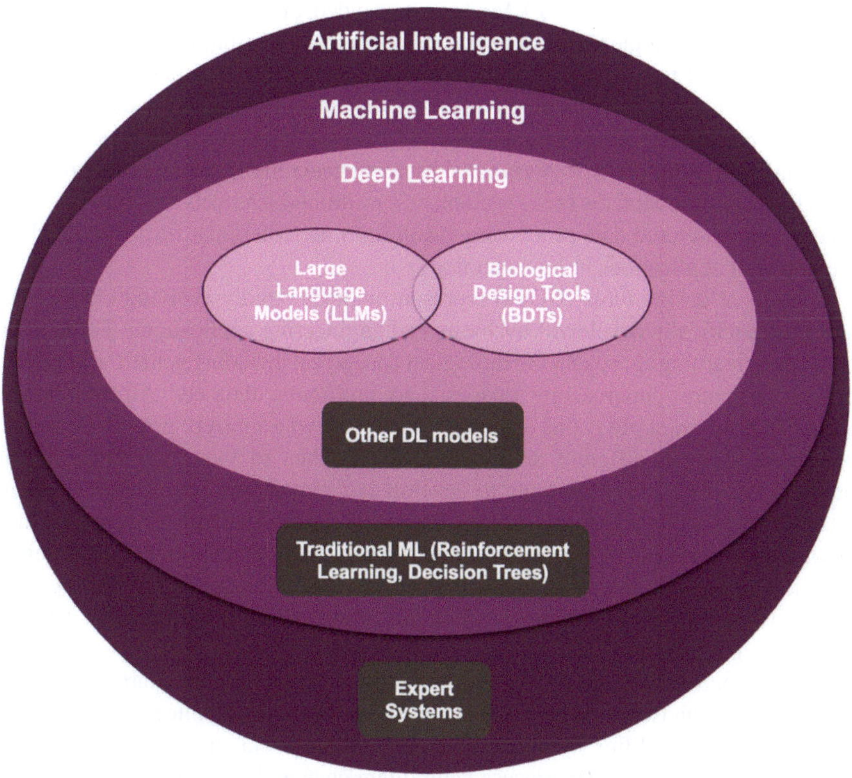

Fig. 2.1 Relationships among artificial intelligence methods applied in biotechnology

biological problems—supervised learning excels at prediction tasks when abundant labeled data exists, unsupervised learning helps discover underlying patterns in complex biological datasets, and reinforcement learning optimizes sequential decision-making processes like experimental design.

DL, a subset of ML that has driven many recent biotechnology breakthroughs, employs neural networks with multiple processing layers to learn hierarchical representations of data. These networks, inspired by biological neural systems, consist of input and output layers separated by numerous hidden layers that progressively transform information [3, 19, 21]. The architecture of these networks continues to evolve rapidly, with transformers, graph neural networks, and diffusion models representing just a few recent innovations that have significantly expanded capabilities. Transformer architectures, which underlie modern language models, have proven remarkably adaptable to biological sequence data, enabling new approaches to protein and genetic sequence analysis that treat biological information as a form of language [22].

Among DL model types, LLMs exemplify the rapid evolution occurring within AI subfields. These deep learning systems, trained on vast natural language datasets, interpret and generate text based on patterns learned from billions of examples [19, 23]. The capabilities of LLMs have expanded dramatically in recent years, progressing from simple text completion to sophisticated reasoning, code generation, and multilingual translation. Models like OpenAI's GPT series and Meta's Llama have become accessible through web interfaces, enabling researchers without specialized AI expertise to leverage natural language processing for biological research tasks including literature review, experimental protocol generation, and data interpretation [24]. Even as this chapter is being written, new multimodal models that combine text, image, and structured data processing capabilities are emerging, further expanding the potential applications in biotechnology.

Biological design tools (BDTs) represent another rapidly evolving class of AI systems specifically developed for biological engineering applications. These tools apply deep learning approaches to biological data types, including genetic sequences, protein structures, microscopy images, and multiomic datasets. AlphaFold and RoseTTAFold exemplify this category, having revolutionized protein structure prediction through different architectural approaches to the same fundamental problem [12, 25]. BDTs continue to develop at a remarkable pace—AlphaFold 3, released just months before this writing, expanded capabilities from protein structure prediction to modeling interactions between diverse biomolecules including proteins, nucleic acids, small molecules, and ions [12]. New BDTs emerge regularly, targeting different aspects of biological design including protein function optimization, genetic circuit design, and metabolic pathway engineering.

The distinction between LLMs and BDTs illustrates important characteristics of the broader AI landscape. While both employ neural network architectures trained on large datasets, they differ significantly in their training data, output modalities, and accessibility profiles. LLMs process natural language and generate textual outputs, making them accessible to anyone with basic language skills. BDTs, conversely, process specialized biological data types and generate outputs (such as protein structural coordinates or genetic sequences) that require domain expertise to interpret and apply meaningfully [19, 26]. This highlights how AI systems vary not only in their technical implementation but in their interfaces with human users and integration into scientific workflows.

Beyond specific techniques, the field of AI encompasses broader research directions including:

1. Symbolic AI approaches that use logic and knowledge representation to reason about biological systems
2. Hybrid systems combining neural networks with explicit knowledge bases or rule systems
3. Neuroevolutionary methods that use principles from evolutionary biology to optimize AI architectures
4. Federated learning approaches that enable collaborative model training while maintaining data privacy
5. Causal inference methods that move beyond correlation to understand mechanistic relationships in biological systems

Importantly, these approaches are not merely theoretical—they represent active research areas with potential applications in biotechnology. For instance, symbolic approaches may help incorporate existing biological knowledge into predictive models, while causal inference methods could improve understanding of gene regulatory networks.

The evolutionary pace of AI approaches requires a framework that emphasizes fundamental principles rather than specific implementations. Current AI systems lie at varying points along multiple technological continuums, and future systems will further expand the ranges of these continuums. As such, frameworks for understanding these AI systems should consider the ranges of key characteristics including:

1. Data requirements: From data-hungry approaches requiring millions of examples (e.g., deep convolutional neural networks [27]) to few-shot learning methods that can generalize from limited examples (e.g., meta-learning [28])
2. Transparency: From inherently interpretable "glass box" models (e.g., decision trees) to complex "black box" systems where decision-making processes remain opaque (e.g., deep neural networks). The introduction of explainable AI (XAI) techniques has also introduced methods to approximate "black box" algorithms with simpler "glass box" models to provide more post-hoc interpretability [29].
3. Autonomy: From systems requiring extensive human guidance (e.g., rules-based automation systems such as robotic labs) to those capable of independent learning and adaptation (e.g., reinforcement learning agents)
4. Generality: From narrow systems designed for specific tasks (e.g., DeepMind's AlphaGo) to more flexible systems capable of addressing diverse problems (e.g., OpenAI's GPT series LLM)

5. Computational demands: From lightweight algorithms deployable on edge devices (e.g., Apple's Siri) to resource-intensive models requiring specialized hardware (e.g., DeepMind's AlphaFold)

The integration of AI with biotechnology creates particularly fertile ground for innovation, as methods from each field inspire advances in the other. Neural networks drew initial inspiration from biological neural systems, while more recent biologically-inspired approaches include reservoir computing (modeled after brain dynamics) and genetic algorithms (inspired by natural selection). Simultaneously, AI techniques are being applied to biological problems in increasingly sophisticated ways—from designing novel proteins with specific functions to optimizing genetic circuits for synthetic biology applications. This reciprocal relationship continues to evolve as researchers develop new computational approaches tailored to the unique challenges of biological systems.

The rapid pace of innovation in AI methodologies highlights the importance of focusing governance efforts on capabilities and impacts rather than specific technologies. Attempting to regulate particular AI techniques will likely prove futile as the field continuously evolves, with new approaches emerging and existing ones being refined or combined in novel ways. Instead, governance frameworks must address the capabilities that these systems enable in biotechnology contexts—such as the ability to design novel biological entities or accelerate experimental cycles—while remaining adaptable to technological shifts.

2.1.2 What Is Convergence?

Convergence represents a fundamental paradigm shift in scientific and technological development—moving beyond interdisciplinary collaboration toward the deep integration of previously distinct fields to create novel capabilities and insights. While interdisciplinary research brings together multiple disciplines to address common problems while maintaining distinct methodological approaches, convergence creates new conceptual frameworks and methodologies at the intersection of fields [30, 31]. The AI-biotechnology convergence manifests across multiple scales, from molecular-level processes to ecosystem-wide applications, creating bidirectional flows of knowledge, techniques, and organizing principles between computational and biological domains.

Bidirectional Knowledge Transfer. The AI-biotechnology interface operates through reciprocal channels of influence, with each domain transforming the capabilities and conceptual frameworks of the other. This bidirectional relationship extends beyond superficial application of techniques from one field to another, creating fundamentally new approaches to solving complex problems.

From a biotechnology-to-AI perspective, biological systems provide sophisticated design patterns and operational principles that inspire computational advances. Neural networks represent the most widely recognized biomimetic computational approach, drawing inspiration from neuronal connectivity patterns and signal processing in biological brains. However, this influence extends far beyond this well-known example. Genetic algorithms implement selection, recombination, and mutation principles from evolutionary biology to optimize solutions in complex search spaces [32]. These algorithms have proven effective in addressing optimization problems resistant to traditional approaches, particularly in cases involving high-dimensional spaces with numerous local optima.

More recent biologically-inspired computational approaches incorporate increasingly sophisticated biological principles. Artificial immune systems model computational security based on biological immune response mechanisms, with anomaly detection systems inspired by how T-cells distinguish between self and non-self entities [33]. Swarm intelligence algorithms draw from collective behaviors observed in insect colonies, bird flocks, and fish schools to solve distributed optimization problems [34]. These approaches leverage principles of local interaction, indirect communication through environmental modification (stigmergy), and emergent coordination without centralized control.

At the cellular level, biological computation provides models for parallel processing systems. Membrane computing (P-systems) draws inspiration from cellular compartmentalization and membrane permeability to develop computational models with hierarchical structure and local processing capabilities [35]. DNA computing demonstrates how biological molecules can perform computational operations through hybridization, ligation, and enzymatic modification, offering potential for massive parallelism and energy efficiency [36].

From an AI-to-biotechnology perspective, computational approaches have fundamentally transformed biological understanding and engineering capabilities. Deep learning has revolutionized protein structure prediction through systems like AlphaFold, which accurately models three-dimensional protein conformations by learning patterns from existing protein structure databases [37]. This capability dramatically accelerates drug discovery by enabling in silico screening of candidate molecules against protein targets with unprecedented accuracy.

Beyond structure prediction, AI systems enhance biological design through optimization of sequence-function relationships. Evolution-guided machine learning approaches, like those implemented in ProGen, generate functional protein sequences by learning statistical patterns from millions of natural proteins [38]. These systems can design synthetic proteins with specific properties while maintaining biophysical constraints necessary for proper folding and function. Similar approaches have been applied to genetic circuit design, metabolic pathway optimization, and synthetic genome development.

Computational Representation of Biological Systems. A critical aspect of AI-biotechnology convergence involves developing computational frameworks that accurately capture the complexity of biological systems. Traditional computational models often struggled with the inherent messiness of biology—its stochasticity, redundancy, and context-dependence. Modern AI approaches embrace these characteristics rather than attempting to simplify them away.

Non-negative matrix factorization (NMF) exemplifies how biological visual processing principles have transformed computational approaches to pattern recognition. Unlike principal component analysis (PCA), which represents data through holistic, often difficult-to-interpret components, NMF learns parts-based representations that align with how visual cortex neurons process information. When applied to facial recognition, NMF identifies interpretable features like eyes, nose, and mouth, mirroring how biological visual systems process such information [39]. This approach has proven valuable in biological data analysis, particularly for deconvolving complex tissue samples into constituent cell types based on gene expression patterns.

The stochastic nature of biological systems has inspired probabilistic computational frameworks that embrace uncertainty rather than attempting to eliminate it. Bayesian neural networks incorporate uncertainty quantification into deep learning models, providing not just predictions but confidence estimates for those predictions. This capability proves particularly valuable when applying AI to biological problems where data may be sparse, noisy, or incomplete. Stochastic optimization methods like simulated annealing draw from physical annealing processes to escape local optima in complex search spaces, a challenge common in protein folding and molecular design problems.

Afek et al. demonstrated how biological decision-making processes can inspire algorithmic innovations that transcend traditional computational approaches [40, 41]. By studying how fruit flies resolve the sensory organ precursor (SOP) selection problem—determining which cells become sensory organs through distributed consensus—they developed a maximal independent set (MIS) selection algorithm with remarkable efficiency. Unlike conventional approaches requiring complete knowledge of network topology, their biologically-inspired algorithm functions with limited local information, demonstrating how biological systems solve complex coordination problems under constraints. This work exemplifies how seemingly "inefficient" biological processes actually represent sophisticated solutions to distributed computing challenges.

2.1.3 Emerging Integration Paradigms

The convergence of AI and biotechnology manifests through several paradigms that represent distinct integration pathways, each with unique implications for research, applications, and governance.

AI-Augmented Biological Design. The design-build-test-learn (DBTL) cycle fundamental to synthetic biology has been transformed by AI integration at multiple stages. In the design phase, generative models create novel biological entities optimized for specific functions. During building, laboratory automation systems directed by AI algorithms implement designs with minimal human intervention [42]. Testing phases employ machine learning for high-throughput phenotypic analysis and characterization. Finally, learning phases utilize reinforcement learning approaches to optimize experimental strategy based on previous outcomes, creating a closed-loop system for biological discovery and engineering.

This paradigm has accelerated developments across multiple biological domains. In protein engineering, directed evolution approaches traditionally required extensive laboratory selection over many generations. AI-augmented approaches like ProSPR combine computational prediction with targeted experimentation, achieving similar results with orders of magnitude less laboratory work [43]. For metabolic engineering, tools like BioWM implement differentiable genome-scale metabolic models that enable end-to-end optimization of pathways for chemical production [44].

Cell reprogramming represents another domain transformed by AI-augmented design. Traditional approaches to converting one cell type to another required extensive trial-and-error identification of transcription factors. Systems like Geneformer leverage large language models trained on single-cell RNA sequencing data to predict optimal transcription factor combinations for specific cell type conversions, dramatically reducing experimental cycles [45].

Biological Computing Substrates. While silicon-based computing has dominated technological development, biological substrates offer alternative computational paradigms with unique advantages in certain contexts. DNA-based data storage demonstrates remarkable information density—one gram of DNA can theoretically store 215 petabytes of data, far exceeding conventional storage media [46]. Recent advances combine DNA synthesis technologies with machine learning algorithms that optimize encoding schemes for error resistance and retrieval efficiency.

Biological circuits constructed from genetic components represent another convergence pathway, where cells function as computational units. These systems can implement logical operations (AND, OR, NOT gates), memory storage, and signal processing functions within living cells. Machine learning approaches have enhanced the design of these circuits by modeling the complex interactions between genetic components and predicting system behavior under various conditions [47].

Neuroelectronic interfaces represent a more direct integration pathway, creating hybrid systems where biological neural tissue interfaces with electronic computing systems. These interfaces enable bidirectional communication between neural tissue and computational hardware, with applications ranging from neuroprosthetics to biological computing platforms. Advanced machine learning algorithms process the complex, noisy signals from neural tissue, extracting meaningful information and providing appropriate feedback [48].

Distributed Biological Intelligence. Natural biological systems demonstrate sophisticated emergent intelligence without centralized control structures. Slime molds solve complex optimization problems like finding shortest paths through mazes despite lacking a central nervous system [49]. Ant colonies implement distributed algorithms for resource allocation and path optimization through simple local interactions and pheromone signaling. Immune systems demonstrate adaptive learning capabilities, recognizing and responding to novel pathogens while maintaining self-tolerance.

The principles underlying these distributed biological intelligence systems have inspired computational approaches that challenge traditional centralized AI paradigms. Cellular automata models implement computational processes through local interactions between simple units, generating complex emergent behaviors [50]. Edge computing architectures distribute AI processing across networks of devices rather than centralizing computation, mirroring how biological systems operate without centralized coordination. Federated learning enables collaborative model training while maintaining data privacy, conceptually similar to how immune systems share information about pathogens without centralizing all immune cells.

Recent experimental systems have demonstrated direct integration of biological and computational intelligence. Organoid intelligence approaches cultivate brain organoids (three-dimensional neural tissue cultures) interfaced with sensing and computing systems [51]. These hybrid systems leverage the inherent information processing capabilities of biological neural networks while providing computational support for tasks where silicon excels.

Synthetic Systems at the AI-Biology Interface. The deepest form of AI-biotechnology convergence involves creating systems that seamlessly integrate computational and biological principles, transcending traditional boundaries between silicon and carbon-based processes.

Engineered living materials represent one manifestation of this integration, combining the programmability of computational systems with the self-assembly and adaptive properties of biological materials. These materials incorporate living cells within structural matrices, enabling functionalities like self-healing, environmental sensing, and adaptive response. Machine learning algorithms optimize both the biological components and the surrounding matrices, predicting how these complex systems will behave under various conditions [52].

Whole-cell computational models represent another synthetic approach, creating detailed in silico representations of entire cells that capture their dynamic behavior. The *E. coli* whole-cell model integrates thousands of known biochemical parameters with machine learning approaches that predict unknown interactions, creating a computational twin of the bacterial cell [53]. These models enable computational experiments that would be impractical or impossible in physical systems, accelerating understanding of fundamental biological processes.

Synthetic morphogenesis systems demonstrate how computational principles can guide biological development and pattern formation. By encoding computational rules within genetic circuits, researchers have programmed cells to form specific three-dimensional structures through coordinated growth and differentiation [54]. These systems implement biological versions of cellular automata, where local cell–cell interactions follow computational rules that generate complex emergent structures.

The concept of a "living laboratory" represents perhaps the most ambitious convergence paradigm—biological systems engineered to conduct experimental processes autonomously. Microfluidic devices containing engineered cells can implement complex experimental protocols, with cellular sensors detecting outcomes and genetic circuits processing this information to determine subsequent steps. Machine learning algorithms optimize these systems' design and operation, creating biological experimentation platforms that function with minimal human intervention.

2.1.4 Toward Fully Integrated Systems

The trajectories of AI and biotechnology convergence point toward computer systems with distinctly biological characteristics: (1) distributed coordination without centralized control; (2) robust adaptation to failure and environmental change; (3) efficient resource utilization through component repurposing; and (4) leveraging stochasticity as a feature rather than a bug. These characteristics represent fundamental shifts from traditional computational paradigms that emphasized determinism, hierarchy, specialization, and noise reduction.

Several active research areas demonstrate progress toward such integrated systems. Neuromorphic computing implements neural processing principles directly in hardware architecture, creating systems that process information through massively parallel, low-power operations similar to biological brains [55]. These systems excel at tasks requiring real-time adaptation to noisy, unpredictable environments—precisely the conditions where traditional computing architectures struggle.

Self-healing computational systems incorporate mechanisms for detecting and repairing damage, inspired by biological repair processes. These systems implement redundancy, modularity, and adaptive reconfiguration to maintain functionality despite component failures [56]. Machine learning approaches enhance these capabilities by predicting failure modes and optimizing response strategies based on system state and available resources.

Developmental computing represents another convergence pathway, where computational systems undergo structured growth and differentiation processes analogous to biological development. Rather than being explicitly programmed with final configurations, these systems contain "genomic" instructions that guide their self-organization through interaction with environmental conditions [57]. This approach enables adaptation to operating environments not anticipated during initial design, similar to how biological development accommodates environmental variation.

The emergence of engineered biological communication networks demonstrates how convergence enables novel modes of information transmission. These systems use biological molecules as information carriers, with engineered cells functioning as network nodes that process and relay signals. Machine learning algorithms optimize signal encoding, transmission protocols, and network topology to maximize information throughput while minimizing resource requirements [58].

The AI-biotechnology convergence represents not merely the application of one field's tools to the other's problems, but the creation of fundamentally new approaches that transcend traditional boundaries between computational and biological systems. This convergence operates across multiple scales and through various integration paradigms, from AI-augmented biological design to fully integrated synthetic systems. Understanding these diverse manifestations is essential for developing governance frameworks that address specific convergence patterns rather than attempting to regulate "AI" and "biotechnology" as monolithic entities.

To effectively assess and manage the governance challenges of AI-biotechnology convergence, we must first distinguish between two related, but distinct, terms: risk and uncertainty. Risk refers to the negative consequences of an event with respect to something of human value [59]. Mathematically, risk is generally characterized as a function of the probability of an event's occurrence and the severity of its consequences [60–62]. Uncertainty, on the other hand, refers to a lack of complete knowledge of the occurrence and/or consequences of an event [63].

2.2 Taxonomy of Uncertainty

Uncertainty in complex systems like those created through AI-biotechnology convergence manifests in multiple forms, each presenting distinct challenges for assessment and governance. Knight established a foundational distinction between risk (where probabilities are known) and uncertainty (where probabilities cannot be determined), but modern understanding requires more nuanced categorization [64]. Walker et al. proposed a comprehensive uncertainty matrix that identifies three dimensions: location (where uncertainty manifests in the system), level (how uncertainty is expressed), and nature (why uncertainty exists) [65].

The location dimension encompasses uncertainties in system context (boundaries and external factors), model structure (relationships between variables), inputs (observed or measured exogenous data), parameter values (constants in models), and model outcomes (accumulated uncertainties through the system). For AI-biotechnology systems, structural uncertainties prove particularly challenging as the relationships between components may not follow predictable patterns, especially when AI systems modify biological components based on complex learning processes.

The level dimension classifies uncertainty according to how it can be characterized, ranging from statistical uncertainty (quantifiable through probability distributions) to scenario uncertainty (discrete possibilities without assigned probabilities) to recognized ignorance (awareness of knowledge gaps without ability to characterize). Deep uncertainty, the most challenging category, occurs when stakeholders cannot agree on model structures, probability distributions, or even the variables that should be considered [66]. AI-biotechnology convergence frequently operates in this deep uncertainty domain, where fundamental aspects of system behavior remain unknown and potentially unknowable before deployment.

The nature dimension distinguishes between epistemic uncertainty (resulting from imperfect knowledge that could theoretically be reduced through more information) and aleatory uncertainty (resulting from inherent variability that cannot be reduced through additional knowledge). Biological systems exhibit high aleatory uncertainty through stochastic processes at molecular, cellular, and ecological levels. AI systems introduce complex epistemic uncertainties through their learning processes and emergent behaviors. The convergence of these domains compounds both types of uncertainty, creating challenges for traditional risk assessment approaches.

In the remainder of this section, we describe the unique challenges posed by complex and systemic risks, such as those introduced by the convergence of AI and biotechnology. We then discuss varying perceptions and interpretations of these risks and their implications for risk governance and policy.

2.2.1 *Simple Versus Complex Risk*

Simple risks present well-defined cause-consequence relationships with low uncertainty. In these cases, conventional probabilistic methods can effectively assess and manage risk [67]. Simple risks can be characterized by discrete probability distributions with narrow confidence intervals and clear causal mechanisms. Examples in biotechnology might include the contamination risk of a laboratory culture or the failure rate of a standardized protocol. These risks operate within established paradigms where past evidence provides reliable guidance for future events.

Risk assessment for these simple cases typically follows established frameworks where hazards are identified, exposure pathways characterized, dose–response relationships determined, and risks calculated through formal mathematical models. Stakeholders generally agree on relevant variables, acceptable thresholds, and appropriate management responses. Traditional risk management tools like cost–benefit analysis function effectively when uncertainty remains confined to statistical parameters rather than fundamental system properties [68].

The risks posed by AI-biotechnology convergence, however, rarely conform to these simple patterns [69]. Instead, they manifest as complex systemic risks characterized by deep uncertainty and potential for cascading consequences, and emerge from multiple interacting factors that cannot be isolated or analyzed independently. They operate across organizational, spatial, and temporal scales, with feedback loops that amplify or attenuate effects in non-linear ways. Risk assessment for these systems requires fundamentally different approaches that acknowledge interconnectedness and emergent properties not predictable from component analysis.

Stirling and Gee identified four contexts for risk assessment based on knowledge dimensions [70]: risk (where probabilities and outcomes are well-characterized), uncertainty (where outcomes are known but probabilities remain undefined), ambiguity (where probabilities might be known but outcomes remain contested), and ignorance (where both probabilities and outcomes remain unknown). Most AI-biotechnology convergence scenarios operate in either the ambiguity or ignorance contexts, where traditional risk assessment proves fundamentally inadequate.

2.2.2 Cascading and Systemic Risks

The risks emerging from AI-biotechnology convergence extend beyond isolated events to encompass cascading and systemic effects. Cascading risks involve sequences of connected failures, where initial events trigger subsequent failures through interdependencies between system components [71]. Systemic risks threaten the stability of entire systems rather than individual elements, often through complex causal pathways that evade simple modeling approaches [72].

The international Risk Governance Council (IRGC) framework classifies risks according to complexity, uncertainty, and ambiguity [67]. Simple risks display low values across all dimensions, while complex risks exhibit multiple cause-effect relationships that may prove difficult to identify but remain theoretically determinable. Uncertain risks involve relationships between causes and effects that cannot be established with certainty even with sophisticated modeling techniques. Ambiguous risks involve contested values and priorities even when the underlying technical relationships may be well-established.

AI-biotechnology convergence introduces risks in the most challenging category—those simultaneously exhibiting high complexity, uncertainty, and ambiguity. The complexity emerges from intricate interactions between computational and biological components, each with their own non-linear dynamics. Uncertainty stems from the novel properties of these hybrid systems, with limited empirical data on their behavior under varying conditions. Ambiguity arises from differing stakeholder perspectives on acceptable outcomes, appropriate safeguards, and value hierarchies for weighing potential benefits against risks.

These systemic risks cannot be adequately represented by single points in a probability-severity space. Instead, they manifest as diffuse regions that reflect fundamental uncertainty about both likelihood and consequence dimensions. Even sophisticated uncertainty quantification methods like probability bounds analysis or fuzzy set theory may prove insufficient when dealing with risks characterized by deep uncertainty about system structure and behavior [73].

2.2.3 Risk Perceptions and Interpretations

Perceptions of risks associated with AI-biotechnology convergence vary dramatically across stakeholder groups, reflecting different interpretative frameworks rather than merely differing access to information. Cultural theory of risk suggests that risk perceptions align with cultural worldviews along dimensions of grid (acceptance of social constraints) and group (degree of social cohesion) [74]. These fundamental orientations shape how individuals interpret uncertainties—whether as opportunities for innovation or precursors to catastrophe.

Kahan et al. demonstrated that individuals filter technical information through cultural cognition processes, selectively crediting or discounting expert claims based on their congruence with cultural values [75]. This dynamic appears particularly pronounced for technologies like AI and biotechnology that touch on fundamental questions about human capability, identity, and governance. The uncertain nature of convergence risks amplifies these interpretive divergences, as the absence of definitive evidence allows greater scope for value-based judgments.

Risk assessment professionals often approach uncertainty through quantitative frameworks using expected values, confidence intervals, and sensitivity analyses. However, public risk perceptions frequently incorporate qualitative dimensions like controllability, familiarity, catastrophic potential, and equity of distribution that formal analysis may overlook [76]. Technologies that appear controllable to experts may trigger strong public concern when perceived as involuntary, inequitable, or potentially catastrophic.

The Amplification of Risk Framework explains how risk signals interact with psychological, social, and cultural processes to intensify or attenuate public perceptions [77]. Technologies occupying the intersection of AI and biotechnology possess numerous characteristics associated with risk amplification: potentially irreversible consequences, technological novelty, contested expertise, and distributed governance responsibility. The uncertainty inherent in these technologies provides fertile ground for competing narratives that frame identical technical information through dramatically different interpretive lenses.

These divergent interpretative frameworks frequently lead to ambiguity about appropriate risk management approaches even when stakeholders agree on technical aspects of the underlying science. Proponents of precautionary principles emphasize avoiding irreversible harm under conditions of uncertainty, while proponents of proactionary principles highlight potential opportunity costs of excessive caution [78]. Neither framework provides objective guidance for decision-making—both represent value judgments about the relative importance of different types of error under conditions of fundamental uncertainty.

2.2.4 Implications for Governance

The distinctive characteristics of AI-biotechnology convergence risks—complexity, uncertainty, and ambiguity—demand governance approaches fundamentally different from those developed for simple risks with well-defined parameters. Linear risk assessment frameworks that proceed sequentially from hazard identification to exposure assessment to risk characterization prove inadequate when system boundaries remain contested, causal pathways exhibit non-linear dynamics, and stakeholders disagree about which outcomes warrant concern.

Adaptive governance represents a more suitable approach for managing these complex risks, emphasizing iterative processes that incorporate new information as it emerges rather than attempting comprehensive ex ante regulation [79, 80]. This approach acknowledges that perfect foresight remains impossible when dealing with complex adaptive systems characterized by emergence, self-organization, and path-dependence. Rather than seeking static optimization against predefined criteria, adaptive governance establishes monitoring systems, feedback mechanisms, and response protocols that activate as system behaviors emerge.

Robust decision-making provides complementary strategies for navigating deep uncertainty, focusing on identification of courses of action that perform acceptably across wide ranges of future scenarios rather than optimally under narrow assumptions [66]. This approach emphasizes broad-based resilience rather than targeted risk mitigation, acknowledging that the specific manifestations of systemic risks remain fundamentally unpredictable despite broad agreement about their potential significance.

Participatory risk governance represents another essential component, incorporating diverse stakeholder perspectives in defining problems, identifying alternatives, and evaluating outcomes [81]. This approach acknowledges that technical risk assessment alone cannot resolve fundamentally ambiguous questions about acceptable risk levels, appropriate safeguards, or equitable distribution of benefits and harms. Effective governance requires transparent negotiation among stakeholders with different values, priorities, and interpretive frameworks.

The International Risk Governance Council framework suggests tailoring governance approaches to the risk characteristics—epistemic discourse for complex risks, reflective discourse for uncertain risks, and participative discourse for ambiguous risks [67]. AI-biotechnology convergence risks typically require all three discourse types operating in parallel, with different emphasis depending on the specific technology and application context.

2.2.5 Uncertainty as Opportunity or Threat

The uncertainty surrounding AI-biotechnology convergence generates dramatically different responses across stakeholder groups, with some interpreting this uncertainty as predominantly threatening while others perceive it primarily as opportunity space. These divergent interpretations stem not from different access to technical information but from fundamentally different orientations toward uncertainty itself.

Those perceiving uncertainty as primarily threatening emphasize irreversibility, catastrophic potential, and governance failures. From this perspective, the inability to predict specific outcomes creates unacceptable risks, particularly for technologies that interact with fundamental biological processes with potential for cascading effects. Precautionary frameworks emerge naturally from this orientation, emphasizing the need to establish safety before proceeding with technological development [82]. Proponents argue that the burden of proof should rest with technology developers to demonstrate safety rather than with regulators to demonstrate harm, particularly when potential harms could prove irreversible or broadly distributed.

Those perceiving uncertainty as primarily opportunity-generating emphasize flexibility, innovation potential, and adaptation capabilities. From this perspective, uncertainty creates possibility space for beneficial applications that cannot be anticipated through prospective analysis. Proactionary frameworks emerge from this orientation, emphasizing the need to balance potential risks against opportunity costs of excessive caution [78]. Proponents argue that governance should focus on establishing responsive monitoring and adaptation mechanisms rather than preventing development based on speculative concerns.

Both perspectives operate from the same fundamental recognition of uncertainty, but reach dramatically different conclusions about its implications for governance. This divergence highlights that the central challenges for AI-biotechnology governance are not merely technical but normative—they concern values, priorities, and acceptable trade-offs under conditions of fundamental uncertainty. No amount of technical analysis can fully resolve these normative questions, though improved technical understanding can help clarify the decision space and identify robust governance approaches that accommodate different value perspectives.

In the remainder of this chapter, we highlight both the unique opportunities and the systemic risks associated with the convergence of AI and biotechnology, identifying adaptive governance principles to ensure robust risk governance. We examine how different stakeholder perspectives interpret identical technologies through different frames, leading to divergent assessments of benefits and risks [83]. Rather than attempting to resolve these normative differences through technical analysis alone, we propose governance principles designed to accommodate diverse interpretive frameworks while establishing guardrails against the most concerning potential outcomes.

2.3 Opportunities and Threats of Convergence

Advances in computation and AI provide new levels of predictability and understanding of the most complex biological systems. At the same time, the stochasticity and complexity of biological systems are inspiring deeper AI algorithms that excel at complex decision making. This convergence of AI and biotechnology presents immense opportunities with potential to reshape many fields, from healthcare and agriculture to energy and environmental sustainability. However, it also introduces complex, systemic risks and uncertainties that will require careful consideration and clear, robust governance frameworks. Understanding the potential benefits and risks of convergence is essential to finding a delicate balance to governance of this emerging landscape.

2.3.1 Opportunities

A major promise of AI and biotechnology convergence is the ability to accurately predict the outcomes of engineered biological systems, thus enabling design for purpose [84]. While AI applications in other fields often focus on simply improving efficiency, in the context of biotechnology, AI is fundamentally transforming our ability to solve and understand complex biological problems that would be intractable without it. For example, the development of BDTs AlphaFold and RoseTTAFold has solved a 50-year problem, accurately predicting (to a high degree of certainty) the three-dimensional structures of proteins and other molecules in mere minutes. These structures are crucial for understanding protein function, which can in turn drive the development of targeted therapies, such as cancer treatments [85].

At present, BDTs typically focus on outputting sequences optimized for a single function. However, future developments could enable the engineering of entire organisms designed to perform multiple specialized functions [19]. A particularly significant initiative in this space is the Genome Project-Write (GP-write), launched in 2016, which aims to revolutionize the development of optimized organisms by leveraging whole-genome engineering of human cell lines and other essential organisms. With the aim to drastically reduce the cost of engineering large genomes by over 1,000-fold within the next decade, GP-write aims to push the boundaries of synthetic biology. The project seeks to define the "solution space" for human and other genomes, allowing for the creation of organisms tailored for specific functions, whether for agricultural, medical, or industrial applications [86]. This work is expected to provide the foundation for advances that could lead to more sustainable agriculture, personalized medicine, and next-generation industrial manufacturing.

The convergence of AI and biotechnology is poised to significantly enhance the speed, precision, and scalability of biological experiments and the design process. This synergy is already transforming drug discovery and development, fields traditionally constrained by lengthy timelines and high costs. Conventional pharmaceutical research relies heavily on extensive trial-and-error testing before a viable drug can be brought to market. AI-driven methods, however, can analyze vast biological datasets and efficiently and accurately predict promising compounds, thereby reducing costs and accelerating development. For example, in 2021, Insilico Medicine introduced INS018_055, a drug designed using generative AI to treat idiopathic pulmonary fibrosis. The drug progressed from discovery to Phase I clinical trials in just three years, demonstrating AI's capacity to revolutionize drug development [87]. Similarly, a partnership between NVIDIA and Roche's Genentech uses generative AI to power a "lab in a loop" that optimizes molecular design through iterative experimentation [88]. AI also plays a pivotal role in pandemic preparedness, as exemplified by the GUIDE program from Lawrence Livermore National Laboratory, which uses ML and high-performance computing to design antibodies and antigens in silico, aiming to develop medical countermeasures against emerging public health threats [89].

AI-biotechnology convergence can automate all aspects of the design-build-test-learn cycle, significantly reducing the need for human intervention [19, 90]. Platforms like BioAutomata, which integrate robotic systems with ML algorithms, can autonomously design and conduct experiments to optimize biological processes with minimal input from researchers. After the initial setup, the researcher acts primarily as a hands-off supervisor while the AI-powered platform iteratively refines experiments. In a proof-of-concept test aimed at optimizing the lycopene biosynthetic pathway, BioAutomata outperformed random screening by 77% while testing less than 1% of possible variants [91]. This coupling of AI with automated experimentation platforms could greatly enhance the efficiency of the design-build-test-learn cycle, lowering costs and reducing labor requirements.

In the realm of genome engineering, CRISPR-GPT, introduced in 2024, automates the design of CRISPR gene-editing experiments. This LLM, augmented with domain-specific knowledge of CRISPR, may eventually enable even non-expert users to develop gene-editing experiments, democratizing access to powerful genome-engineering techniques [92]. As tools such as CRISPR-GPT and other scientific LLMs become more sophisticated, they promise to lower the barriers to entry in biotechnology, empowering more researchers to harness the full potential of synthetic biology. LLMs with intuitive user interfaces may allow individuals from a variety of disciplines—not just those with specialized knowledge—to participate in biotechnology research and development. These tools could also be instrumental in the education and training of the next generation of biologists and engineers.

Democratizing the field will enable broader participation in bio-innovation and can foster new interdisciplinary collaborations. In this vein, the DIY biology movement is a grassroots initiative in which lay people engage in biological experimentation outside traditional academic or industrial labs. Seyfried et al. examined the benefits and societal concerns regarding safety and ethics associated with the initiative, concluding that, with appropriate oversight and support, DIY biology has the capacity to enhance public engagement with science and contribute to biotechnological advancements [93]. Further, the initiative's independence from institutional constraints may foster a unique contribution to biotechnological innovation and inspire novel approaches in the field [94]. The democratization of biotechnology through easily accessible LLMs and other web-based AI tools can expand the power of DIY biology and similar initiatives, encouraging out-of-the-box thinking in biotech research and development.

The influence of biology on AI tools driven by convergence could provide valuable guidance in the development of computational systems that can adapt, recover, and self-correct—much like how DNA repairs itself after damage. In biological systems, mechanisms such as DNA repair and cellular regeneration allow organisms to maintain functionality and resilience despite disruption. Taking influence from these natural processes, AI researchers could design algorithms and architectures that mimic these adaptive qualities. For example, AI models could incorporate self-repairing mechanisms to detect and recover from faulty inputs, bad data, or even internal errors, enabling them to "heal" from mistakes without external intervention.

Such bio-inspired AI could lead to more resilient technologies, capable of learning from and correcting errors autonomously.

2.3.2 *Threats*

Many of the risks present in biotechnology and biological research are heightened for policymakers tasked with assessing and regulating AI-biotechnology convergence [90]. The integration of AI, and its associated increase in speed and efficiency, with the consequential nature of biotech development and bioengineering presents a unique combination of scale and speed that threatens to outpace existing technology and R&D regulatory frameworks. Broadly, these threats and uncertainties can be categorized into four groups: (1) Misuse, dual-use, and malicious intent; (2) Errors and the black box problem; (3) Deskilling and reduced barriers to entry; and (4) Failing to capitalize and opportunity costs.

Misuse, dual-use, and malicious intent. An often-discussed threat of AI-biotechnology convergence is the increased speed and precision at which a harmful biological agent could be developed and released by a bad actor. Assessing this threat is complicated by the so-called "dual-use dilemma:" findings of legitimate scientific innovation can often be easily misused to cause societal harm. Atlas and Dando described three aspects of the dual-use dilemma of modern biology [95]: (1) outwardly-seeming civilian facilities that actually function for military or terrorist bioweapon development; (2) other equipment that can be misused for bioweapon development; and (3) knowledge and expertise that could be applied to bioweapon development. AI tools, particularly scientific LLMs and BDTs, cover aspects (2) and (3), potentially providing the tools and knowledge required for bioweapon development.

In their current state, LLMs alone do not necessarily present a major risk of dual-use facilitation of bioweapon development by lay persons, as they are generally comprised of knowledge already available on the internet. While information about how to develop bioweapons may be accessible, to actually implement a viral mutation, for example, an actor would need much more sophisticated biological knowledge [96]. Mitigating misuse or dual-use of LLMs or BDTs by those with specialized knowledge is more difficult. As noted by Epstein, without complete visibility into potential bad actors' actions, we cannot know with certainty how well oversight systems will mitigate misuse of biotechnology [97]. However, the author suggests adopting robust, balanced oversight frameworks that can reasonably secure the buy-in of the research community and the trust of the public at large.

Errors and the black box problem. Further risks stem from the design and training of the AI models and algorithms themselves. One notable concern is the potential for AI models to generate inaccurate outputs [26]. These inaccuracies can arise from various factors, including errors within AI algorithms or training on biased and/or incomplete data sets. Furthering this concern, models may learn from faulty interpretations of training data, perpetuating these mistaken outputs in future iterations. A related issue specific to LLMs is the phenomenon of confabulation, where a model generates plausible-sounding, but incorrect or misleading information [19]. While the risk of such errors of AI models is not unique to biotechnology, the consequences of these errors may yield more irreversible harms than those encountered in other fields.

The "black box" nature of many AI systems poses significant challenges to transparency and accountability. While the complexity of many DL models can enhance pattern discovery, their opacity makes it difficult to interpret model outputs and understand decision-making processes [90]. This opacity also complicates the identification and correction of errors or biases within the models. Furthermore, the lack of interpretability of results may hinder public trust in AI-enabled biotechnology tools and limit their societal acceptance. These training data and algorithmic risks underscore the necessity for rigorous validation and continuous oversight of AI systems to mitigate potential harms and ensure their reliable application.

Deskilling and reduced barriers to entry. An overreliance on any technology can result in a deskilling effect, where practitioners lose the necessary expertise and skills required to manually complete a particular task. As AI tools become more sophisticated, they are beginning to reflect both the domain knowledge and the intuition of expert researchers [90]. As such, experts and non-experts alike may be able to rely exclusively on AI systems for research and experimentation. This potential for reduced expert involvement raises significant concerns regarding the preservation of critical thinking, problem-solving skills, and judgement in scientific research. Without the need to engage deeply with the nuances of a task, researchers may become mere operators of technology, losing the ability to innovate and respond to unexpected challenges.

As outlined earlier, the potential for AI-driven democratization of biotechnology could heighten innovation and provide opportunities for those outside of traditional research organizations to get involved in biotechnology. On the other hand, expanding access to biotechnology and biodesign tools to the general public (who are not necessarily educated in the field) could result in irrational or unintentional misuse. Imagine a scenario where an individual experiences discomfort from mosquito exposure during outdoor activities. The future democratization of gene editing technologies may present scenarios where untrained actors could access and deploy powerful interventions with serious, global consequences. Advanced CRISPR-Cas9 gene drive systems designed for insect genome manipulation could become accessible to the untrained user through AI-based biodesign tools and desktop synthesis devices. Such a system could guide the user through complex biological processes without requiring any understanding of the underlying molecular biology or ecological implications. The resulting modifications could spread exponentially through ecosystems, resulting in potentially severe consequences to global ecology.

Failing to capitalize and opportunity costs. From a strategic perspective, neglecting to leverage AI tools in biotechnology discovery could lead to a competitive disadvantage. In the private sector, innovation-driven companies are positioned to capitalize on advancements, thereby gaining a market edge. For national governments, failure to adopt these technologies could result in delayed discoveries and the forfeiture of economic, national security, and geopolitical advantages [26].

2.3.3 Quadrant of Perspectives

In a very simplified way, we can classify concerns and hopes vis-à-vis the convergence of AI and biotechnology into four perspectives, as borrowed from Thiel and Masters [98]. The stereotypical unspecific pessimist would assume that somehow convergence will be the final step towards the end of human civilization, combining the unaligned machine superintelligence with dangerous enhanced bioengineering capabilities. The unspecific optimist, however, cheers the arrival of unprecedented technical capabilities, where AI unlocks the full potential of biological engineering that will bring the mythological cornucopia. The specific pessimist envisions a dangerous future where people with hostile intentions will use AI to create new forms of bioweapons. The specific optimist, on the other hand, anticipates a future where particular applications of AI in biotechnology will lead to concrete market opportunities and financial gains. More detailed examples of the four personas are presented below.

Unspecific pessimism. The unspecific pessimist approaches AI-biotechnology convergence through a lens of fundamental technological skepticism, grounded in complex systems theory and historical patterns of technological disruption. This perspective views convergence as an inherently destabilizing force that introduces novel systemic risks beyond our governance capacity. Rather than identifying specific failure modes, the unspecific pessimist focuses on emergent properties of the combined technologies—particularly their capacity for autonomous evolution beyond initial design parameters.

Central to this worldview is the concept of technological autonomy, where technologies develop according to internal logics rather than human intentions. Self-improving AI systems paired with biological design capabilities represent a particularly concerning manifestation of this phenomenon, as they could potentially optimize for outcomes orthogonal to human welfare. The unspecific pessimist identifies certain technological characteristics that make convergence particularly concerning: non-human timescales of operation (microseconds for computational processes, generations for biological evolution), autonomous decision-making capabilities, and potential for rapid, self-directed adaptation.

This perspective draws on complexity science concepts of attractor states and irreversibility. Once technological systems cross certain thresholds—such as AI systems capable of designing novel biological entities without human oversight—reversion to previous states becomes impractical or impossible. The mathematics of chaos theory and dynamical systems suggests that minor perturbations in initial conditions can lead to dramatically divergent outcomes in complex systems, making reliable prediction impossible. When such systems operate at the interface of computation and biology, the unspecific pessimist argues that unforeseen consequences become not merely possible but inevitable.

Historical technological transitions inform this perspective, particularly the industrial revolution's profound disruption of social, economic, and ecological systems. The convergence of AI and biotechnology represents a potentially more fundamental transformation operating at the level of living systems themselves. The unspecific pessimist points to historical examples of unforeseen ecological consequences from technological interventions, such as introduced species disrupting native ecosystems or agricultural monocultures creating vulnerability to novel pathogens. AI-biotechnology convergence operates on more fundamental genetic, metabolic, and ecological levels, potentially introducing more profound disruptions.

From a governance standpoint, the unspecific pessimist advocates comprehensive application of precautionary principles, arguing that uncertainty should be interpreted as reason for restraint rather than permission for experimentation. This approach emphasizes strict technology assessment before deployment, moratoriums on high-risk research directions, and international governance frameworks with substantial enforcement mechanisms. The unspecific pessimist would prioritize development of "containment" technologies—methods to limit the spread of engineered biological systems or constrain AI capabilities—over the technologies themselves.

Specific pessimism. The specific pessimist grounds concerns in concrete technological capabilities emerging from AI-biotechnology convergence, particularly regarding novel biosecurity threats. This perspective acknowledges potential benefits while maintaining that specific risk vectors will inevitably emerge and outweigh those benefits. Unlike the unspecific pessimist, this viewpoint identifies particular technological pathways and failure modes rather than expressing general technological skepticism.

A primary concern involves the democratization of biological capabilities through AI-augmented design tools. While traditional bioweapon development required specialized knowledge and infrastructure, AI systems trained on biological data could potentially enable less sophisticated actors to design pathogenic organisms with specific characteristics. For instance, machine learning models trained on chemical structures and bioactivity data can generate novel toxic compounds without being explicitly trained on chemical weapons. Applied to biological design, similar systems could identify viral protein modifications that enhance transmissibility or virulence.

The specific pessimist points to technological developments that lower barriers to biological engineering, such as CRISPR-based gene editing systems, automated DNA synthesis platforms, and machine learning tools for protein design. Developments like OpenCRISPR, which uses large language models to design CRISPR systems with novel functionalities, exemplify how AI can accelerate capability development in biological engineering [14]. Combining these technologies creates potential for rapid design-build-test cycles that could enable malicious actors to develop biological agents resistant to existing countermeasures.

This perspective often emphasizes dual-use research concerns, where legitimate scientific advances simultaneously enable harmful applications. Gain-of-function research demonstrating enhanced transmissibility of influenza serves as a frequently cited example, where scientific knowledge intended to improve pandemic preparedness simultaneously created potential for misuse. AI tools accelerate this dynamic by making it easier to identify potential modifications that enhance pathogen capabilities or generate novel biological functions.

The specific pessimist expresses particular concern about verification challenges in biological arms control. Unlike nuclear technologies with distinctive signatures and specialized infrastructure requirements, biological research operates with increasingly accessible technologies and dual-use materials. AI-enhanced design capabilities could enable development of biological agents through distributed research networks without centralized facilities that might trigger traditional monitoring systems. The reconstruction of horsepox virus using synthetic DNA ordered from commercial providers demonstrated the feasibility of recreating complex viral genomes without specialized infrastructure.

From a governance perspective, the specific pessimist advocates targeted regulatory interventions focused on identified high-risk capabilities. These include enhanced screening of DNA synthesis orders, licensing requirements for advanced biological design software, international monitoring of dual-use research, and development of technical safeguards like genetic barcoding of synthesized DNA. Rather than calling for broad technological restrictions, this perspective seeks to create targeted barriers against misuse while allowing beneficial applications to proceed under appropriate safeguards.

Organizations like the Nuclear Threat Initiative's biological program and certain biosecurity researchers articulate variations of this perspective, focusing on specific technology-enabled threats rather than generalized technological skepticism. They advocate for governance measures tailored to particular risk vectors, often drawing analogies to nuclear nonproliferation regimes while acknowledging the different verification challenges presented by biological technologies.

Unspecific optimism. The unspecific optimist approaches AI-biotechnology convergence through a lens of technological determinism, viewing technological development as an inherently progressive force that naturally produces societal benefits regardless of specific applications. This perspective emphasizes the historical pattern of technological revolutions creating initial disruption followed by broadly distributed benefits. Rather than focusing on particular applications, the unspecific optimist emphasizes emergent capabilities and second-order effects that transform society in ways that initial developers could not have anticipated.

Central to this worldview is the concept of technological solutionism—the belief that technology represents the primary mechanism for addressing societal challenges.

The convergence of AI and biotechnology represents a particularly powerful manifestation of this approach, as it combines the information-processing capabilities of AI with the material transformation capabilities of biotechnology. Together, these technologies could potentially address challenges across multiple domains—healthcare, agriculture, environmental remediation, energy production—through general capability enhancement rather than domain-specific solutions.

This perspective often draws on exponential growth frameworks, particularly regarding computational capabilities and biological information processing. The historical pattern of Moore's Law in computing combined with declining DNA sequencing and synthesis costs suggests accelerating capabilities at the AI-biotechnology interface. When computational systems capable of learning from biological data interface with rapidly advancing biological engineering technologies, the unspecific optimist argues that capabilities will emerge faster than specific applications can be envisioned.

Historical technological transitions inform this perspective, particularly the digital revolution's transformation of information access and communication capabilities. The unspecific optimist points to how technologies initially developed for narrow purposes (like ARPANET for military communication) evolved into broadly transformative systems (the modern internet) with applications far beyond initial design specifications. Similarly, genetic engineering technologies initially developed for research purposes have evolved into sophisticated biotechnology platforms with applications across multiple industries.

From a governance standpoint, the unspecific optimist advocates minimal intervention focused primarily on removing barriers to innovation rather than imposing constraints. This approach emphasizes international technological competition, intellectual property rights protection, and market-based incentives for technology development. The unspecific optimist would prioritize rapid technological progress and deployment, arguing that benefits will naturally outweigh risks as technologies mature and societal adoption increases.

Organizations like various Silicon Valley think tanks and certain technology industry associations often articulate variations of this perspective, emphasizing the transformative potential of emerging technologies while downplaying specific risk concerns. They advocate for governance approaches that maximize innovation potential while addressing harms reactively rather than preemptively restricting technological development based on speculative concerns.

Specific optimism. The specific optimist grounds technological optimism in concrete applications emerging from AI-biotechnology convergence, particularly regarding precision medicine, sustainable biomanufacturing, and targeted environmental interventions. This perspective identifies particular technological pathways and beneficial applications while acknowledging that intentional design, rather than automatic technological determinism, drives positive outcomes.

A primary focus involves healthcare transformation through AI-enhanced precision medicine. Machine learning systems trained on multi-omic patient data (genomic, transcriptomic, proteomic, metabolomic) can identify subtle patterns associated with disease states before clinical symptoms manifest. Combined with biological intervention capabilities like CRISPR-based therapies, these systems enable targeted treatments addressing root causes rather than symptoms. The development of AlphaFold 3, which accurately predicts protein complex structures including interactions with small molecules, exemplifies how AI can accelerate drug discovery by identifying potential therapeutic candidates with unprecedented precision.

The specific optimist points to technological developments enabling sustainable biomanufacturing of chemicals, materials, and food products. Machine learning algorithms optimize metabolic pathways in engineered microorganisms, enabling production of complex molecules from renewable feedstocks rather than petroleum derivatives. Developments like AI-optimized yeast strains producing cannabinoids at industrial scale exemplify how computational design can transform manufacturing processes. Combining these technologies creates potential for circular bioeconomies that reduce environmental impact while meeting human needs.

This perspective often emphasizes specific ecological applications, where AI-guided biological interventions address environmental challenges. Machine learning algorithms analyzing environmental DNA samples can track biodiversity changes and identify early warning signs of ecosystem disturbance. Combined with targeted biological interventions like engineered microorganisms that degrade pollutants or restore soil fertility, these technologies enable precise environmental management. The development of machine learning models that predict coral bleaching vulnerability and identify resilient genotypes demonstrates how computational approaches can guide ecological conservation efforts.

The specific optimist expresses particular enthusiasm about AI-accelerated discovery processes in fundamental biology. Machine learning algorithms analyzing vast datasets can identify patterns invisible to human researchers, generating testable hypotheses about biological mechanisms. The discovery of novel antibiotic candidates through machine learning analysis of molecular structures exemplifies how AI can identify promising intervention points that traditional research methods might overlook.

From a governance perspective, the specific optimist advocates strategic investment and regulatory streamlining focused on high-potential application areas. These include expedited approval pathways for AI-designed therapeutics with strong safety profiles, public–private partnerships for biomanufacturing infrastructure development, and targeted funding for AI-biotechnology applications addressing critical societal challenges. Rather than allowing technologies to develop haphazardly, this perspective seeks to channel innovation toward specific beneficial applications through intentional policy choices.

Organizations like various bioeconomy initiatives and certain academic research consortia articulate variations of this perspective, focusing on specific beneficial applications rather than generalized technological optimism. They advocate for governance measures that accelerate specific high-value applications while maintaining appropriate safety standards, often emphasizing the opportunity costs of regulatory delay in addressing urgent societal challenges like climate change, food security, and disease burden.

Navigating diverse perspectives. Each of these four perspectives provides a unique lens through which to assess the potential risks and opportunities of AI-biotechnology convergence. Policymakers are likely to encounter many of these perspectives in stakeholder groups. However, reconciling these views within a shared governance framework is inherently challenging due to the differences in underlying assumptions of each worldview. In this chapter, we propose addressing this challenge by defining governance principles that can accommodate each perspective, even though the particular interpretations may vary depending on the viewpoint of an individual stakeholder.

The *unspecific pessimist* emphasizes a precautionary approach, preferring stringent oversight or outright bans to prevent catastrophic disasters. The *specific pessimist* instead focuses on specific threats and pushes for concrete regulatory measures to mitigate the particular risks that they perceive as inevitable. On the other hand, the *unspecific optimist* supports a laissez-faire approach, placing trust in humanity and social and economic systems to manage emerging challenges. The *specific optimist* advocates for targeted investment policies in research and innovation to achieve profound technological breakthroughs.

These contrasting viewpoints highlight the difficulty of achieving consensus on governance strategies amongst stakeholder groups. Disagreements will inevitably arise due to the divergence in fundamental views on risk, human nature and capability, and the role of policy. Instead of endorsing or dismissing any one perspective, we advocate for a pluralistic approach that acknowledges the validity of each viewpoint while seeking common ground through shared principles, such as safety, equity, sustainability, and growth. Policymakers should adopt an inquisitive and open-minded approach, recognizing that individual and cultural perspectives may evolve as our knowledge and experience with new technologies grows. By doing so, we can foster a governance approach that remains flexible and inclusive, capable of accommodating a range of worldviews while ensuring that the benefits of convergence are realized in a responsible manner.

2.4 Tipping Points and Cascading Dynamics Applied to Convergence

Until now, we have approached the topic of convergence between artificial intelligence and biotechnology by explaining possible opportunities and threats, particularly highlighting the role of uncertainty. We proposed four perspectives, ranging from pessimistic to optimistic, to provide an overview of the emerging challenges for policymakers and governance. In the next section, we will adopt a systemic approach to consider the critical "crossroads" that could cause rapid escalation crises as convergence develops, and the measures that could be activated to ensure resilience. The study of system risk has significantly evolved over the past decade, a trend expected to accelerate as new technologies become associated with emerging vulnerabilities [99]. The increasing complexity of reality, characterized by the interaction of various factors within systems, heightens the unpredictability of their evolutionary trajectories and how risk is understood or addressed [72]. More recently, convergence can be associated with the characteristics of high-impact, low-probability (HILP) events, particularly due to the high uncertainty in their development [100]. These events can be challenging because comparable precursors could be distant in time or occurred before significant socio-technological breakthroughs, which could compromise the capacity to respond, or represent a new combination of previously known risks and vulnerabilities into a "perfect storm." In other words, when manifested, HILPs impact an operational reality that could be significantly different from the last known precursor, in terms of both exposure and vulnerability. As often happens, the era of AI is compared to the industrial revolution in terms of potential impact. But how much is similar and how much is different? What is predictable based on experience and what is not?

This section explores how the convergence of AI and biotechnology can trigger shifts within complex systems, beginning with an examination of "tipping points"—critical thresholds beyond which change becomes rapid or irreversible. Next, we discuss cascading dynamics that may unfold when such tipping points are crossed, highlighting system interdependence and risk amplification across domains. Finally, we introduce the metaphor of "A Fence or an Ambulance" to frame how decision-makers can balance proactive preventive measures and reactive response in managing complex, systemic risks.

2.4.1 The Concept of "Tipping Points" Applied to Convergence

The concept of panarchies provides a useful framework for understanding the dynamic interactions and evolution within complex systems, including the convergence of biotechnology and AI. Originally developed to describe ecological systems, it explains how systems evolve through adaptive cycles consisting of phases: growth (exploitation), conservation, release, and reorganization [101]. These cycles reflect the accumulation and transformation of resources, followed by periods of crisis or "creative destruction," leading to significant reorganization. The approach has limitations in both theory and application [102], but it offers some insights that can be leveraged to understand how AI and biotechnology convergence might progress through similar adaptive cycles. For example, the rapid growth phase could be fueled by breakthroughs in AI algorithms or biotechnological innovations. However, this growth may lead to a conservation phase where the system becomes rigid and more vulnerable to disturbances.

As the system continues to evolve in the cycles of accumulation and releases, certain "tipping points" may emerge, representing critical thresholds beyond which irreversible changes occur. According to [103], this concept introduces new challenges in how we think about the future, suggesting that we are not merely moving toward the future, but that it is rapidly advancing toward us. While "tipping points" often imply a point of no return—though sometimes they refer to the destabilization of a system that could eventually recover—the future remains inherently unpredictable. To consider a world beyond the tipping point is to face an uncertain reality that lies beyond our collective experience.

In the context of AI-biotechnology convergence, tipping points could manifest as sudden, widespread failures in AI-driven healthcare systems due to flawed algorithms or unintended consequences of gene editing technologies, potentially leading to ecological or public health crises. Lu et al. discusses existing tipping points in healthcare, such as the "care debt" from deferred surgical treatments and worsening medical conditions due to cancelled elective procedures [104]. Convergence between AI and biotechnology could influence this tipping point. On one hand, it could optimize care delivery, transforming healthcare systems to be more resilient and capable of managing cascading effects in the face of future disruptions. On the other hand, it could introduce significant challenges, such as data-sharing issues. If a tipping point is reached—where an unforeseen interaction between AI and biotechnology triggers a systemic failure—the system could enter a release phase, exposing vulnerabilities and unfolding a crisis. As Kinzig et al. suggest, once these thresholds are crossed, cascading effects generate consequent regime shifts across scales and domains, where small perturbations can lead to disproportionate impacts [105]. Holling emphasized that these tipping points and the subsequent cascading effects are not isolated events but are deeply interconnected across different scales—temporal, spatial, and organizational [101]. For instance, a failure in AI-driven diagnostic tools could lead to widespread misdiagnoses, overwhelming healthcare facilities, and creating broader public health emergencies. These cascading effects underscore the importance of understanding systemic interactions and potential points of failure within the AI-biotechnology convergence.

There might also be positive tipping points, for example, by discovering an inexpensive and highly effective treatment against cancer or even aging which underlies a lot of different diseases [106, 107]. Such synthetic biology-enabled tipping points could not only be reached in public health but also in the field of environmental remediation, e.g. halting ecosystem destruction or mitigating climate change via large scale bioremediation [94].

2.4.2 Cascading Dynamics in AI and Biotechnology Convergence

Existing theories on system dynamics highlight the need to consider the further evolution of complex networks and operational fields across multiple aspects of crisis management, organizational resilience, and adaptation [99]. Once a tipping point is crossed, addressing a crisis as it unfolds and progresses becomes increasingly difficult, as the crisis accelerates along its own pathways.

As AI and biotechnology continue to converge, these cascading dynamics are likely to intensify, particularly if the systems involved lack sufficient resilience. The integration of AI into biotechnological processes, such as synthetic biology or personalized medicine, presents both opportunities and risks. For example, an AI-driven error in genomic editing could have far-reaching consequences, potentially triggering a cascade of failures that impact health, the environment, and society at large. The COVID-19 pandemic highlighted the need for robust systemic risk monitoring and management in global health infrastructures, a challenge exacerbated by the introduction of AI into medical and diagnostic systems. Research on global systemic risk and resilience for novel coronaviruses [108] and its post-pandemic implications [109] underscores how interconnected networks of AI-driven decision-making systems can either enhance or destabilize healthcare resilience. AI-driven bio surveillance and diagnostic systems must be designed to withstand disruptions, ensuring that their integration into biotechnology does not lead to cascading failures in pandemic response and healthcare infrastructure. The interdependencies between technological systems and societal functions underscore the urgency of better defining operational thresholds and addressing the uncertainties involved.

Traditional risk assessment approaches, such as risk registers, may be inadequate for addressing these complex dynamics. Current policies and planning efforts often focus narrowly on specific hazards, driven by threat-oriented scenarios [110]. Such approaches, based on historical or perceived data, fail to consider the broader, interconnected nature of risks, which can be particularly problematic in situations of heightened uncertainty. Cascading dynamics at the intersection of biotechnology and AI are especially concerning given the speed and scale at which AI can operate, reducing the predictability of potential recombination. For example, a possible tipping point could arise from the application of convergence between biotechnology and AI to the military field, and more at large to new forms of warfare [111]. Once a first step is triggered, any control measure will become harder and the potential application could increase too fast to keep track of them, or having a clear outline of their implications. Other examples may include (1) the commercialization of certain tools or techniques derived by convergence to private individuals that are not subject to any compliance or for which the forms of controls and oversight may be limited, and (2) wide accessibility of bioengineered systems to the public where users do not have the requisite background knowledge to use these tools responsibly.

The tight coupling of AI-driven systems with biotechnological applications means that any disruption generated by sudden variations or unexpected scenarios can spread rapidly, potentially outpacing human oversight and escalating into a major crisis, affecting not only the immediate system but also interconnected global networks. Although these changes can be positive, following Perrow's theory of normal accidents, it must be considered that the complexity and interconnections within technological systems render certain failures inevitable [112].

2.4.3 A Fence or an Ambulance

Government leaders must navigate these complexities with a balanced approach, emphasizing both preventive measures to avert potential tipping points and reactive measures to address incidents when they occur. This balanced approach can help ensure that AI and biotechnology do not suffer from reactionary fear-based oversight, as typified by GMO scare seen in the EU, which led to reduced crop yields and hindered agricultural innovation [113–115].

Joseph Malins' 1895 poem "A Fence or an Ambulance" provides a compelling metaphor for this dual approach. In the poem, a community debates whether to build a fence at the top of a cliff to prevent accidents or to place an ambulance at the bottom to treat those who fall. This metaphor aptly illustrates the necessity of proactive planning and preparedness (the fence) alongside robust response mechanisms (the ambulance) [116]. The below sections present philosophies for managing the convergence of AI and biotechnology through this lens, highlighting both preventive and reactive principles.

The Fence: Preventive Measures. In the context of AI and biotechnology convergence, "the fence" symbolizes preventive measures designed to mitigate risks before they materialize. Preventive measures are crucial for ensuring the ethical and safe development of these technologies, minimizing the likelihood of adverse outcomes.

One of the foremost preventive measures is the establishment of robust ethical standards and regulatory frameworks. International ethical standards should guide the development and application of AI and biotechnology, ensuring that these technologies are used responsibly and for the benefit of humanity. Regulatory frameworks must be adaptive and able to keep pace with rapid technological advancements while safeguarding against potential harms [117]. Specific measures include adopting the National Institute of Standards and Technology (NIST) AI Risk Management Framework (RMF) to manage risk and increase trustworthiness of AI systems, as well as adopting the Organization for Economic Co-operation and Development (OECD) AI Principles, the United Nations Educational, Scientific, and Cultural Organization (UNESCO) Recommendations on AI Ethics, and the Universal Guidelines for AI (UGAI), to ensure transparency, accountability, and protection of human rights [118–120].

Preventive oversight of research and development (R&D) activities is essential to identify and address potential risks early in the innovation process. Establishing oversight bodies that monitor R&D projects can help ensure that safety protocols are adhered to and that ethical considerations are integrated into every stage of development. These oversight bodies should follow existing laws or policies where already applicable, such as sector-specific privacy laws and oversight requirements. Where law or policy is lacking, they should follow the Blueprint for an AI Bill of Rights to inform policy decisions [120]. Funding should be directed towards research that prioritizes safety and addresses societal needs, promoting responsible innovation [121].

Education and training programs are vital for equipping stakeholders with the knowledge and skills needed to responsibly develop and utilize AI and biotechnology. This includes training for researchers, policymakers, and the general public on the risks and benefits associated with these technologies and promoting the inclusion of ethical AI practices and human rights considerations in curricula and professional development programs. By fostering a culture of awareness and responsibility, governments and agencies can reduce societal fear, prevent misuse of these technologies, and ensure that the convergence of AI and biotechnology is guided by informed and ethical decision-making.

Finally, introducing a dedicated ethics and society review (ESR) process can systematically evaluate the ethical and societal impacts of AI and biotechnology research and provide structured oversight. This process should be a requirement for funding and project approval, involving iterative feedback and collaboration with researchers to address ethical challenges and propose mitigation strategies [122].

The Ambulance: Reactive Measures. While preventive measures are crucial for averting potential risks, it is equally important to have robust reactive measures in place to address incidents when they occur. The metaphor of "the ambulance" in Malins' poem represents these reactive strategies, emphasizing the need for preparedness and rapid response to mitigate the impact of unforeseen events [116].

A comprehensive crisis management and response framework is essential for dealing with emergencies arising from AI and biotechnology convergence. This includes establishing rapid response teams equipped to handle biosecurity threats, AI malfunctions, and other critical incidents. These teams should be trained in crisis management, possess the necessary technical expertise, and have access to advanced tools and resources. Developing and regularly updating response protocols for a variety of scenarios ensures that agencies are prepared to act swiftly and effectively in the event of a crisis [123].

Effective incident reporting and analysis systems are fundamental to understanding and learning from failures. Creating a centralized database for reporting incidents related to AI and biotechnology allows for the collection and analysis of data, helping to identify patterns and underlying causes. This data-driven approach enables refining of preventive measures and improvement of reactive strategies. Transparency in reporting and a culture of continuous improvement are vital for building trust and enhancing resilience [122]

In the aftermath of an incident, remediation and recovery programs are essential to mitigate the damage and restore normalcy. These programs should focus on addressing the immediate impacts, such as containing biohazards, restoring system functionalities, and providing support to affected individuals and communities. Long-term recovery plans should also be in place to manage the consequences of incidents, such as environmental cleanup, health monitoring, and infrastructure rebuilding. Ensuring adequate resources and support for these programs is crucial for effective recovery [121].

Integrating Preventive and Reactive Measures. A balanced approach that integrates both preventive and reactive measures is necessary for managing the convergence of AI and biotechnology effectively. Preventive measures reduce the likelihood of incidents, while reactive measures ensure preparedness for addressing them when they occur. By combining these strategies, policymakers can create a comprehensive framework that enhances resilience and promotes the safe and beneficial use of these technologies.

Consider the case of dual-use research of concern (DURC). DURC refers to research that, while intended for beneficial purposes, can also be misused to pose significant threats to public health, safety, or security. The convergence of AI and biotechnology presents several DURC scenarios, such as the development of synthetic pathogens or the enhancement of human capabilities through genetic modification. In managing DURC, both preventive and reactive measures are essential. Preventive measures include establishing strict ethical guidelines, conducting thorough risk assessments, and implementing robust oversight mechanisms. Reactive measures involve creating rapid response protocols for incidents, ensuring effective communication channels for reporting DURC-related issues, and developing remediation plans to address potential consequences [122].

A further principle, continuous risk assessment is vital for adapting to the dynamic nature of convergence. Predicting outcomes in these complex systems is challenging, so it is important to prepare for surprises that will inevitably arise due to the many possibilities created by convergence. This preparation involves understanding potential scenarios, planning responses, and determining appropriate actions—whether through regulation, observation, support, or restriction. Ongoing evaluation of risks and benefits can keep policies up to date with emerging threats and opportunities. This involves regular monitoring of technological advancements, conducting scenario analyses, and updating risk management strategies accordingly, with the NIST AI RMF being adapted for this continuous risk assessment [118]. Engaging with experts from various fields and fostering a collaborative approach to risk assessment ensures a comprehensive understanding of the evolving landscape [123].

2.5 Governance Principles

Ultimately, the impacts of AI-biotechnology convergence will hinge on the implementation of thoughtful regulations, robust safety measures instituted at the earliest stages of research and development [124], and responsible research and development practices. Effective governance should establish appropriate safeguards to mitigate potential catastrophic and cascading risks. Rather than constructing rigid frameworks that may stifle innovation, adaptable policies can evolve alongside technological advancements. This adaptive approach can ensure that regulations remain relevant and effective in the face of rapid change. Just as automobiles are equipped with brakes not to limit speed, but to enable safe acceleration, well-designed governance mechanisms can facilitate swift and secure progress in AI-biotechnology convergence. We propose the below 10 principles for adaptive and effective governance of this convergence landscape:

1. Establish and monitor tipping points, metrics, and lead indicators for critical transitions: Governing the convergence of AI and biotechnology requires careful monitoring of critical transitions that could significantly impact ecosystems, societies, or economies. Establishing quantifiable metrics and leading indicators enables early risk detection and informed intervention strategies to mitigate risks and equally facilitate development of convergent AI and biotechnology necessary to unlock new capabilities and benefits for society [125]. Given the unpredictable and complex nature of AI-synthetic biology technologies, ongoing assessment is essential for adaptive governance. This ongoing monitoring is crucial for developing resilient systems that can withstand disruptions.
2. Identify decision-makers and authority for response: Effective governance requires precise delineation of decision-making authority and responsibility allocation across relevant stakeholders. A robust governance framework must specify jurisdictional boundaries, escalation pathways, and accountability mechanisms, particularly for novel scenarios emerging from AI-biotechnology convergence. This ensures accountability, facilitates timely responses, and reduces confusion during critical moments of decision-making.
3. Allocate sufficient resources to support research, development, and regulation: The convergence of AI and biotechnology requires adequate resource allocation for continued research, development, and regulatory oversight. Policymakers must ensure that funding and other resources are directed toward not only scientific advancement but also the creation of effective regulatory mechanisms. This will allow for innovation while safeguarding societal interests and preventing harmful applications.
4. Adopt evidence-based, proportionate, and adaptable policy outcomes and decision-making frameworks: Governance in rapidly evolving fields like AI and biotechnology should rely on evidence-based approaches, ensuring that policies are grounded in the latest scientific knowledge. While policymakers and experts should use latest evidence to try to predict probable outcomes, adaptable solutions that can handle inevitable surprises are preferable. Furthermore, policies must be proportionate to the risks and adaptable as new information emerges. This approach avoids overregulation that could stifle innovation while ensuring that governance measures remain relevant in the face of ongoing technological advancements.

5. Leverage AI to address challenges and opportunities presented by AI-biotechnology convergence: AI itself offers tools for addressing the complexities associated with its convergence with biotechnology. Policymakers should harness the capabilities of AI for monitoring, regulating, and understanding the impacts of AI in biotechnology. For instance, AI can be used for predictive modeling of ecological or societal impacts, helping stakeholders make more informed decisions. Further, one can imagine a future where AI models can "repair" the structure of their own learning algorithms from human-introduced error, similar to the self-repair mechanisms of DNA. The self-referential use of AI can be a key component in managing the risks and maximizing the benefits of the convergence.
6. Invest in comprehensive, interdisciplinary education and training programs: A successful governance framework for the AI-biotechnology convergence requires a workforce that is well-versed in both fields. Investment in interdisciplinary education and training programs is essential to equip researchers, regulators, and policymakers with the necessary skills to navigate the challenges posed by these emerging technologies. Such programs should integrate knowledge from biology, computer science, ethics, law, and policy, ensuring a holistic understanding of the complexities involved.
7. Foster open collaboration, dialogue, and knowledge-sharing among stakeholders: Collaboration across multiple sectors—public, private, academic, and civil society—is crucial for addressing the challenges of this convergence. Open dialogue and knowledge-sharing among stakeholders promote a comprehensive understanding of risks, benefits, and societal impacts. Transparency and inclusiveness in discussions will lead to more balanced decision-making processes and help build public trust in the governance framework. Establishing formal agreements and frameworks for collaboration ensures that information flows smoothly and that joint initiatives are undertaken to enhance global security [117]. Further, engaging in bilateral and multilateral international partnerships can help address cross-border biosecurity threats and ethical dilemmas, creating a more cohesive and resilient international community.
8. Promote public engagement, awareness, and participatory decision-making processes: Engaging the public in discussions about AI and biotechnology fosters accountability and ensures that governance frameworks reflect societal values. Public awareness campaigns, coupled with participatory decision-making processes, allow for more inclusive policy development, ensuring that diverse perspectives are considered. Given the profound implications of these technologies, public trust and involvement are critical for their responsible governance [126].
9. Develop and promote harmonized ethical frameworks, guidelines, and best practices: Ethical considerations are paramount in the convergence of AI and biotechnology, where issues such as data privacy, biosecurity, and human enhancement may arise. Harmonized ethical frameworks and guidelines are necessary to ensure that global efforts are aligned in addressing these concerns.

Best practices must be established and promoted to guide researchers, developers, and regulators in ethically responsible decision-making. Ensuring that technologies respect human rights and privacy is paramount. This can include implementing robust data protection measures, conducting ethical reviews of research projects, and developing policies that prioritize the welfare of individuals and communities. These ethical considerations include protection of personal data and public health but should also extend to the fair distribution of benefits derived from AI and biotechnology convergence. This includes advocating for access to new treatments, technologies, and information for all populations, including those most vulnerable [117].

10. Invest in long-term, multidisciplinary, and international research and foresight initiatives: Given the transformative potential of AI and biotechnology, it is essential to invest in long-term, multidisciplinary research and foresight initiatives. These initiatives will enable policymakers to anticipate future developments and challenges, preparing for potential scenarios before they materialize. International collaboration is key, as the global nature of these technologies and their potential effects requires coordinated efforts to address shared risks and opportunities. Agencies such as NATO can lead these collaborative efforts by harmonizing efforts across countries, promoting best practices, ethical standards, and regulatory frameworks [119, 120].

2.6 Discussion

The convergence of artificial intelligence and biotechnology creates a governance challenge fundamentally different from those posed by either technology independently. This challenge stems not merely from technical complexity but from deep uncertainty about both potential benefits and risks, combined with divergent stakeholder interpretations of that uncertainty. Throughout this chapter, we have examined how this convergence manifests across multiple domains—from protein structure prediction to automated biological design to distributed biological intelligence—and how these manifestations generate both unprecedented capabilities and novel concerns.

Four distinct interpretive frameworks dominate stakeholder responses to AI-biotechnology convergence, characterized along the axes of specificity and outlook (pessimistic versus optimistic) (Sect. 2.3.3). These divergent perspectives do not simply reflect different access to technical information—they represent fundamentally different orientations toward technological uncertainty itself. When the specific outcomes of technological development remain indeterminate, stakeholders filter available information through distinct cognitive frameworks that emphasize either opportunity space or threat potential. These frameworks derive from deeper values concerning risk tolerance, institutional trust, and appropriate distribution of decision-making authority.

Traditional governance approaches falter when confronted with this combination of technical uncertainty and interpretive divergence. Linear risk assessment frameworks that proceed from hazard identification to risk characterization assume agreement on what constitutes a hazard and how severity should be evaluated. Conventional cost–benefit analysis presupposes quantifiable outcomes that can be compared on common metrics. Both approaches break down when dealing with deep uncertainty and contested values.

More productive governance approaches acknowledge these fundamental challenges rather than attempting to resolve them through technical analysis alone. Robust decision-making identifies courses of action that perform acceptably across multiple future scenarios and diverse value frameworks rather than optimally under narrow assumptions. This approach emphasizes identifying areas of overlapping consensus among stakeholders with different interpretive frameworks—interventions that unspecific pessimists view as essential safeguards, specific pessimists recognize as addressing concrete threats, unspecific optimists can accept as minimally restrictive, and specific optimists value for enabling beneficial applications.

1. Such overlapping consensus emerges more readily through structured deliberative processes that explicitly acknowledge interpretive differences rather than attempting to flatten them into unified risk assessments. When stakeholders understand that their divergent responses to uncertainty stem from different but equally legitimate frames rather than factual disagreements, more productive dialogue becomes possible. This deliberative approach requires: Separating empirical claims from normative judgments about appropriate responses to uncertainty
2. Distinguishing between specific technological applications and general technological trajectories
3. Making explicit the values that inform different prioritizations of opportunity versus risk
4. Identifying potential actions that accommodate multiple interpretive frameworks

Practical implementation of this approach involves creating multi-stakeholder forums explicitly designed to incorporate diverse perspectives rather than achieving technical consensus. These forums must include voices from all four interpretive quadrants, with facilitators trained to identify when disagreements stem from different factual claims versus different normative judgments about uncertainty. Rather than seeking unanimous agreement, these processes should identify tiered intervention options that respect multiple interpretive frames.

Specific governance mechanisms flowing from this approach might include:

1. Differentiated oversight regimes for distinct technology applications rather than blanket policies for "AI" or "biotechnology"
2. Targeted safety research addressing specific risk vectors identified by both pessimistic and optimistic stakeholders
3. Horizon scanning systems incorporating diverse interpretive perspectives to identify emerging concerns

4. Periodic reassessment triggers built into regulatory frameworks to accommodate evolving understanding
5. Capability-based governance that focuses on functional impacts rather than specific technological implementations

Ultimately, governing AI-biotechnology convergence requires abandoning the notion that technical analysis alone can resolve fundamentally normative questions about appropriate responses to uncertainty. The challenges stem not from insufficient information but from legitimate differences in how stakeholders interpret ambiguity, value potential benefits versus risks, and assign decision-making authority. By acknowledging these differences explicitly and designing governance processes that accommodate multiple interpretive frames, we can develop more resilient approaches to managing technological uncertainty that neither stifle beneficial innovation nor ignore legitimate concerns.

Acknowledgements The authors wish to thank the NATO Science for Peace and Security Programme, the U.S. Army Corps of Engineers, Universidad de los Andes, the University of Virginia, the Commonwealth Center for Advanced Logistics Systems, the Office of Naval Research, and the Air Force Research Laboratory for their funding support. G.P. was supported by HE AGILE (AGnostic risk management for high Impact Low probability Events), funded by HORIZON EUROPE Research and Innovation Action (Project number: 101121356; Call: HORIZON-CL3-2022-DRS-01; Topic: HORIZON-CL3-2022-DRS-01-02), ESA EO4MULTIHAZARDS (Earth Observation for High-Impact Multi-Hazards Science), funded by the European Space Agency and launched as part of the joint ESA-European Commission Earth System Science Initiative, and CIRCLE (Community and Infrastructure Resilience to Climate-geological Long-term Effects), supported by New Frontiers in Research Fund—International (NFRFI-2023-00548), UKDRI ES/Z000122). M.S. was supported by funding from the European Union for "Bioindustry 4.0" (101094287) and deCYPher (101081782).

References

1. McCarthy J, Minsky ML, Rochester N, Shannon CE (2006) A proposal for the Dartmouth summer research project on artificial intelligence, August 31, 1955. AI Mag 27(4):12
2. Haenlein M, Kaplan A (2019) A brief history of artificial intelligence: on the past, present, and future of artificial intelligence. Calif Manag Rev 61(4):5–14
3. Holzinger A, Keiblinger K, Holub P, Zatloukal K, Müller H (2023) AI for life: trends in artificial intelligence for biotechnology. New Biotechnol 74:16–24
4. Cao Y, Li S, Liu Y, Yan Z, Dai Y, Yu PS, Sun L (2023) A comprehensive survey of AI-generated content (AIGC): a history of generative AI from GAN to ChatGPT. Cornell University
5. Nah FF, Zheng R, Cai J, Siau K, Chen L (2023) Generative AI and ChatGPT: applications, challenges, and AI-human collaboration. J Inf Technol Case Appl Res 25(3):277–304
6. Kendrew JC, Bodo G, Dintzis HM, Parrish RG, Wyckoff H, Phillips DC (1958) A three-dimensional model of the myoglobin molecule obtained by X-ray analysis. Nature 181(4610):662–666
7. Kendrew JC, Dickerson RE, Strandberg BE, Hart RG, Davies DR, Phillips DC, Shore VC (1960) Structure of myoglobin: a three-dimensional Fourier synthesis at 2 Å resolution. Nature 185(4711):422–427

8. Strasser BJ (2009) Collecting, comparing, and computing sequences: the making of Margaret O. Dayhoff's Atlas of protein sequence and structure, 1954–1965. J Hist Biol 43(4):623–660
9. Maxam AM, Gilbert W (1977) A new method for sequencing DNA. Proc Natl Acad Sci 74(2):560–564
10. Felsenstein J (1981) Evolutionary trees from DNA sequences: a maximum likelihood approach. J Mol Evol 17(6):368–376
11. Venter JC, Adams MD, Myers EW, Li PW, Mural RJ, Sutton GG, Smith HO, Yandell M, Evans CA, Holt RA et al (2001) The sequence of the human genome. Science 291(5507):1304–1351
12. Abramson J, Adler J, Dunger J, Evans R, Green T, Pritzel A, Ronneberger O, Willmore L, Ballard AJ, Bambrick J et al (2024) Accurate structure prediction of biomolecular interactions with AlphaFold 3. Nature 630(8016):493–500
13. Bhat AA, Nisar S, Mukherjee S, Saha N, Yarravarapu N, Lone SN, Masoodi T, Chauhan R, Maacha S, Bagga P et al (2022) Integration of CRISPR/Cas9 with artificial intelligence for improved cancer therapeutics. J Transl Med 20(1)
14. Ruffolo JA, Nayfach S, Gallagher J, Bhatnagar A, Beazer J, Hussain R, Russ J, Yip J, Hill E, Pacesa M et al (2024) Design of highly functional genome editors by modeling the universe of CRISPR-Cas sequences. Cold Spring Harbor Laboratory
15. Häse F, Roch LM, Aspuru-Guzik A (2019) Next-generation experimentation with self-driving laboratories. Trends Chem 1(3):282–291
16. Martin HG, Radivojevic T, Zucker J, Bouchard K, Sustarich J, Peisert S, Arnold D, Hillson N, Babnigg G, Marti JM et al (2023) Perspectives for self-driving labs in synthetic biology. Curr Opin Biotechnol 79:102881
17. Ligo AK, Rand K, Bassett J, Galaitsi SE, Trump BD, Jayabalasingham B et al (2021) Comparing the emergence of technical and social sciences research in artificial intelligence. Front Comput Sci 3:653235
18. Johnson J, Lucas, McNerney, Olson, Lipinski, Weber (2020) National artificial intelligence initiative act of 2020. In: HR 6216
19. Donkor SA, Walsh ME, Titus AJ (2024) Computing in the life sciences: from early algorithms to modern AI
20. Helm JM, Swiergosz AM, Haeberle HS, Karnuta JM, Schaffer JL, Krebs VE, Spitzer AI, Ramkumar PN (2020) Machine learning and artificial intelligence: definitions, applications, and future directions. Curr Rev Musculoskelet Med 13(1):69–76
21. Lee BD, Gitter A, Greene CS, Raschka S, Maguire F, Titus AJ, Kessler MD, Lee AJ, Chevrette MG, Stewart PA et al (2022) Ten quick tips for deep learning in biology. PLoS Comput Biol 18(3):e1009803
22. Rives A, Meier J, Sercu T, Goyal S, Lin Z, Liu J et al (2021) Biological structure and function emerge from scaling unsupervised learning to 250 million protein sequences. Proc Natl Acad Sci 118(15):e2016239118
23. Wang L, Ma C, Feng X, Zhang Z, Yang H, Zhang J, Chen Z, Tang J, Chen X, Lin Y et al (2024) A survey on large language model based autonomous agents. Front Comput Sci 18(6)
24. Thirunavukarasu AJ, Ting DSJ, Elangovan K, Gutierrez L, Tan TF, Ting DSW (2023) Large language models in medicine. Nat Med 29(8):1930–1940
25. Baek M, DiMaio F, Anishchenko I, Dauparas J, Ovchinnikov S, Lee GR, Wang J, Cong Q, Kinch LN, Schaeffer RD et al (2021) Accurate prediction of protein structures and interactions using a three-track neural network. Science 373(6557):871–876
26. NSCEB White Paper Series on AIxBio (2024) White paper 1: introduction to AI and Biotech
27. Rather IH, Kumar S, Gandomi AH (2024) Breaking the data barrier: a review of deep learning techniques for democratizing AI with small datasets. Artif Intell Rev 57(9)
28. Song Y, Wang T, Cai P, Mondal SK, Sahoo JP (2023) A comprehensive survey of few-shot learning: evolution, applications, challenges, and opportunities. ACM Comput Surv 55(13s):1–40
29. Rai A (2019) Explainable AI: from black box to glass box. J Acad Mark Sci 48(1):137–141
30. National Science Foundation (2019) NSF's 10 BIG IDEAS—SPECIAL REPORT. U.S. National Science Foundation

31. Sharp PA (2013) The third revolution: convergence of life, physical, and engineering sciences
32. Holland JH (1992) Genetic algorithms. Sci Am 267(1):66–73
33. Forrest TL, Ware RE, Howard T, Jaffee BD, Denning SM (1994) Novel mechanisms of brequinar sodium immunosuppression on T cell activation. Transplantation 58(8):920–926
34. Bonabeau E, Theraulaz G, Deneubourg JL, Aron S, Camazine S (1997) Self-organization in social insects. Trends Ecol Evol 12(5):188–193
35. Păun G (2000) Computing with membranes. J Comput Syst Sci 61(1):108–143
36. Adleman LM (1994) Molecular computation of solutions to combinatorial problems. Science 266(5187):1021–1024
37. Jumper J, Evans R, Pritzel A, Green T, Figurnov M, Ronneberger O et al (2021) Highly accurate protein structure prediction with AlphaFold. Nature 596(7873):583–589
38. Madani A, Krause B, Greene ER, Subramanian S, Mohr BP, Holton JM et al (2023) Large language models generate functional protein sequences across diverse families. Nat Biotechnol 41(8):1099–1106
39. Lee DD, Seung HS (1999) Learning the parts of objects by non-negative matrix factorization. Nature 401(6755):788–791
40. Afek Y, Alon N, Barad O, Hornstein E, Barkai N, Bar-Joseph Z (2011) A biological solution to a fundamental distributed computing problem. Science 331(6014):183–185
41. Afek Y, Alon N, Bar-Joseph Z, Cornejo A, Haeupler B, Kuhn F (2012) Beeping a maximal independent set. Distrib Comput 26(4):195–208
42. Strelzoff A, Trump BD, Cummings CL, Smith M, Galaitsi SE, Stoddard K et al (2024) Human intuition and algorithmic efficiency must be balanced to enhance data mesh resilience. Commun ACM 67(5):48–51
43. Dhar MS et al (2021) Genomic characterization and epidemiology of an emerging SARS-CoV-2 variant in Delhi, India. Science 374(6570):995–999
44. Schroeder WL, Kuil T, van Maris AJ, Olson DG, Lynd LR, Maranas CD (2023) A detailed genome-scale metabolic model of Clostridium thermocellum investigates sources of pyrophosphate for driving glycolysis. Metab Eng 77:306–322
45. Theodoris CV, Xiao L, Chopra A, Chaffin MD, Al Sayed ZR, Hill MC et al (2023) Transfer learning enables predictions in network biology. Nature 618(7965):616–624
46. Church GM, Gao Y, Kosuri S (2012) Next-generation digital information storage in DNA. Science 337(6102):1628–1628
47. Nielsen AA, Der BS, Shin J, Vaidyanathan P, Paralanov V, Strychalski EA et al (2016) Genetic circuit design automation. Science 352(6281):aac7341
48. Vassanelli S, Mahmud M (2016) Trends and challenges in neuroengineering: toward "intelligent" neuroprostheses through brain-"brain inspired systems" communication. Front Neurosci 10:438
49. Nakagaki T, Yamada H, Tóth Á (2000) Maze-solving by an amoeboid organism. Nature 407(6803):470–470
50. Wolfram S (2018) Cellular automata and complexity: collected papers. CRC Press
51. Smirnova L, Caffo BS, Gracias DH, Huang Q, Morales Pantoja IE, Tang B et al (2023) Organoid intelligence (OI): the new frontier in biocomputing and intelligence-in-a-dish. Front Sci 1:1017235
52. Gilbert C, Ellis T (2018) Biological engineered living materials: growing functional materials with genetically programmable properties. ACS Synth Biol 8(1):1–15
53. Karr JR, Sanghvi JC, Macklin DN, Gutschow MV, Jacobs JM, Bolival B et al (2012) A whole-cell computational model predicts phenotype from genotype. Cell 150(2):389–401
54. Davies JA, Cachat E (2016) Synthetic biology meets tissue engineering. Biochem Soc Trans 44(3):696–701
55. Schuman CD, Kulkarni SR, Parsa M, Mitchell JP, Date P, Kay B (2022) Opportunities for neuromorphic computing algorithms and applications. Nat Comput Sci 2(1):10–19
56. Ghosh D, Sharman R, Rao HR, Upadhyaya S (2007) Self-healing systems—survey and synthesis. Decis Support Syst 42(4):2164–2185

57. Doursat R (2013) Bridging the mind-brain gap by morphogenetic "neuron flocking": the dynamic self-organization of neural activity into mental shapes. In: AAAI fall symposia
58. Bi Q, Goodman KE, Kaminsky J, Lessler J (2019) What is machine learning? A primer for the epidemiologist. Am J Epidemiol 188(12):2222–2239
59. Aven T, Ben-Haim Y, Andersen HB, Cox T, López Droguett E, Greenberg M et al (2018) Society for risk analysis glossary
60. Kaplan S, Garrick BJ (1981) On the quantitative definition of risk. Risk Anal 1(1):11–27
61. Lowrance WW (1976) Of acceptable risk: science and the determination of safety
62. Galaitsi SE, Keisler JM, Trump BD, Linkov I (2021) The need to reconcile concepts that characterize systems facing threats. Risk Anal 41(1):3–15
63. Whipple CG (1986) Dealing with uncertainty about risk in risk management. In: Hazards: technology and fairness. National Academies Press (US)
64. Knight FH (1921) Risk, uncertainty and profit, vol 31. Houghton Mifflin
65. Walker WE, Harremoës P, Rotmans J, Van Der Sluijs JP, Van Asselt MB, Janssen P, Krayer von Krauss MP (2003) Defining uncertainty: a conceptual basis for uncertainty management in model-based decision support. Integr Assess 4(1):5–17
66. Walker WE, Lempert RJ, Kwakkel JH (2012) Deep uncertainty. Delft Univ Technol 1(2):1
67. Renn O, Klinke A, Van Asselt M (2011) Coping with complexity, uncertainty and ambiguity in risk governance: a synthesis. Ambio 40(2):231–246
68. Morgan MG, Henrion M, Small M (1990) Uncertainty: a guide to dealing with uncertainty in quantitative risk and policy analysis. Cambridge University Press
69. Trump BD, Cummings CL, Kuzma J, Linkov I (eds) (2020) Synthetic biology 2020: frontiers in risk analysis and governance. Springer, Cham
70. Stirling A, Gee D (2002) Science, precaution, and practice. Public Health Rep 117(6):521
71. Pescaroli G, Alexander D (2016) Critical infrastructure, panarchies and the vulnerability paths of cascading disasters. Nat Hazards 82:175–192
72. Helbing D (2013) Globally networked risks and how to respond. Nature 497(7447):51–59
73. Ferson S (2014) Model uncertainty in risk analysis. Université de Technologie de Compiègne
74. Douglas M, Wildavsky A (1982) How can we know the risks we face? Why risk selection is a social process. Risk Anal 2(2):49–58
75. Kahan DM, Braman D, Slovic P, Gastil J, Cohen G (2009) Cultural cognition of the risks and benefits of nanotechnology. Nat Nanotechnol 4(2):87–90
76. Slovic P (1987) Perception of risk. Science 236(4799):280–285
77. Kasperson RE, Renn O, Slovic P, Brown HS, Emel J, Goble R, Ratick S (1988) The social amplification of risk: a conceptual framework. Risk Anal 8(2):177–187
78. Fuller S, Lipinska V (2014) The proactionary imperative: a foundation for transhumanism. Springer
79. Chaffin BC, Gosnell H, Cosens BA (2014) A decade of adaptive governance scholarship: synthesis and future directions. Ecol Soc 19(3)
80. Greer SL, Trump B (2019) Regulation and regime: the comparative politics of adaptive regulation in synthetic biology. Policy Sci 52(4):505–524
81. Renn O, Schweizer PJ (2009) Inclusive risk governance: concepts and application to environmental policy making. Environ Policy Gov 19(3):174–185
82. Kriebel D, Tickner J, Epstein P, Lemons J, Levins R, Loechler EL, Stoto M (2001) The precautionary principle in environmental science. Environ Health Perspect 109(9):871–876
83. Keisler JM, Trump BD, Wells E, Linkov I (2021) Emergent technologies, divergent frames: differences in regulator vs. developer views on innovation. Eur J Futures Res 9:1–6
84. Eslami M, Adler A, Caceres RS, Dunn JG, Kelley-Loughnane N, Varaljay VA, Garcia Martin H (2022) Artificial intelligence for synthetic biology: the opportunities and challenges of adapting and applying AI principles to synbio. Commun ACM
85. Staff at the National Security Commission on Emerging Biotechnology (2024) White paper 2: emerging technologies. In: NSCEB white paper series on AIxBio
86. Boeke JD, Church G, Hessel A, Kelley NJ, Arkin A, Cai Y, Carlson R, Chakravarti A, Cornish VW, Holt L, Isaacs FJ, Kuiken T, Lajoie M, Lessor T, Lunshof J, Maurano MT, Mitchell LA, Rine J, Rosser S, Yang L (2016) The genome project-write. Science 353(6295):126–127

87. Arnold C (2023) Inside the nascent industry of AI-designed drugs. Nat Med 29(6):1292–1295
88. Kelleher R (2024) NVIDIA collaborates with GenenTech to accelerate drug discovery. NVIDIA Blog
89. Desautels TA, Arrildt KT, Zemla AT, Lau EY, Zhu F, Ricci D, Cronin S, Zost SJ, Binshtein E, Scheaffer SM, Dadonaite B, Petersen BK, Engdahl TB, Chen E, Handal LS, Hall L, Goforth JW, Vashchenko D, Nguyen S, Faissol DM (2024) Computationally restoring the potency of a clinical antibody against Omicron. Nature 629(8013):878–885
90. Vindman C, Trump B, Cummings C, Smith M, Titus AJ, Oye K, Prado V, Turmus E, Linkov I (2024) The convergence of AI and synthetic biology: the looming deluge
91. HamediRad M, Chao R, Weisberg S, Lian J, Sinha S, Zhao H (2019) Towards a fully automated algorithm driven platform for biosystems design. Nat Commun 10(1)
92. Huang K, Qu Y, Cousins H, Johnson WA, Yin D, Shah M, Zhou D, Altman R, Wang M, Cong L (2024) CRISPR-GPT: an LLM agent for automated design of gene-editing experiments
93. Seyfried G, Pei L, Schmidt M (2014) European do-it-yourself (DIY) biology: beyond the hope, hype and horror. BioEssays 36(6):548–551
94. de Lorenzo V, Schmidt M (2017) The do-it-yourself movement as a source of innovation in biotechnology—and much more. Microb Biotechnol 10(3):517–519
95. Atlas RM, Dando M (2006) The dual-use dilemma for the life sciences: perspectives, conundrums, and global solutions. Biosecur Bioterror 4(3):276–286
96. NSCEB White Paper Series on AIxBio (2024) White paper 3: risks of AIxBio
97. Epstein GL (2012) Preventing biological weapon development through the governance of life science research. Biosecur Bioterror 10(1):17–37
98. Thiel P, Masters B (2014) Zero to one: notes on startups, or how to build the future. Crown Currency
99. Pescaroli G, Suppasri A, Galbusera L (2024) Progressing the research on systemic risk, cascading disasters, and compound events. Prog Disaster Sci 22:100319
100. Pescaroli G, Dryhurst S, Karagiannis GM (2025) Bridging gaps in research and practice for early warning systems: new datasets for public response. Front Commun 10:1451800
101. Holling CS (2001) Understanding the complexity of economic, ecological, and social systems. Ecosystems 4(5):390–405
102. Allen CR, Angeler DG, Garmestani AS, Gunderson LH, Holling CS (2014) Panarchy: theory and application. Ecosystems 17(4):578–589
103. Nuttall M (2012) Tipping points and the human world: living with change and thinking about the future. Ambio 41(1):96–105
104. Lu AC, Cannesson M, Kamdar N (2020) The tipping point of medical technology: implications for the postpandemic era. Anesth Analg 131(2):335–339
105. Kinzig AP, Ryan P, Etienne M, Allison H, Elmqvist T, Walker BH (2006) Resilience and regime shifts: assessing cascading effects. Ecol Soc 11(1):1–24
106. Marino N, Putignano G, Cappilli S, Chersoni E, Santuccione A, Calabrese G, Bischof E, Vanhaelen Q, Zhavoronkov A, Scarano B, Mazzotta AD, Santus E (2023) Towards AI-driven longevity research: an overview. Front Aging 4
107. Bernal M, Batista E, Martínez-Ballesté A et al (2024) Artificial intelligence for the study of human ageing: a systematic literature review. Appl Intell 54:11949–11977
108. Wu DD, Mitchell J, Lambert JH (2021) Global systemic risk and resilience for novel coronavirus and Covid-19. Risk Anal 41(5):701–704
109. Wu DD, Mitchell J, Lambert JH (2022) Global systemic risk and resilience for novel coronavirus in postpandemic era. Risk Anal 42(1):1–4
110. Pescaroli G, Guida K, Reynolds J, Pulwarty RS, Linkov I, Alexander DE (2023) Managing systemic risk in emergency management, organizational resilience and climate change adaptation. Disaster Prev Manag 32(1):234–251
111. Linkov I, Baiardi F, Florin MV, Greer S, Lambert JH, Pollock M, Trump BD (2019) Applying resilience to hybrid threats. IEEE Secur Priv 17(5):78–83
112. Perrow C (1999) Normal accidents, with a new 'afterword'. Princeton University Press

113. Brookes G (2022) Genetically modified crops and sustainability: 25 years since their widespread introduction. Genetic literacy project
114. McDivitt P (2023) Does GMO corn increase crop yields? More than 20 years of data confirm it does. Genetic literacy project
115. Maina J (2022) GMOs could shrink Europe's climate footprint, study suggests. Alliance Sci
116. Malins J (1985) The fence or the ambulance
117. United Nations (2012) Guiding principles on business and human rights. United Nations Human Rights Office
118. NIST (2023) Artificial intelligence risk management framework (AI RMF 1.0). National Institute of Standards and Technology
119. CAIDP (2023) Comments to the Office of Science and Technology Policy on the national priorities for artificial intelligence. Center for AI and Digital Policy
120. OSTP (2023) National artificial intelligence research and development strategic plan 2023 update. Networking and information technology research and development program
121. World Health Organization (2021) Ethics and governance of artificial intelligence for health: WHO guidance executive summary. World Health Organization
122. Bernstein MS, Levi M, Magnus D, Rajala B, Satz D, Waeiss C (2021) Ethics and society review of artificial intelligence research
123. Stanley-Lockman Z, Christie EH (2021) An artificial intelligence strategy for NATO. NATO Rev
124. Trump BD, Keisler JM, Galaitsi SE, Palma-Oliveira JM, Linkov I (2020) Safety-by-design as a governance problem. Nano Today 35:100989
125. Linkov I, Trump BD, Keisler JM, Cegan J, Foran CM, Collier ZA, Kuklja MM (2020) Signals and metrics identifying partnerships for innovation. IEEE Eng Manag Rev 48(2):39–46
126. Trump B, Cummings C, Klasa K, Galaitsi S, Linkov I (2023) Governing biotechnology to provide safety and security and address ethical, legal, and social implications. Front Genet 13:1052371

Chapter 3
Safeguarding the Responsible Use and Equitable Benefit-Sharing of Biotechnology in an Age of Scientific Modernization

Tatyana Novossiolova, Beth Ellinport, Cindy Vindman, Thomas Binz, Holly Jarman, Chris J. Myers, Mariam Natsvlishvili, and María Juliana Rolón Rojas

Abstract Biotechnology modernization, driven by advances in synthetic biology, AI, and automation, offers transformative solutions in health, agriculture, and sustainability but also holds a potential to fuel a biological arms race. This chapter maps approaches and tools for mitigating the potential of misuse and malign exploitation of biotechnology knowledge, materials, and infrastructure without hindering legitimate access to novel scientific and technological advances. In particular, it demonstrates how research security can support equitable biotechnology modernization through strengthened international coordination and inclusive governance. The chapter, first, identifies a set of structural prerequisites for the governance of biotechnology-AI convergence that impact states' rate of capability adoption. Second, it examines

T. Novossiolova (✉)
Center for the Study of Democracy, Sofia, Bulgaria
e-mail: tatyana.novossiolova@csd.eu

B. Ellinport · C. Vindman
U.S. Army Corp of Engineers, Engineer Research and Development Center, Concord, USA

T. Binz
Federal Office of Public Health, Liebefeld, Switzerland

H. Jarman
Department of Health Management and Policy, University of Michigan School of Public Health, Ann Arbor, USA

C. J. Myers
Department of Electrical, Computer, and Energy Engineering, University of Colorado Boulder, Boulder, USA

M. Natsvlishvili
Georgian Biosafety Association, Tbilisi, Georgia

M. J. R. Rojas
Faculty of Health Sciences, University of Santander, Bucaramanga, Colombia

C. L. Cummings et al. (eds.), *Biotechnology and AI: Technological Convergence and Information Hazards*, NATO Science for Peace and Security Series A: Chemistry and Biology, https://doi.org/10.1007/978-3-032-05246-9_3

research and innovation security as part of the norms of responsible use of biotechnology. And third, it outlines practical options for integrating research and innovation security within biotechnology governance to promote the collaborative nature of scientific and technological progress while mitigating risks of misuse and malign exploitation.

Keywords Biotechnology modernization · Responsible governance · Innovation security

3.1 Introduction

Biotechnology modernization is defined here as the pursuit of cutting-edge capabilities and products enabled by the convergence of life sciences and related fields with advances in engineering and computing. This includes technologies such as artificial intelligence (AI), machine learning, and robotics. While progress is rapid, it remains uneven across nations. This divergence is driven by a combination of factors. These include differences in national investment in innovation and development, state and quality of existing research infrastructure, access to data and advanced technologies, and differing risk cultures. Some countries aggressively pursue early adoption while others prioritize risk management over potential benefits. The emergence of an equitable bioeconomy, where the benefits of biotechnological advances are broadly shared, holds tremendous promise to mitigate pressing health, societal, and environmental concerns. Biotechnical growth not only enhances capabilities, it also offers opportunities to boost biomedical preparedness and health security, reduce reliance on petrochemical supply, leverage bioremediation capacity, and establish sustainable circular manufacturing systems.

The field of synthetic biology is expanding at a rapid pace providing benefits in the form of disease treatment methods, vaccines and therapeutics, agricultural products and resilient crops, biofuels, eco-friendly waste treatment, and biodegradable materials for use in manufacturing and construction [1–5]. Advances in computing and machine learning, alongside the advent of high-power AI tools, promise to further revolutionize this field enabling data analytics and modelling for experiment design on an unprecedented scale [6]. But beyond Big Data processing, AI technologies are also transforming the physical research environments. Examples include the installation of cloud laboratories that function with a high-degree of automation and can perform in networks for real-time data sharing [7].

Harnessing the potential of synthetic biology-AI convergence, however, is far from straightforward. Several immediate challenges persist. First, there is limited availability of high-quality datasets for training algorithms, which creates bias and undermines the accuracy of results. Second, there is a lack of established synergies between life science and AI research and innovation communities (e.g. lack of access to high-performance computing and data centers at life science departments) which hinders interdisciplinary collaboration. Third, the convergence raises serious ethical and security concerns, including the risk of misuse [8].

As biotechnology capabilities advance, worsening divergence can lead to a future global landscape characterized by a widening gap between leaders and laggards where early adopters gain significant technological, economic, and political advantage in international agenda-setting [9]. Without international coordination to harmonize governance and manage risks, unchecked divergence may exacerbate global inequities, hinder cross-border collaboration, and disrupt geopolitical balances. This includes the potential to fuel new forms of arms race despite existing robust international norms against the development and use of biological weapons. Addressing key gaps and needs will require the development of viable governance mechanisms that make it possible to safeguard biotechnology innovation and capabilities from misuse without restricting legitimate access to knowledge, data, infrastructure, and expertise. This includes advancing effective action across government, business, and academia to counter activities that ultimately undermine the ethos and integrity of responsible science, promoting inclusive strategies for biotechnology innovation and equitable benefit-sharing, and integrating foresight and security risk management in research and innovation to ensure that international collaborations are not exploited for illicit or other nefarious purposes.

This chapter maps approaches and tools for mitigating the potential of misuse and malign exploitation of biotechnology knowledge, materials, and infrastructure without hindering legitimate access to novel scientific and technological advances. The chapter, first, identifies a set of structural prerequisites for the governance of biotechnology-AI convergence that impact states' rate of capability adoption. Second, it examines research and innovation security as part of the norms of responsible use of biotechnology. And third, it outlines practical options for integrating research and innovation security within biotechnology governance to promote the collaborative nature of scientific and technological progress while mitigating risks of misuse and malign exploitation. The conclusion addresses how research security can support equitable biotechnology modernization through strengthened international coordination and inclusive governance.

3.2 Navigating the Biotechnology-AI Convergence: Governance Prerequisites

Biotechnology is an economic driver that offers sustainable solutions in areas such as agriculture, veterinary science, medicine, pharmaceuticals and fine chemicals production [10, 11]. However, in order to take advantage of these benefits, countries need to make sure that certain prerequisites are in place. These include a national governance framework aligned with international standards and good practices; appropriate budget allocation and funding arrangements that enable the translation of research findings into viable products and applications; sustainable infrastructure; and sufficient and well-trained workforce.

3.2.1 National Governance Frameworks

Effective national governance provides suitable conditions for the implementation of biotechnology research and innovation programs and helps ensure that these programs meet local priorities [12, 13]. Relevant regulatory, policy, and self-governance initiatives cover multiple areas such as adoption of technological advances (e.g. use of particular technologies, experimental methods etc.); certification and licensing of entities, facilities, and persons; protection of intellectual property; management of research involving human and animal subjects; clinical trials; drug and product safety; occupational health and safety; management of activities involving hazardous substances and hazardous waste; mitigation of negative societal and/or environmental impacts; countering the misuse or other forms of illicit diversion of infrastructure, materials, or personnel, etc.—all of which are vital for the smooth functioning of the research and innovation cycle.

The use of AI tools in biotechnology research adds another layer of complexity especially when it comes to the management, protection, and use of personal data in research and healthcare settings [8, 14]. There are not common international standards for personal data sharing and use, including for the utilization of such datasets for algorithm training. In addition, the lack of universal guidelines on AI governance is likely to lead to significant discrepancies regarding the types of approved AI systems in different countries, the monitoring and audit procedures for ensuring the safety and reliability of approved AI systems, and the level and quality of training that the deployers and end-users of AI systems receive.

Governance requires foresight so that it can accommodate technological change [15]. The development and implementation of effective biotechnology governance involves a multitude of stakeholders spanning government agencies, academia, industry, private research foundations, and publishers across different disciplines and sectors [16]. Historically, many of these stakeholders thus far have had only little interaction with their counterparts in the field of AI development. Understanding how the integration of AI tools with biotechnology innovation can impact research and development practices is essential for updating the existing and developing new relevant regulatory and policy initiatives [17]. This in turn requires fostering synergies between quite disparate professional communities. Transparency in decision-making and the participation of civil society are essential to gain public trust.

To support international research and product development, governance frameworks must promote regulatory interoperability. Shared standards for biosafety, data ethics, AI auditing can prevent gaps that enable misuse and reduce barriers to collaboration [18]. Biosafety, biosecurity, and ethics governance frameworks require corresponding structures such as containment facilities, hazardous waste systems, and institutional ethics review boards to ensure responsible research practices, particularly as technologies become more powerful and more distributed. These structures must be scalable and resilient, capable of withstanding external threats, and adapting to continuous innovation.

3.2.2 Funding and Investment

Robust and sustainable financial support is essential for biotechnology research and development across all stages from basic science to commercialization. Early-stage R&D often relies on public funding and grant schemes, while late-stage development and market entry typically attract private sector investment [19, 20]. By and large, government agencies and private foundations account for the lion share of funding for early-stage biomedical research. By contrast, private equity firms specialize in investing around late-stage development, providing capital for expansion with the aim of selling for profit. In addition, big pharma partnerships are increasingly important for financing biotech start-ups in later stages, helping to bridge funding gaps. This creates a revolving door where companies at times have been battling with a lack of private investment but then begin to face a lack of government funds.

Tools such as targeted project financing, seed funding, and venture capital are some of the instruments that can speed up industry's uptake of cutting-edge scientific advances so that these are converted into products [21]. Fostering synergies between academia and business, providing tax incentives for companies and institutions that invest in biotechnology, and promoting public–private partnerships can facilitate knowledge transfer and infrastructure sharing which in turn helps reduce costs and enables efficient allocation of resources.

At the same time, AI development, increasingly integrated with biotechnology, requires a distinct set of financial considerations. The development of high-performance AI tools for biotechnology demands significant investment in computing infrastructure, data resources, and talent. Without adequate financial support many institutions, especially in middle- and low-income countries, lack access to the tools necessary to participate in global research.

International finance mechanisms should support technology transfer and innovation cooperation [22]. This may include funding for human capital development through fellowships, training programs, and career support for scientists. Financial support should address regional disparities in access to research infrastructure to ensure innovation environments are globally interconnected.

3.2.3 Sustainable Infrastructure

Infrastructure is the backbone of biotechnological development encompassing institutional systems comprising digital and physical assets. At the physical level, this includes, for example, well-equipped and maintained research and laboratory facilities providing reliable and safe settings for innovation, production facilities for industrial biotechnology, and modernized health, agricultural, and manufacturing systems that both rely on and drive the adoption of advanced technologies. Investment in infrastructure attracts talent and international collaborations and enables the implementation of large-scale research and innovation programs.

Digital infrastructure is equally essential [23]. High-performance computing systems, cloud-based data storage, and secure-data sharing platforms can facilitate AI integration in biotechnology. However, access to these resources is often limited, particularly in lower resourced institutions and countries. Harnessing big data analytics in biotechnology requires synergies between life sciences and AI research communities. Many life scientists do not have readily available access to analytical tools for large, complex biological datasets or technologies for AI-enhanced discovery. This digital divide impedes both innovation and equitable modernization.

The development of high-precision AI tools for modelling and forecasting require good-quality and diverse datasets so that algorithms generate more accurate and less repetitive outputs. Interoperable systems, shared infrastructure, and standardized data protocols are needed to facilitate collaboration and model training across countries and borders.

3.2.4 Well-Trained Workforce

Trained human capital is essential to drive progress in biotechnology [24]. Education and continuous training programs for scientists, technicians, and laboratory personnel need to prioritize the acquisition of transferable skills, as well as cross-disciplinary cooperation. The latter is particularly important for promoting innovation in fields such as synthetic biology which combines concepts and tools from different disciplines including biology, chemistry, engineering, computing, and material science. Life science education should be practice-oriented and directed toward the application of relevant knowledge for resolving pressing health and societal concerns. It should also encompass grounding in safe, secure, and ethical research conduct.

Professional training programs should encourage the set-up of national and international networks of experts for knowledge and experience sharing and advocacy. Education and training are also useful tools for connecting the life science and AI professional communities including by fostering joint projects and peer-learning. Strengthening digital fluency and data literacy among life scientists is especially important for ensuring they can responsibly and effectively apply AI tools for modelling and decision support. At the same time, AI professionals who support biotechnology research must be equipped with an understanding of biological systems and laboratory research to build context awareness.

Efforts to strengthen the biotechnology workforce must also address disparities in access to quality education and training resources. International cooperation, through remote learning, fellowships, and institutional partnerships can support equitable capacity building to foster a more inclusive global innovation environment [25, 26].

Whereas the four prerequisites described above are not exhaustive, they are interdependent and their development is essential for promoting the adoption of biotechnology capabilities and ensuring that these are utilized for beneficial purposes that advance health, societal betterment, and environment protection. Enforcing appropriate regulatory arrangements guarantees safe, secure, and ethical research practice, adequate protection of intellectual property, and fair competition and funding allocation. For its own part, developing and implementing a robust national policy and regulatory framework requires resources including institutional and technical capacity (Box 1). In the absence of a long-term strategy for biotechnology oversight systems, countries can face significant challenges to keeping these systems fully staffed and operational which, in turn, can create regulatory loopholes and enforcement weaknesses raising the risk of accidents, facility malfunction, and malign exploitation of infrastructure.

Box 1: Sustainability of biotechnology governance systems in emerging economies

Colombia is one of the most biodiverse countries in the world hosting close to 10% of the global biodiversity [27]. With its varied ecosystems ranging from the Amazon jungles to the Andean mountains, it harbors an incredible diversity of flora and fauna species. This biological wealth is not only a natural treasure but also an invaluable source of genetic resources and ecosystem services. Colombia is an emerging economy and like other South American countries, relies on its rich natural resources for development and revenue. Agriculture and industries driven by raw material extraction account for a significant share of Colombia's GDP and the main exports include oil, coal, gold, and coffee.

Colombia hosted the negotiations within the framework of the Convention on Biological Diversity which led to the Cartagena Protocol on Biosafety. Formally adopted in 2000, the Cartagena Protocol addresses the safe development, handling, transport, and use of living modified organisms (LMOs). As part of the national implementation of the Protocol, Colombia initiated the development of a regulatory framework that should designate national competent authorities. In this framework, the Ministry of Environment and Sustainable Development, the Ministry of Agriculture and Rural Development, and the Ministry of Health and Social Protection shared assigned responsibilities for managing, assessing, approving, and monitoring the use of genetically modified organisms (GMOs) in the country.

To enhance the national implementation efforts in support of the Cartagena Protocol, Colombia launched a World Bank-supported project to strengthen the regulatory framework and build technical and scientific capacity in biosafety to harness the benefits of GMOs without compromising its rich biodiversity [28]. This project delivered the following outputs:

- Strengthening the legislative framework and operational mechanisms of the national biosafety system: Regulations were published, nine institutional biosafety groups were formed, and regional workshops were held across the country to promote awareness of the Cartagena Protocol and related national regulatory requirements. These activities resulted in an improved institutional capacity for risk assessment and management related to GMOs among Colombian stakeholders.
- Development of Institutional Technical Capacity Nuclei: A series of training activities were implemented including inter-institutional seminars, workshops, and internships which culminated in the establishment of a postgraduate program in biosafety.
- Establishment of a national Biosafety Clearing House (BCH): The Biosafety Clearing House under the Cartagena Protocol facilitates the exchange of good practices among states parties. The BCH Colombia was set-up as an online resource modelled on the Protocol's BCH to provide competent authorities and the general public with accurate and timely information so that they are kept up-to-date with relevant developments.
- Creation of the Inter-Institutional Laboratory for GMO Detection and Monitoring: The laboratory was established at the Colombian Institute for Agriculture (ICA) headquarters in Tibaitatá, Cundinamarca, in partnership with the National Institute of Drug and Food Surveillance (INVIMA), and the Alexander von Humboldt Biological Resources Research Institute (IAvH). Besides carrying out research projects, the laboratory published a procedural manual for GMO detection.

This initial investment enabled Colombia to establish an operational foundation for implementing the provisions of the Cartagena Protocol. Local health and agricultural sectors have operated in a coordinated manner, consolidating regulatory and technical infrastructure for the oversight of activities involving GMOs.

By contrast, the environmental sector has faced structural challenges in consolidating its role as a leading national authority in this field. This is largely due to the fact that the environmental sector is not driven by immediate economic outputs, but is instead oriented toward the protection of Colombia's natural heritage which requires distinct institutional capacities [29].

3.3 Research and Innovation Security as Elements of the Responsible Use of Biotechnology

Biotechnology modernization has far-reaching implications for society, and public trust is an important factor for the adoption of technologies and products [30]. Accidental release or deliberate misuse of pathogens, genetically modified organisms or critical technology may cause great harm to society, agriculture and the environment with local, regional or even global consequences. All life science research, independently on its geographical location, and its political or economic context, must be used responsibly.

Fundamental principles that both underpin and guide the responsible use of biotechnology modernization include: (1) respect for human life and human autonomy; (2) do no harm; (3) beneficence; and (4) justice. Traditionally, these principles are associated with research ethics and have their origins in pivotal international instruments including the WMA Declaration of Helsinki—Ethical Principles for Medical Research Involving Human Subjects [31]. Here, the intention is to use these four principles in much broader terms and demonstrate how they translate into requirements for biotechnology innovation. The discussion of these principles should be considered within the framework of responsible research and innovation (RRI) which aims to align technological innovation with broader social values [32–34].

International law guarantees the right to life of all human beings. Life science activities, regardless of whether they involve human subjects need to comply with this rule and, where applicable, ensure that appropriate procedures for informed consent and voluntary participation are applied. Research and innovation in biotechnology should not pose a threat to human life and if such a risk exists—as is the case with research involving dangerous pathogens, for instance—sufficient precautions and risk mitigation measures need to be undertaken. Life science stakeholders need to guarantee human autonomy throughout the research and innovation cycle. Key areas include personal data management, human oversight whenever AI tools are used, and provision of timely, accurate, and accessible information on issues of major public interest to prevent the effects of misinformation and disinformation and counter self-threatening behavior and practices.

The principle of nonmaleficence stipulates that life science activities should not cause harm to humans, animals, plants, the environment, or infrastructure. The implications of this principle are far-reaching and concern both the processes of research and innovation in biotechnology and the outcomes of these processes. Life science stakeholders need to ensure that their activities are carried out in ways that prevent suffering and any forms of damage to social or natural ecosystems. Relevant actions range from ethical treatment of research animals to regular maintenance of laboratory equipment to prevent the accidental release of pathogens and control on the release of living modified or genetically engineered organisms to reduce the risk of environmental damage or ecological disaster. Stakeholders should also ensure that the results, products, and services generated in biotechnology are safe and do not lead to negative side effects. Vaccines, medicines, food, foodstuffs, and diagnostic methods need to meet established quality standards and certification procedures before they are placed on the market.

The principle of beneficence goes a step further and stipulates that biotechnology research and innovation are expected to generate positive results and benefits. These include, for example, advancing knowledge and product-development, and promoting human and animal wellbeing, the protection of the environment, and economic development. Biotechnology should benefit both stakeholders and end-users. Scientists and their institutions gain recognition and credit through research publications and patents. Businesses and private companies generate profits by providing products and services. National regulations and government policies shape and at times, direct biotechnology research and innovation. Equally, the effective implementation of certain policies—e.g. providing accessible healthcare, ensuring affordable medical supplies and countermeasures in quantities that meet public demand, and tackling health crises—is dependent on the progress of biotechnology and local capacity to mobilize resources, expertise and, infrastructure to respond to societal needs. As the primary end-users of biotechnology results, the general public should be duly informed about the value that scientific advances bring to everyday life and how any potential risks are addressed to mitigate potential negative impacts.

The justice principle concerns the distribution of risks and benefits of biotechnology within society and across countries. It stipulates that no social group should disproportionately bear the costs of biotechnology research and innovation, or enjoy the benefits accrued therefore at the expense of other groups. Related issues include underrepresentation in research which could exclude certain groups of the population and restrict their access to products and services; exploiting loopholes and discrepancies in governance for outsourcing potentially hazardous activities; and unfair collaboration practices resulting in inequitable sharing of benefits. Biotechnology stakeholders should adopt rules and mechanisms that prevent unethical conduct and uphold public interest and welfare along the entire research and innovation cycle. To be effective, such mechanisms should be anticipatory, inclusive, deliberative, and responsive. They should take into account local socio-economic, political, and cultural dynamics and provide sufficient opportunities for cross-sectoral dialogue and systematic public engagement.

The responsible use of biotechnology modernization requires a fine balance between research security and research integrity. Research security refers to the approaches, mechanisms, and practices to protect the research ecosystem against the risk of undesirable transfer of critical technology, malign influence, and ethical or integrity violations by third countries [35]. The primary goal of research security is to protect the research ecosystem and guarantee legitimate national and economic interests. Research integrity concerns the values, norms, and principles that constitute good scientific practice (e.g. freedom of scientific research, openness, honesty, accountability, etc.) and regulate international research collaboration (e.g. reciprocity, equity, non-discrimination, etc.) [36].

The implementation of research and innovation security is multi-layered, comprising strategies and tools for protecting collaborations against misuse and malicious exploitation. It is the shared responsibility of multiple stakeholders in government, academia, and industry. Research and innovation security covers different types of collaborative activities including funding and investment; knowledge, material, and technology transfer and sharing; student and staff exchange; and physical security and cyber-security.

Knowledge sharing is a core aspect of science that drives innovation and plays a critical role in states' acquisition of biotechnology capabilities. But nefarious actors can hijack legitimate data sharing mechanisms for harmful purposes including biological weapon development, bioterrorism, or competitive sabotage. Whereas making research data sets openly available for reuse (e.g. in line with data sharing standards such as the FAIR framework—findable, accessible, interoperable, reusable) allows replication and validation of findings, AI tools for data mining can provide rogue actors with insights into the development of biological technologies for malicious purposes [37, 38].

Scientists and policymakers must carefully navigate the fine line between protecting sensitive information, materials, and infrastructure and fostering an environment which is conducive to progress and innovation [39]. Excessive secrecy can stifle collaboration, slow down scientific advancement, breed uncertainty among institutions and nations, and undermine the accuracy of threats assessments. At the same time, a failure to uphold appropriate security standards and implement due diligence practices renders entities vulnerable to attacks from ill-intended actors.

3.4 Balancing Security and Access in Biotechnology Research and Innovation

Research and innovation security policies and measures should be flexible and proportionate. Government agencies should develop and promote clear guidelines, in order to foster a shared understanding of research security among stakeholders in academia and business. Risk assessment is at the core of research security. Adopting a risk-based approach makes it possible to design graded systems and mechanisms for anticipating and mitigating risks without restricting legitimate collaborations. A risk assessment helps prioritize sensitive areas of research and innovation that are likely to be targeted. It also helps design appropriate screening protocols and procedures for materials, technology, and infrastructure that are considered high-risk. Conducting a risk assessment allows institutions and organizations to review what relevant internal policies already exist and how these are implemented, in order to decide if additional precautions or improvements are necessary. Burdensome administrative procedures may discourage scientists from pursuing certain research projects which is why it is essential that research security requirements are streamlined within already established institutional processes. Recurring training that is adapted to the needs and professional duties of the staff with responsibilities for research and innovation security helps professionals be up-to-date with pertinent regulatory developments and emerging risks.

Practicing due diligence empowers scientists, academic institutions, and business to protect research and technology assets against misappropriation, diversion, and misuse. Screening schemes such as know-your-collaborator and know-your customer are useful tools for building relationships with persons and entities with whom there is no prior interaction. They allow collecting information about prospective funders, collaborators, and clients, as well as the purpose for which any materials, data, or equipment will be used. In this way, scientists and their institutions can identify potential red flags early on and take appropriate risk mitigation measures to ensure their assets and interests are adequately protected. The red flags applied for screening should be regularly updated so that they are compatible with the fast-evolving nature of security concerns (Box 2). They should reflect past institutional experience with suspicious situations including security-related incidents and near-incidents to promote situational awareness among scientists and administrative staff and inform capacity building efforts for monitoring and reporting.

Box 2: Indicative list of red flags [40]

Regarding persons/institutions

- New partner/collaborator/client about whom little information is readily available.
- Difficult to locate in databases; unconvincing online presence/incomplete website.
- Inconsistent contact/affiliation/billing details.
- Located in or has ties with a country/entity of concern.
- Showing lack of/limited evidence of technical knowledge/experience with the research area or technology about which the person/entity is willing to work in/make use of.

Regarding funding/contracting conditions

- An offer to pay an excessive price for items/products.
- Cash payment offered even for large purchases.
- Funder refuses to accept a clause that does not allow military end-use.
- Unexplained/unusually large budget for research that does not contain concrete outputs.
- Funding from private donor with no expectations or requirements beyond access to research findings.

Regarding the shipment of orders/materials/technology

- Unusual shipping/packaging/labelling arrangements are requested, including collection by the end-user's personal vehicle.

- Delivery to a post box office address/incomplete or forwarding address.
- The cost of delivery significantly exceeds the product cost.

Regarding the stated purpose of the collaboration/end-use of the material/technology

- The person/entity is vague about/reluctant to disclose how the technology/item/material will be used.
- Unusual requirements for excessive confidentiality.
- Indications that the intended use is inconsistent with the stated purpose.
- No obvious need for requesting the technology/item in question.

Besides screening mechanisms, additional approaches and tools that scientists and their institutions can use to enhance research and innovation security include material transfer agreements, personnel reliability requirements, and cyber-biosecurity frameworks.

A material transfer agreement (MTA) is a legal contract governing the transfer of biological, or other laboratory materials from one institution to another, in order to ensure that the recipient of the materials agrees to use them for a specific purpose and to follow certain conditions [41]. An MTA defines the scope of the transfer, e.g. nature and quantity of the material being transferred, as well as any limitations on the use or distribution of the material. It also assigns rights and responsibilities of both the provider and recipient of the material, such as confidentiality obligations, intellectual property rights and liability for any harm resulting from the use of the material.

Specific issues that a generic MTA should cover include term, amendment and termination conditions; notifications and communication; policy about the publication of information about or information generated using the biological material; requirements for transfer and transport; specifications for the final disposal or retention of the material, and/or biosafety and biosecurity requirements for the recipient. The goal of MTAs is to ensure that rights of all parties are guaranteed and that none of the parties should utilize the transferred biological material in a way that harms the interests of other parties.

Institutional personnel reliability programs seek to prevent 'insider threats'. The insider threat refers to a situation in which one or more individuals with legitimate access to biological agents, expertise, data, or equipment exploit their position to assist or otherwise contribute to or facilitate the misuse, diversion, or misappropriation of biological assets. ISO 35001:2019 on biological risk management for laboratories and other related organizations lays down requirements for strengthening personnel reliability, in order to mitigate the insider threat and promote an institutional culture of trust and transparency [42]. The main purpose of personnel reliability programs is to confirm the identity and credentials of employees; ensure that employees do not suffer from a health condition that may negatively impact on their performance and result in biosafety and/or biosecurity risks; and carry out a security assessment of employees' circumstances.

Personnel reliability programs should be graded taking into account the role and responsibilities of staff within the facility. For example, researchers working within facilities where activities with high-risk biological agents and toxins are performed might be required to undergo a thorough background check covering employment history, financial circumstances, criminal history, and physical and mental health records. Other staff within these facilities might also be subject to background screening procedures that are not so extensive. It is important that staff employed on temporary or fixed-term contracts, or visiting research staff including students are covered by institutional personnel reliability programs. This should also include individuals accessing research facilities to perform specific tasks, e.g. maintenance workers, cleaners, and suppliers.

Personnel reliability is inter-related with physical security at laboratories and research facilities. Graded access, door and freezer locks, identity checks at entrance, escorted visits, real-time video surveillance, and on-site security presence are essential to protect sensitive biological materials and data against misuse or misappropriation. Visiting researchers should always undergo an induction training prior to the start of their appointment within facilities to ensure that they are well -acquainted with the applicable biosafety and biosecurity procedures and protocols.

Cyber-biosecurity refers to the need to protect both digital and physical systems within research facilities and laboratories against unauthorized access, damage, or manipulation [43, 44]. As the digital and physical domains are closely integrated, so is their security. Cybersecurity within laboratories extends beyond information security and includes measures and procedures that ensure the operation of all institutional networked electronic systems, e.g. for access control, video surveillance, material inventory, etc.

Cyber-hygiene is a core aspect of cybersecurity. A cyberattack through an institutional account or inbox, or shared online infrastructure can result in an indefinite disruption of operations, loss of data, and reputational damage. Internal cybersecurity policies should cover the use of institutional devices such as laptops, tablets, and mobile phones during international travel and work overseas, as well as requirements for remote connection and access to institutional servers to guarantee that sensitive information is not released or stolen. The integration of AI with biotechnology also reinforces the importance of cyberbiosecurity, as AI tools are vulnerable to manipulation and hacking [45].

3.5 Conclusion

The COVID-19 pandemic has demonstrated that no country in the world can face emerging biological risks alone. Collaboration is at the heart of scientific research and innovation. It is also a critical factor for establishing and sustaining effective governance. A vibrant biotechnology ecosystem requires a fully operational national

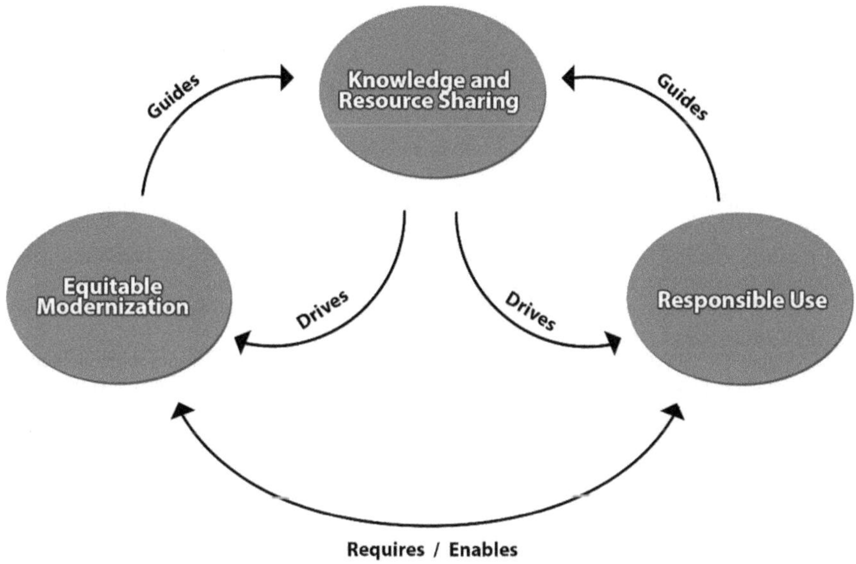

Fig. 3.1 Framework for beneficial biotechnology modernization—role of research and innovation security

policy, regulatory, and institutional framework; financial backing; robust and resilient infrastructure; and well-trained workforce. International cooperation and assistance enables countries to access funding and investment opportunities, resources, and capacities that may not be otherwise readily available.

Respect for the principles of responsible use of biotechnology modernization is essential to guarantee that collaborative activities including resource sharing and capacity building result in fair and equitable access to benefits. Research and innovation security offers approaches and tools that help achieve the goals of responsible use of biotechnology modernization in a fragmented and fast-evolving international geopolitical landscape. The framework of research and innovation security aims to enhance trust and transparency among biotechnology stakeholders across government, industry, and academia, and safeguard the exchange and transfer of data, expertise, materials, and equipment against misuse and diversion (Fig. 3.1).

Research and innovation security plays an important role in reassuring the public that biological assets are used only for peaceful and beneficial purposes. Informing the public of the purpose of security measures which different stakeholders undertake helps to resolve misconceptions and counter the spread of misleading and false narratives that aim to erode the ethos of international cooperation in biotechnology. Recent examples of mis- and disinformation campaigns related to biological threats and unsubstantiated allegations of bioweapon development breed uncertainty about the positive role that biotechnology plays in society, foster a negative perception of biomedical laboratories and the activities conducted there, and undermine public trust in science [46, 47].

References

1. Voigt C (2020) Synthetic biology 2020–2030: six commercially-available products that are changing our world. Nat Commun 11:6379
2. Wang L et al (2023) Synthetic biology: a powerful booster for future agriculture. Adv Agrochem 1:1
3. Liu Z et al (2022) Yeast synthetic biology advances biofuel production. Curr Opin Microbiol 65
4. Yan X et al (2023) Applications of synthetic biology in medical and pharmaceutical fields. Signal Transduct Target Ther 8:199
5. Zhao N et al (2023) Synthetic biology-inspired cell engineering in diagnosis, treatment and drug development. Signal Transduct Target Ther 8:112
6. Vindman C et al (2024) The convergence of AI and synthetic biology: the looming deluge
7. National Academies of Sciences, Engineering, and Medicine (2025) The age of AI in the life sciences: benefits and biosecurity considerations. National Sciences Press, Washington DC
8. O'Brien J, Nelson C (2020) Assessing the risks posed by the convergence of artificial intelligence and biotechnology. Health Secur 18(3):219–227
9. Trump B et al (2023) The worsening divergence of biotechnology: the importance of risk culture. Front Bioeng Biotechnol 11
10. Robinson D, Nadal D (2023) Synthetic biology: a game changer for economic sustainability, security and resilience. OECD Blog
11. Titus AJ, van Opstal E, Rozo M (2020) Biotechnology in defense of economic and national security. Health Secur 18(4):310–312
12. Dietz T et al (2023) Towards effective national and international governance for a sustainable bioeconomy: a global expert perspective. EFB Bioecon J 3
13. National Academies of Sciences, Engineering, and Medicine (2017) Preparing for future products of biotechnology. National Academies Press
14. Zaidan E, Ibrahim IA (2024) AI governance in a complex and rapidly changing regulatory landscape: a global perspective. Human Soc Sci Commun 11:1121
15. Keisler JM, Trump BD, Wells E, Linkov I (2021) Emergent technologies, divergent frames: differences in regulator vs. developer views on innovation. Eur J Futures Res 9:1–6
16. Novossiolova T et al (2021) The vital importance of a web of prevention for effective biosafety and biosecurity in the twenty-first century. One Health Outlook 3
17. Zakaria S (2024) Regulating a revolution: as biotechnology and AI converge and advance at high speed, developing robust oversight is challenging-but not impossible. Biologist 71(4)
18. de Lorenzo V, Schmidt M (2018) Biological standards for the knowledge-based bioeconomy: what is at stake. New Biotechnol 40:170–180
19. Linkov I, Trump BD, Keisler JM, Cegan J, Foran CM, Collier ZA, Kuklja MM (2020) Signals and metrics identifying partnerships for innovation. IEEE Eng Manag Rev 48(2):39–46
20. Héder M (2017) From NASA to EU: the evolution of the TRL scale in public sector innovation. Innov J 22(2):1–23
21. Bains W, Wooder S (2014) Funding biotech start-ups in a post-VC world. J Commer Biotechnol 20(1)
22. Panday N et al (2022) Beyond technology transfer: innovation cooperation to advance sustainable development in developing countries. WIREs Energy Environ 11:2
23. Dhindsa A et al (2021) A novel model for building digital infrastructure for biodiversity studies. J Phys 1714
24. Pascoli DU, Aui A, Frank J, Therasme O, Dixon K, Gustafson R et al (2022) The US bioeconomy at the intersection of technology, policy, and education. Biofuels Bioprod Biorefin 16(1):9–26
25. Vera-Choqqueccota S et al (2025) Reducing education inequalities through cloud-enabled live-cell biotechnology. Trends Biotechnol 43(1)
26. Danies G (2023) Does biotech education need new teaching methodologies? Res Direct Biotechnol Des 1

27. Convention on Biological Diversity (2025) Colombia—country profile
28. GEF Trust Fund (2003–2008) Capacity building for the implementation of the cartagena protocol
29. Colombia (2020) Fourth national report on the implementation of the Cartagena protocol on biosafety, BCH-NR4-CO-251831-1
30. Trump BD, Cummings CL, Kuzma J, Linkov I (2020) Synthetic biology 2020: frontiers in risk analysis and governance. Springer, Cham
31. World Medical Association (2023) Declaration of Helsinki—ethical principles for medical research involving human subjects
32. Mashiah D et al (2023) Responsible research and innovation in Europe: empirical evidence from regional planning initiatives in Austria, Norway, and Spain. Eur Plan Stud 31(9)
33. ERA4HEALTH Partnership (2025) Responsible research and innovation
34. DiEuliis D, Giordano J (2022) The need for modernization of biosecurity in the post-COVID world. Msphere 7(2):e00025-e122
35. Council Recommendation on Research Security (2024)
36. EC-OECD STIP Compass (2025) Research security
37. Wilkinson MD et al (2016) The FAIR guiding principles for scientific data management and stewardship. Sci Data 3(1)
38. Wheeler N (2025) Responsible AI in biotechnology: balancing discovery, innovation and biosecurity risks. Front Bioeng Biotechnol 13
39. Nieuwenweg AC, Trump BD, Klasa K, Bleijs DA, Oye KA (2021) Emerging biotechnology and information hazards. In: Emerging threats of synthetic biology and biotechnology: addressing security and resilience issues, pp 131–140
40. European Commission (2021) Commission Recommendation (EU) 2021/1700 on internal compliance programmes for controls of research involving dual-use items under Regulation (EU) 2021/821
41. World Health Organization (2024) Laboratory biosecurity guidance
42. ISO (2019) ISO 35001:2019 biorisk management for laboratories and other related organisations
43. Richardson L et al (2019) Cyberbiosecurity: a call for cooperation in a new threat landscape. Front Bioeng Biotechnol 7:99
44. AL-Eitan L et al (2025) Addressing cyberbiosecurity challenges in the modern era of biotechnology and artificial intelligence. Glob Biosecur 7(1)
45. Laith AE, Jaouni H, Mihyar A (2025) Addressing cyberbiosecurity challenges in the modern era of biotechnology and artificial intelligence. Glob Biosecur
46. Wright L, Pandey A (2018) Russia claims US tested bioweapons in Georgia. DW
47. Field M (2022) Russian media spreading disinformation about US bioweapons as troops mass near Ukraine. Bull Atom Sci

Chapter 4
Information Hazards and Biotechnology

Nils Braun, Gerald L. Epstein, Kevin Esvelt, Rickard Knutsson, Michael G. Montague, Kenneth A. Oye, Anca-Georgiana Rusu, Lia Vizzotti, and Alla Yemets

Abstract Information hazards are defined as risks that arise from the dissemination of true information that may enable some agent to cause harm. The working group focused on cases where information potentially enabled significant irreversible harm but for which the possibility also existed of limiting or delaying access to that critical information. Discussion centered on key actors including funders, researchers and institutional review boards, regulators, and publishers. The group discussed each with reference to its authority and resources for mitigating information hazards, sources of uncertainty with respect to information on which decisions are based, and the extent that adaptive policies might improve handling of uncertainty. Several perverse points emerged: (1) Imposing limitations on access to information may alert potential malevolent actors to the existence of the very information that poses a

N. Braun
Secrétariat général de la Défense et de la Sécurité nationale, Paris, France

G. L. Epstein (✉)
RAND Corporation, Arlington, VA, USA
e-mail: gepstein@alum.mit.edu

K. Esvelt
MIT Media Lab, Cambridge, MA, USA

R. Knutsson
Swedish Veterinary Institute, Uppsala, Sweden

M. G. Montague
Archimedes Network, Woodbine, MD, USA

K. A. Oye
Massachusetts Institute of Technology, Cambridge, MA, USA

A.-G. Rusu
Université Paris Dauphine—PSL, Paris, France

L. Vizzotti
World BioHazTec Colombia, Bogotá, Colombia

A. Yemets
Institute of Food Biotechnology and Genomics, National Academy of Sciences of Ukraine, Kyiv, Ukraine

C. L. Cummings et al. (eds.), *Biotechnology and AI: Technological Convergence and Information Hazards*, NATO Science for Peace and Security Series A: Chemistry and Biology, https://doi.org/10.1007/978-3-032-05246-9_4

hazard, (2) Security benefits of controlling information may be offset by adverse side effects, including slowing scientific advance, limiting development of applications, impeding oversight and self-policing, and empowering information controllers to advance their interests, (3) Deliberations on information hazards were complicated by the fact that reporting those very discussions could itself have constituted an information hazard. The good news is the bad news. In other domains, imposition of controls has garnered support only in the face of manifest disasters. In biosecurity, we fortunately have not yet had such a disaster. However, given the challenges involved with managing hazardous information, even if future events were to create the political will to implement information controls, it is not clear what those controls would be.

Keywords Information hazards · Biosecurity · Dual-Use Research of Concern (DURC)

4.1 Definitions and Assumptions

An information hazard has been defined as "a risk that arises from the dissemination or the potential dissemination of (true) information that may cause harm or enable some agent to cause harm" [1]. The information hazards working group identified several challenges at the outset and premised discussion on the following assumptions.

First. The sources of information hazards vary significantly. Some hazards may arise when novel results of a life science experiment reveal a previously unknown vulnerability that can be exploited by malefactors to cause disease or other harm. Some hazards may arise when descriptions of research methods could facilitate the development of a biological agent already known to a potential attacker. In some instances, publication of research results may call attention to hazards already available to a potential attacker but that may have not been recognized by the potential attacker.

Second. The danger that might result from the misuse of information depends entirely on who is misusing it and how they will do so. Consequently any analysis of the danger that might result from revealing that information is fundamentally based on assumptions regarding the likelihood that such actors will discover the information, the likelihood that they have the technical skills to make use of the information (i.e., the chemical structure of a novel toxin that could evade existing countermeasures would be useless to someone who had no means of synthesizing it), and the likelihood that they have (or are inspired by the information to create) plans to use the information for harm. A worst-case analysis of the potential consequences of an information hazard must assume that there are people interested in and capable of utilizing it for harm, that the information is certain to get to them, and that they are certain to use it—but those assumptions may not actually be true.

Third. Blocking access to information by malefactors may also limit access of those who would use information for legitimate purposes. The working group recognized that limiting scientists' access to information may affect the pace and direction of future research. The working group also assessed how limiting access to information on hazards might itself cause harm by affecting the ability of biosafety officers, regulatory decision makers, and the national security community to evaluate and manage risks. A decision to restrict access to information therefore has an ethical component to it, as it involves weighing not only the harms of sharing the information against the harms of withholding it, but also the benefits that the information may confer to some against the harms it might facilitate to others.

Fourth. Managing information hazards is particularly challenging when the legal authority, political support, and technical feasibility of controlling access to information are problematic. Accordingly, the working group evaluated the legal foundations and technical methods associated with existing and potential policies and identified external factors that bore on resistance to or acceptance of information controls.

Finally. One is always at the mercy of the still-unknown. Information thought to be hazardous might (at least in principle) be rendered harmless by a future discovery that can mitigate the harms that that information could have generated. At the same time, new discoveries or technological possibilities can make information dangerous that previously had not been thought of as harmful. However, the fact that science and technology are constantly evolving should not be a rationale to do nothing.

4.2 What Information Hazards Are Appropriate Foci for Controls?

The working group began by differentiating among classes of information hazards using two dimensions:

- the severity of potential harm that might result from access to information; and
- the extent to which access to the information can be controlled or delayed.

The group used these dimensions to define the cells in Table 4.1, and it focused on cases in the northwest quadrant of the table. Cases in that quadrant combine potential significant harms associated with access to information with the possibility of limiting access to critical information.

The working group reviewed the extensive existing literature on estimating potential harms, with emphasis on information on what features of organisms map into infectivity, virulence, and robustness that are in turn elements of pandemic potential. Information hazards are of course not limited to pandemic threats, but triggering or facilitating a man-made pandemic provides a very clear pathway by which misuse

Table 4.1 Dimensions of harm and controllability of information hazards

	Information access Controllable or delayable	Information access Not controllable or delayable
High harms	Prime info hazard cases Focus on control or delay of release	Desperate futility Focus on response and resilience
Low harms	Monitor possible changes in harms and controllability/delayability	Monitor possible changes in harms and controllability/delayability

of information might lead to global harm. The working group considered how information that created a perception of offensive advantage could be particularly destabilizing by creating incentives for and fears of preemptive attacks. Members of the working group divided on the feasibility of differentiating operationally between offensive and defensive implications of release of information. Some argued that since key defenses against the use of an infectious agent as a weapon are nonbiological (e.g., personal protective equipment, airflow sterilization), hazardous biological information would be more useful to those seeking to develop weapons than those working to defend against them.

The working group then turned to the formidable challenge of withholding discoveries about the natural world. The security benefits of restricting communication should be viewed in terms of delaying the widespread availability of that information, rather than preventing it. Given the possibility of rediscovery, information controls can only buy time. It is possible that the scientific insight leading to a surprising discovery is so novel that others are unlikely to replicate it, or that the technology newly applied to making a discovery is uniquely available to a single research group. But neither situation provides any assurance that the first group to make a potentially hazardous discovery will be so far ahead of the second that it can effectively suppress awareness of the result. Consequently, considering the control of information hazards as a choice between keeping a secret or not is a false dichotomy. Rather, it is a question of delaying broad access at the cost of sacrificing the first-mover advantages that would accrue from revealing it first oneself, or seizing first-mover advantage at the cost of sacrificing a delay in broad access to the information hazard.

The group turned to the implications of widespread adoption of artificial intelligence in accessing, filtering, and applying information relevant to the design of organisms with pandemic producing potential and other hazardous properties. The group concluded that AI would have the effect of reducing the practical possibility of delaying the diffusion of information but could also have the effect of increasing the possibility of segmenting information flows, blocking access by malevolent actors and enabling access by benevolent actors. At present, conventional profiling methods are used to identify potentially malevolent and benevolent actors. AI pattern recognition methods using diverse sources of digitized information should improve the performance of conventional profiling, with respect to the following classes of potential recipients and appropriate treatment of each class:

- To tag those who are most likely to use information for good reasons and green-light information access. We note that tagging the large number of responsible researchers and biosecurity professionals and enabling information access is the least problematic class.
- To tag potential malefactors and red-light access. We note that the very small number of malevolent biosecurity actions will make identification of predictors of malevolence difficult.
- To tag those likely to screw up and provide information conditioned on training and oversight. We note that the number of accidents will facilitate identification of predictors of accident proneness and that training programs to reduce risks exist.

Utilizing AI to target information flows in this way presumes the ability to "narrow-cast" information to specific groups or even specific users, as compared to the classical "broad brush" approach of publishing research in a widely available scientific journal. Of course, the degree to which the redistribution of information can be controlled complicates any effort to segment its initial distribution.

Moreover, the discussion of AI did not extend to consideration of practical issues such as empirical difficulties in reliably profiling users or securing distribution channels for scientific knowledge in ways that would differ from existing practice. Neither did the working group address the issue of whether generative AI systems could discover or synthesize hazardous information, and if so, how to control its distribution.

4.3 Policies—Existing Oversight and Other Methods of Control

The working group focused on United States policies and regulations because they are widely known worldwide. Other countries and international organizations also have relevant guidance, policies, and regulations with the same objectives, but a dedicated review would have been necessary to address the full spectrum of global approaches.

The 2024 US Government Policy for Oversight of Dual Use Research of Concern and Pathogens with Enhanced Pandemic Potential defines essential elements of oversight [2]. It calls for assessing the risks posed by certain types of proposed research projects and for preparing mitigation plans to deal with them. The Implementation Plan for that Policy offers a menu of mitigation options, including modifying the design or conduct of the research; determining the mode, venue, and context of research communication; voluntary redaction; security classification (which is often legally or practically not available); or even not performing the research at all.

Any researcher who discovers information that he or she believes could cause danger if disseminated has always been fully empowered not to share that information. Policy challenges only arise when information restrictions involve gatekeeping other people—particularly those who are under no obligation to follow the precedents that some may seek to impose. Moreover, different stakeholders have different interests; some may seek dissemination whereas others seek to block it, and no mechanisms have been institutionalized to mediate such conflicts.

Unlike classified plans and other human-created information, which in principle could be kept secret indefinitely if no one with access to them divulged them, nature makes the results of scientific research available to anyone capable of conducting that research. Anyone making a discovery can effectively veto (either knowingly or otherwise) the decisions of those who have previously discovered—but withheld—the same results. Indeed, discoveries are often made when new technologies or previous discoveries permit accessing or characterizing natural phenomena in ways that had previously not been done—and once these new discoveries are made, it is not unusual for them to be made independently and nearly simultaneously [3]. On the other hand, some members of the working group questioned the relevance of this point to information hazards. They suggested that the most serious information hazards are not true facts about nature, but rather conceptual engineering insights regarding how scientific facts and principles can be exploited for harm. They argued that very few individuals combine the requisite scientific knowledge and security orientation, thereby limiting the likelihood of independent discoveries or realizations that create prime information hazards and increasing the importance of not revealing those that do.

Participants also agreed that it is too simple to view options for managing hazardous information as a binary choice of either sharing or withholding. The security benefits of withholding information and the protective or other benefits of sharing may both be attainable if communications regarding that information are creatively shaped to convey the beneficial aspects while not highlighting the dangerous ones. Shaping communications in this way is a matter of giving different emphasis to different aspects of the information; it does not imply generating and sharing misinformation. Some participants emphasized the role of responsible communications in lessening opportunities for misuse, suggesting that organizations should offer educational initiatives on communications for scientists.

Similarly, one way of permitting the beneficial application of information while lessening the probability of its harmful use could be by distracting those most likely to misuse it through the provision of additional information that makes it harder for those parties to find or recognize the hazardous aspects. Like information suppression, distraction is not a robust approach. The security world is replete with examples of the failures of so-called "security by obscurity." But given that there may be no completely effective way of controlling access to information, distraction offers a potential tool.[1]

[1] The shaping and distraction strategies discussed in the working group echo one of the approaches described by Sam Weiss Evans and Megan Palmer for addressing "anomalies" in regulatory systems—things that cannot necessarily be adequately addressed within the current regulatory framework. In particular, Weiss Evans and Palmer suggest that anomalies that cannot be handled by modifying the anomaly itself or modifying the regulatory system might be addressable by "diversion," in which a "decoy" activity is established to distract attention away from the problem [4].

The working group concluded that if the availability of certain information to malefactors is the primary determinant of whether they commit serious harm, then the temptation to do more than simply restrict access to that information will be great. Specifically, the idea of adding information to distract, confuse, or mislead potential misusers of information will have appeal. The group recognized, however, that if the information added to confuse potential misusers of information is untruthful, distraction becomes "disinformation" or "data poisoning." This approach raises serious practical and ethical issues. Any resort to such an option would require intensive discussion and debate, and working group members wanted to make clear that they were not recommending disinformation as a tool to manage information hazards.

Finally, the group discussed what can be termed "the backfire problem," or "Streisand effect." In some cases, imposing limitations on access to information may alert potential malevolent actors to the existence of the very information that poses a hazard. For example, knowledge of a decision to withhold the details of a toolkit that might be used to produce novel pathogens could be a flashing arrow pointing at the toolkit. Even if details on methods and results are withheld, the knowledge that something has been withheld at all can be revealing. In this way, efforts to suppress information can backfire, highlighting the very areas of danger that are objects of concern.

4.4 What Tradeoffs Are Generated by Controls?

The imposition of controls on information access has three significant sets of adverse side effects.

4.4.1 Limiting General Scientific Advance and the Development of Specific Applications

Novel research results may be the purest example of an information hazard. So long as a research result has not been rediscovered by others, there would be no other sources by which the information might be derived. At the same time, suppression of novel research findings has substantial potential for slowing scientific progress. Scientific research has always flourished in an environment of open participation in and open communication of research results. In recognition of this fact, leading scientific editors and authors issued a statement in 2003, not long after the 9/11 terrorist attacks and the October 2001 anthrax mailings in the United States, that emphasized the importance of scientific openness [5–7]. Nevertheless, statement signatories recognized that "on occasions an editor may conclude that the potential harm of publication outweighs the societal potential benefits. Under such circumstances, the paper should be modified, or not be published." This statement was an important acknowledgement of the risks of information hazards, and the fact that mitigating them might be warranted notwithstanding the constraining effect such controls may have on science. However, in the two decades that have elapsed since its publication, only limited progress has been made in operationalizing this conclusion. On a number of occasions, draft (or even published) manuscripts containing potentially hazardous information have been flagged and reviewed, but the only generally applicable conclusion from these cases appears to be that the publication stage is not the time to address such concerns. Potential information hazards are much more appropriately addressed at the proposal stage, before the potentially hazardous information has been generated and disseminated. Indeed, such advance review is one of the primary motivations for U.S. government policies governing so-called dual use research of concern. On the other hand, developing a systematic and complete way of identifying dual use research of concern before the research results are in hand poses its own challenges, which can include some of the same ethical and practical considerations of gatekeeping science.

4.4.2 Impairing Oversight and Self-policing

Biosecurity systems rely on formal oversight led by biosecurity and biosafety professionals and on informal self-policing by individual members of the research community. Institutional Biosafety Committees (IBCs) are formally tasked with ensuring that research is in compliance with the NIH Guidelines for Research Involving Recombinant or Synthetic Nucleic Acid Molecules [8], and in many institutions they oversee biosafety matters pertaining to research that is not covered by those guidelines. Individual researchers play a critical but less formally defined role, policing their own work for potential safety and security issues. In order for these oversight mechanisms to function, biosafety professionals charged with institutional oversight and individual researchers engaging in self-policing must know what is dangerous and what is safe. But limits on access to information on the design and production of hazardous organisms, including those with pandemic potential, may deny biosafety professionals and self-policing researchers the information that they need to mitigate risks. As noted in Part II above, the application of artificial intelligence could, at least in theory, permit segmentation of information flows in a manner that would differentiate between malevolent and benevolent actor, thereby improving the terms of the tradeoff between controlling information and crippling oversight.

4.4.3 Empowering Self-interested Controllers of Information Access

The implementation of controls over sensitive information requires empowering individuals and organizations with the authority to impose controls. Self-interested individuals operating in bad faith may find it easy to use information controls to advance their narrow economic, political and scientific interests. Information on research results can affect decisions on whether further research should be funded. Information on biosecurity risks can affect decisions on whether research on countermeasures and methods of detection should be funded. Information on biosafety and biosecurity hazards can affect decisions on the regulation of hazards that in turn may create barriers to entry by potential competitors. Information on tools with dual use potential can be used to justify restrictions on access to tools that in turn advantage some firms at the expense of others. Federal rules requiring the disclosure of possible conflicts of interest and recusal from decisions in which individuals have a material self-interest are intended to address this tradeoff, but not all actors in a position to influence whether information is shared or not may be subject to such rules, and not all potential conflicts of interest or commitment involve financial matters that are most directly addressed by these rules.

4.5 When, How and by Whom Should Decisions Be Made?

At the outset of its deliberations, the group acknowledged the existence of significant uncertainty associated with both the extent of potential harms that might result from access to information, and with the degree to which access to that information could be controlled. In specifying the roles of actors at various stages, the group focused on management of uncertainty as a central feature of the design problem. From start to end, the actors include funders, researchers and institutional review boards, governmental regulators, and publishers of scientific journals. The group discussed each of these sets of actors with reference to the following points.

- What authority and resources do these actors have for mitigating information hazards? To what extent do external advisory committees and external referees provide independent guidance and legitimation on the management of hazards?
- What are sources of uncertainty with respect to critical information on which decisions are based? To what extent can adaptive policies with provisions for information gathering and assessment and updating of policies in light of emerging information improve handling of uncertainty?
- What possible harms are irreversible? To what extent would the selective application of precautionary policies be appropriate, particularly in circumstances with the potential for global pandemic hazards with mass casualties? Information that is placed in the public domain will remain there forever.

Funders. Funders are in a unique position. On the one hand, the decisions that they reach on what research will or will not be funded are of fundamental significance. On the other hand, early-stage decisions rest on a relatively weak base for anticipating information hazards associated with research outcomes and assessing the potential controllability of information flows. If the results of a research project, and its implications, were known before the research was conducted, there would be no need to conduct it. Working group discussion focused on how funders have used their superpower—their ability to impose conditions on funding that contribute to the management of information hazards and the mitigation of uncertainty. Specific examples include NIH requirements for oversight at the level of research institutions and guidelines for researcher conduct.

As a limited check on potential conflicts of interest, the NIH relies on an extensive array of external advisory boards including the now-superseded Recombinant DNA Advisory Committee (RAC) with its assessment of individual projects; its replacement, the Novel and Exceptional Technologies Research Advisory Committee (NExTRAC), with its focus on horizon scanning; and most critically the National Science Advisory Board for Biosecurity (NSABB), with its mandate to address the security implications (including information hazards) of life science research. Unlike most life science researchers, NSABB members have some access to classified information, should such information have a direct bearing on the matters that come before it.

Each of these advisory groups added to the adaptive capacity of NIH in distinctive ways. Through its review of projects submitted for funding, the RAC gathered concrete information on extant areas of research. With its emphasis on horizon scanning, the NExTRAC has developed a capacity for gathering information on emerging technologies in advance of their realization. With its mandate, the NSABB has been presented with information hazard cases of critical significance, often at a late stage after a controversy has developed. The independence and power of these advisory boards and committees is limited by their dependence on staff for managing membership appointments, selecting witnesses and experts, evaluating conflicts of interest, and writing reports. Moreover, all of these boards are advisory only; none of them have any legal authority to control the content of any scientific manuscript.

Researchers and Institutional Review Boards. Researchers and Institutional Review Boards are tasked by funders with responsibilities to assess and manage biosecurity and biosafety risks, with some attention to information hazards, but they often lack the resources to perform those tasks.

Researchers concentrate on designing, funding, conducting and publishing research. In most instances, their attention to biosecurity matters and potential information hazards is purely instrumental, a matter of satisfying requirements and recommendations of regulators and funders in order to proceed with their research.

Institutional review boards are typically led and supported by biosafety and biosecurity professionals and composed of scientists and engineers. Biosecurity professionals complain that they do not have the time or staff to fulfill their duties. Scientists and engineers typically care more about their research and teaching than matters of biosafety and biosecurity. These problems are compounded by the reluctance of institutions to incentivize faculty with a limited self interest in biosecurity and biosafety matters to devote time and energy to their IBC or Institutional Review Entity.

Finally, researchers, biosecurity professionals and members of review boards may lack information on novel biosecurity and safety developments that are most likely to pose information hazard problems, or even on information hazards that may have been dealt with previously by others. For obvious reasons, there is no Journal of Results Too Sensitive to Publish. Well-intentioned researchers will not report research that may pose a hazard to their IBCs if they are not aware of the hazard. IBCs will not know how to handle safety and security problems and to manage potential information hazards if they are not informed on the hazards. However, sharing this information among IBCs—whose collective memberships are not necessarily vetted or controlled—may create precisely the information hazard that such sharing would be intended to preclude.

Regulators. The roles of regulators in addressing basic biosecurity and biosafety problems are well established. Their role in managing information hazard problems, however, is less well developed. After discussing domestic regulations and international agreements that govern movements of select agents and insurers' requirements on safe and secure handling of materials as a condition of coverage, the group turned to problems of information hazards.

Regulatory agencies and national security organizations have dealt explicitly with aspects of information hazards. Most obviously, information hazards are the raison d'etre for classification systems. Determining what information constitutes possible threats to national security, how such information should be protected, setting specific limits on who has access to information, and deciding on when information controls can be released are elements of classification. However, in the United States (and likely at least some other countries), security classification generally cannot be imposed on information that has not been developed by or for the government.

Group discussion turned to the special case of patent secrecy orders, one exception to the rule stated immediately above.[2] Under the Invention Secrecy Act of 1951, patent applications are reviewed to determine where disclosure or publication might be detrimental to national security [9]. Secrecy orders may block the award of patents, mandate that patents awarded be kept secret, and bar disclosure of information associated with an invention deemed sensitive. Secrecy orders may be applied to patents based on privately funded non-classified research as well as patents based on classified research with governmental funding. Secrecy orders may bar communications by third parties that know of an invention as well as by inventors. Patent secrecy orders highlight tradeoffs across benefits and risks of controlling information. The patent system exists to incentivize creativity by rewarding inventors with payments from licensing and to enable sharing of the benefits of creativity by placing new, nonobvious and useful inventions into registries that are accessible to the public. The

[2] A second exception in the United States would be information classified as Restricted Data under the Atomic Energy Act. However, this exception is likely not applicable to the type of life science information hazard discussed here.

imposition of a secrecy order on a patent denies inventors revenues from licensing and blocks public awareness of potentially useful knowledge.

Publishers. Publishers of scientific journals face the clearest tradeoffs across benefits and costs of restricting access to information. The working group considered the most prominent recent case when the NSABB deliberated on whether the editors of Science and Nature should publish papers on the results of research on development of strains of H5N1 influenza virus that were transmissible via aerosol from mammal to mammal [10–12]. Discussion centered on how much specific information on methods and results was necessary for the scientific and medical communities to derive benefit from the research, as compared to what might enable malefactors to create such enhanced strains. The group also noted that rather than deciding on the release of this information at the publication stage, after it had already been created and shared in detail among reviewers, and when the fact of (if not the details of) this information was widely known, it would have been preferable to address these issues during earlier deliberations by NIH on funding the research and by review boards that approved the research with safeguards. To strengthen early incorporation of security awareness, the working group discussed the possibility of improving the security culture of NIH and life science researchers at the request for proposal and project review stages.

4.6 Conclusions

Deliberations of the working group on information hazards were impeded by several issues.

First, working group discussion of these issues was complicated by the existence of possible information hazards. For example, the group considered the need for precautionary controls limiting release of information on possible global pandemic producing pathogens with the potential for mass casualties. Our discussion hit a wall when the working group member raising this point explained that he could not provide more details on specific pathogens and scenarios for development and release without creating a serious information hazard. That limitation in turn restricted analysis of the plausibility of pandemic potential and of the feasibility of controlling essential information on methods and organisms. The working group noted this issue without providing a solution to the problem.

Second, working group discussion of the need for adaptive self-correction was impeded by a basic evidentiary issue. The very notion of self-correction rests on the ability to identify errors in forecasting and in resulting policy choices. That in turn rests on the ability to generate counterfactuals on what would have happened if other decisions based on other forecasts had been made. We lack a firm basis for projecting counterfactual forecasts on the dire consequences of not controlling or delaying information. We lack a firm basis for counterfactual forecasts on the pace and direction of scientific advance if a paper had been published in full form. We lack a firm basis for counterfactual forecasts on the effectiveness of biosafety officers if controls on full access to information on potentially dangerous activities had not been in place.

Third, the legitimation, effectiveness, and underlying legal authorities of controls on information are context dependent. In other spheres, preparations for disasters including controls on behavior and information have garnered popular support only in the face of manifest disasters. In the realm of information hazards in biosecurity, we do not have disasters to prompt decisive action. No malefactor has started a wildfire pandemic. Setting aside the still uncertain origins of COVID-19, our pandemics have been of natural origins, and our concerns over information hazards remain largely hypothetical. Without the impetus of a disaster clearly traceable to disclosure of scientific information, the working group concluded that the implementation of coercive information control policies for scientific publications—even if they were legally permissible—would be unlikely.

Members of the group recognized that controlling access to scientific information that might be used to facilitate harm is extremely challenging, for all the reasons described above. Nevertheless, they found that there could be some utility in shaping communications by wording them in ways that convey the beneficial aspects of potentially concerning information while not highlighting dangerous ones. They also recognized that researchers and organizations have a responsibility to communicate research outcomes responsibly. In addition, distraction, or providing potentially concerning information in a context that makes it more difficult to find or derive, could be useful as well. Finally, the application of information segmentation, including the use of AI techniques to differentiate among potential recipients of the information in ways that allow messages to be shaped for each, is worth exploring. At the same time, datasets that would allow artificially intelligent systems to differentiate "legitimate users" from "illegitimate" ones are very sparse, and with any luck will remain that way.

References

1. Bostrom N (2011) Information hazards: a typology of potential harms from knowledge. Rev Contemp Philos 10:44–79
2. Executive Office of the President (2024) United States Government Policy for oversight of dual use research of concern and pathogens with enhanced pandemic potential. Washington, DC
3. Merton RK (1963) Resistance to the systematic study of multiple discoveries in science. Eur J Sociol/Archives Européennes de Sociologie 4(2):237–282
4. Evans SW, Palmer MJ (2017) Anomaly handling and the politics of gene drives. J Responsible Innov
5. Atlas R, Campbell P, Cozzarelli NR, Curfman G, Enquist L, Fink G, Flanagin A, Fletcher J, George E, Hammes G, Heyman D (2003) Statement on scientific publication and security. Science 299:1149
6. Atlas R, Campbell P, Cozzarelli NR, Curfman G, Enquist L, Fink G, Flanagin A, Fletcher J, George E, Hammes G, Heyman D (2003) Statement on the consideration of biodefence and biosecurity. Nature 421:771
7. Journal Editors and Authors Group (2003) Uncensored exchange of scientific results. PNAS 100(4):1464

8. National Institutes of Health (2024) NIH guidelines for research involving recombinant or synthetic nucleic acid. Bethesda, MD
9. U.S. Code of Federal Regulations (n.d.) Title 37, Sections 5.1–5.5
10. National Science Advisory Board for Biosecurity (2012) Findings and recommendations. Washington, DC. https://www.ncbi.nlm.nih.gov/books/NBK206979/
11. Imai M, Watanabe T, Hatta M et al (2012) Experimental adaptation of an influenza H5 HA confers respiratory droplet transmission to a reassortant H5 HA/H1N1 virus in ferrets. Nature 486:420–428
12. Herfst S, Schrauwen EJA, Linster M et al (2012) Airborne transmission of influenza A/H5N1 virus between ferrets. Science 336(6088):1534–1541

Chapter 5
Risk Culture in a Time of Biotechnological Convergence: Implications for Regulatory Divergence and Bioeconomic Development

Benjamin D. Trump, Felicia Wu, and Christopher L. Cummings

Abstract This chapter examines how differing risk cultures shape the regulatory approaches that nations adopt when confronting emerging biotechnologies, with a specific focus on synthetic biology and its convergence with artificial intelligence. Despite confronting identical technological uncertainties and evidence, nations interpret technological risk through distinct cultural, political, and institutional lenses, which lead to significant variations in regulation, commercialization pathways, and international competitiveness. Through comparative analysis of regulatory frameworks in the United States, European Union, Singapore, and China, this chapter identifies how specific elements of risk culture—including historical path dependency, styles of regulatory dispute resolution, and governmental power centralization—influence biotechnology governance. These variations generate consequences for bioeconomic development, commercial innovation, and international competitiveness. Nations with adaptive regulatory frameworks that maintain appropriate safeguards while enabling technological progress may secure first-mover advantages in the emerging bioeconomy, while those with excessively rigid or permissive approaches may face either innovation deficits or unmitigated risks. The chapter concludes by exploring how divergent risk cultures will shape the future trajectory of synthetic biology development amid increasing technological convergence with artificial intelligence, with implications for global leadership in the bioeconomy.

Keywords Biotechnology convergence · Risk culture · Regulatory frameworks

---

B. D. Trump
Tor Intelligence, LLC, Apex, NC, USA
e-mail: ben@torintelligence.com

F. Wu
Michigan State University, East Lansing, MI, USA

C. L. Cummings (✉)
US Army Engineer Research and Development Center, Vicksburg, MS, USA
e-mail: Christopher.l.cummings@usace.army.mil

C. L. Cummings et al. (eds.), *Biotechnology and AI: Technological Convergence and Information Hazards*, NATO Science for Peace and Security Series A: Chemistry and Biology, https://doi.org/10.1007/978-3-032-05246-9_5

5.1 Introduction

The development of synthetic biology marks a pivotal advancement in biotechnology, enabling scientists to design and construct biological systems with novel or enhanced functions through the application of engineering principles. Now synthetic biologists are able to create not just novel proteins, but also synthetic cells with the capability of replication [1]. This field represents not merely an incremental improvement over previous genetic engineering capabilities but a fundamental shift toward programmatic control of biological systems [2]. The convergence of synthetic biology with artificial intelligence further accelerates these capabilities, enhancing the design and prediction of biological functions with unprecedented precision [3]. Despite the shared scientific basis across national borders, significant variations exist in how different nations regulate these technologies, commercialize resulting products, and position themselves within the international bioeconomy.

These variations stem at least in part from differing risk cultures—the intertwined institutional, political, and social factors that shape how societies perceive, evaluate, and respond to technological risks [4]. Risk culture encompasses not only formal regulatory frameworks but also the historical pathways of technological governance, styles of dispute resolution, degrees of power centralization, and underlying societal values regarding technological progress [5]. From these differences, nations can vary in their interpretation of risk and benefit of a technology or its applications—even with similar availability of risk information and uncertainty as the global community. While synthetic biology presents similar technical challenges and uncertainties across jurisdictions, nations' responses to these uncertainties diverge substantially due to these deeply embedded differences in risk culture.

This chapter examines how these divergent risk cultures influence the trajectory of synthetic biology development amid increasing technological convergence with artificial intelligence. Nations that establish adaptive regulatory frameworks that maintain appropriate safeguards while enabling technological progress may secure first-mover advantages in the emerging bioeconomy, while those with excessively rigid or permissive approaches may face either innovation deficits or unmitigated risks. As synthetic biology capabilities advance, these divergent approaches to risk will increasingly shape the global distribution of biotechnological power, influence patterns of international trade, and determine which nations lead the development of the bioeconomy.

5.2 Synthetic Biology and Technological Convergence

Synthetic biology represents a transformative approach to biotechnology that applies engineering principles to biological systems, enabling the design and construction of novel biological components and systems with pre-determined functions. This field evolved from earlier genetic engineering practices through several distinct phases of development. The first phase, emerging in the early 2000s, focused on constructing simple genetic circuits and standardizing biological parts. This period witnessed foundational developments such as the creation of a synthetic oscillatory network in *E. coli* [6] and the establishment of the Registry of Standard Biological Parts to catalog and standardize DNA components for biological engineering. The second phase, from approximately 2008–2014, expanded these capabilities through improved DNA synthesis techniques, pathway engineering, and the development of minimal genomes [7]. A landmark achievement during this period was the creation of the first synthetic bacterial genome by the J. Craig Venter Institute in 2010 [8]. The current phase has witnessed accelerating capabilities through CRISPR-based genome editing, cell-free systems, and the integration of computational design tools.

The convergence of synthetic biology with artificial intelligence represents a critical technological development that transforms both fields [9]. Machine learning algorithms now enhance the design of synthetic biological systems by predicting protein structures, optimizing metabolic pathways, and identifying novel enzymatic functions. Deep learning techniques analyze vast biological datasets to identify patterns often invisible to human researchers, enabling the prediction of complex biological phenomena. This convergence accelerates the development cycle of synthetic biology products by reducing experimental iterations and improving design precision. The application of reinforcement learning algorithms to biological design problems enables computational exploration of design spaces far beyond what human engineers could manually evaluate. Furthermore, integrating automated laboratory systems with AI-directed experimental design creates continuous design-build-test cycles that operate with minimal human intervention, significantly accelerating research productivity.

Applications of synthetic biology span numerous sectors with profound economic implications. In medicine, synthetic biology enables the production of complex therapeutic molecules that would be challenging or impossible to synthesize through traditional means, as demonstrated by the artemisinin precursor production pioneered by Keasling and colleagues [10]. The biofuel sector utilizes synthetic biology to engineer microorganisms that produce advanced biofuels from renewable feedstocks, potentially reducing dependency on fossil fuels [11]. Agricultural applications include crop improvements for enhanced nutrition, drought resistance, and disease prevention without conventional transgenic approaches [12]. The materials sector has witnessed the development of programmable biomaterials with tunable properties for applications in construction, textiles, and consumer products [13]. Environmental applications include engineered organisms for bioremediation of pollutants and biosensors for environmental monitoring [14]. These diverse applications collectively form the foundation of an emerging bioeconomy.

The economic significance of synthetic biology continues to grow, with global private investment into the billions, and many governments committing to a decade or more of infrastructure development and basic research support [15]. Major corporations across pharmaceuticals, agriculture, materials, and energy sectors have established synthetic biology divisions or partnerships. Countries including the United

States, China, the United Kingdom, and Singapore have established national bioeconomy strategies that explicitly include synthetic biology as a priority area [16, 17]. The development of biofoundries—facilities that integrate automation, computation, and biological engineering for rapid prototyping—has accelerated commercialization timelines. However, these economic developments proceed against a backdrop of regulatory uncertainty shaped by differing risk cultures across nations.

5.3 Risk Culture: Theoretical Framework

Risk culture encompasses the constellation of institutional arrangements, historical experiences, political structures, and social values that collectively shape how societies perceive and respond to technological risks [5]. This concept extends beyond formal regulatory frameworks to include the underlying factors that influence risk perception, regulatory formation, and technological governance [18].

Several key dimensions characterize risk culture within a given society. Historical path dependency refers to the ways previous regulatory experiences constrain current regulatory options and shape perceptions of new technologies [19]. Societies with histories of technological controversies often develop risk-averse regulatory approaches, while those with successful technological innovation histories may adopt more permissive frameworks. Styles of regulatory dispute resolution represent another critical dimension, with some societies relying on adversarial legalism and formal judicial processes while others emphasize cooperative arrangements and informal negotiation [20]. The degree of power centralization influences regulatory responsiveness and flexibility, where centralized systems may implement decisive changes rapidly while decentralized systems require more extensive consensus-building but potentially offer greater stability.

The manifestation of these risk culture dimensions produces distinct approaches to technological governance across jurisdictions. The precautionary approach emphasizes potential harms from new technologies and places the burden of proof on proponents to demonstrate safety before permitting deployment. This approach prioritizes avoiding potential harms over realizing potential benefits when faced with scientific uncertainty [21]. The cost–benefit approach attempts to quantify both potential risks and benefits, permitting technologies when expected benefits outweigh potential risks. This approach accepts that uncertainties exist but does not view uncertainty itself as sufficient justification for prohibition. The proactionary approach emphasizes technological advancement and places the burden of proof on those seeking restrictions to demonstrate harm [22]. This approach prioritizes realizing potential benefits and views excessive precaution as imposing "opportunity costs" through foregone innovation. The adaptive governance approach emphasizes iterative regulation that evolves with technological development and improved risk understanding. This approach utilizes staged deployment, careful monitoring, and regulatory learning to manage emerging risks while enabling innovation.

These theoretical frameworks help explain regulatory divergence in synthetic biology governance across different jurisdictions. For example, the European Union's approach to genetically modified organisms and, by extension, synthetic biology reflects a strongly precautionary orientation shaped by previous controversies. The United States follows a more cost–benefit oriented approach with significant emphasis on commercial development potential. Singapore demonstrates adaptive governance characteristics through its flexible regulatory frameworks that emphasize both innovation and sufficient safeguards. China exemplifies an approach that combines elements of proactionary advancement in priority sectors with centralized control mechanisms. Each approach emerges from the unique constellation of historical experiences, political institutions, and cultural values that constitute these nations' risk cultures. These differences manifest in concrete regulatory frameworks, commercial development pathways, and international strategies that significantly impact the trajectory of synthetic biology development.

Indeed, on opposite sides of the Atlantic, risk analysis had its birth and development in very different cultural contexts in the 1960s–1970s. In Britain, anthropologist Mary Douglas first posited her cultural theory of risk in her book Natural Symbols: Explorations in Cosmology. In this work, Douglas introduces the group-grid theory, with the group representing how an individual fits or does not within a social group, and the grid representing an individual's social role within a network of a society's privileges, claims, and obligations [23]. This theory later formed the basis for grouping risk cultures into hierarchist, individualist, egalitarian, and fatalist worldviews. From this start of cultural theory, risk perception, risk assessment, risk management, and risk communication were embedded within social structures.

Meanwhile, in the United States, risk assessment had its birth following the founding of the US Environmental Protection Agency (USEPA) in 1970, after the 1969 fire on the Cuyahoga River in Cleveland, Ohio, awakened public and policy awareness of the risks of pollution [24]. This area of Cleveland was heavily populated with industries that, while responsible for the pollution that caused the river fire, also generated wealth and employment opportunities for the local population—so how to balance these population risks and benefits, and what should be considered an acceptable level of risk? After a decade of studying and formalizing how to consider these questions, specifically on how technology affects the environment, the US National Academy of Sciences published a seminal work Risk Assessment in the Federal Government: Managing the Process in 1983 [25]. It codified the four steps of risk assessment: hazard identification, dose–response assessment (or hazard characterization), exposure assessment, and risk characterization. A follow-up National Academies compendium to this 1983 work, Science and Decisions: Advancing Risk Assessment, includes the step of problem formulation before these four risk assessment steps, and lists risk management as the final step while providing recommendations for such management with a view to risk-based policy decision-making [26]. In comparison with Douglas's Cultural Theory, risk assessment development in the US was more focused around a culmination of events that necessitated regulatory involvement, and a quantitative method of bounding and then assessing risks for management purposes.

5.3.1 Cultural Theory of Risk and Institutional Variations

Cultural theory of risk (often shortened as "cultural theory"), pioneered by Douglas and political scientist Aaron Wildavsky, provides a foundational framework for understanding regulatory divergence across national boundaries [5]. Douglas and Wildavsky argued that risk perception and management are fundamentally social processes shaped by cultural values and institutional structures rather than purely objective assessments of hazard probabilities. The theory identifies four cultural worldviews—hierarchist, individualist, egalitarian, and fatalist—that influence risk perception and policy preferences. Each worldview entails distinct assumptions about nature, human behavior, and proper social ordering that shape attitudes toward technological risk. Jasanoff extends this analysis through the concept of "civic epistemologies," describing how national political cultures determine what constitutes credible scientific evidence and how such evidence should inform policy decisions [27]. Nations maintain distinctive approaches to assessing expertise, demonstrating knowledge, and establishing objectivity that produce divergent interpretations of identical scientific evidence.

These cultural differences manifest in institutional arrangements that structure risk governance processes. The European Union's distinctive approach to biotechnology regulation emerged not from greater risk aversion in general, but from specific institutional transformations; including the politicization of food safety following the BSE crisis and the strengthening of environmental ministries relative to agriculture ministries [28–30]. Meanwhile, American institutional structures maintained greater decision-making authority within science-based agencies with closer industry relationships. Sheila Jasanoff characterizes this difference as "communities of reason" versus "communities of interests," with European regulations emerging from negotiated consensus across diverse stakeholders while American approaches privilege technical expertise within narrower policy communities [5]. These institutional variations subsequently entrench themselves through positive feedback mechanisms, creating path dependencies that shape responses to new technologies.

National approaches to synthetic biology governance thus reflect deeper cultural assumptions about technological development rather than simply differing technical assessments. Parthasarathy demonstrates how patent decisions regarding biotechnology in the United States and Europe reflect fundamentally different understandings of the relationship between science, commerce, and public values [31]. American patent decisions emphasized promoting innovation and commercialization, in keeping with an individualistic culture of risk; while European deliberations incorporated broader social and ethical considerations, in keeping with an egalitarian culture of risk. These differences extended beyond mere policy preferences to encompass distinct conceptions of how technological governance should operate in democratic societies. Similar variations appear in approaches to synthetic biology, where identical technical uncertainties generate divergent regulatory responses shaped by cultural understandings of appropriate governance models. These cultural differences prove remarkably durable despite globalization forces and international harmonization efforts, reflecting their embeddedness in national institutions and political cultures.

5.3.2 Risk Culture and Societal Trust Dynamics

The relationship between institutional trust and risk regulation represents another crucial dimension of risk culture that shapes synthetic biology governance. Löfstedt's work demonstrates how varying levels of trust in regulatory institutions correlate with different approaches to technological risk management [32, 33]. Societies with high institutional trust can maintain more flexible regulatory frameworks and delegate greater discretion to expert agencies, while those with trust deficits require more prescriptive rules and transparency mechanisms. These trust relationships develop through historical experiences with regulatory successes and failures, creating feedback loops that reinforce institutional arrangements. The European Union's approach to biotechnology regulation reflects partial reactions to trust deficits following the BSE crisis, which revealed regulatory failures and conflicts of interest within food safety institutions. In contrast, American biotechnology regulation, which had its early roots in the decade leading to the 1986 Coordinated Framework for the Regulation of Biotechnology [34], developed during periods of relatively higher trust in scientific agencies, allowing more discretionary approaches to risk assessment that persist despite subsequent trust erosion.

Trust dynamics operate not only between citizens and regulatory institutions but also across institutional networks that implement risk governance. Renn and Roco identify how differences in inter-institutional trust relationships shape emerging technology governance [35]. European regulatory systems emphasize formalized coordination mechanisms and explicit delineation of responsibilities across agencies, reflecting lower inter-institutional trust and stronger demarcation of authority boundaries. American governance networks operate through more informal coordination with overlapping jurisdictions and negotiated responsibilities, requiring greater inter-agency trust but potentially creating regulatory gaps for novel technologies. Singapore exemplifies strong vertical trust relationships that enable coordinated governance across research funding, regulatory oversight, and economic development agencies, facilitating integrated approaches to synthetic biology [36]. These varying trust architectures influence how effectively governance systems can address emerging risks that cross traditional regulatory boundaries.

Public risk perception further shapes regulatory approaches through political pressure on governance institutions. Slovic illustrates how public risk perceptions incorporate factors beyond statistical risk calculations, including voluntariness, familiarity, catastrophic potential, and equity considerations [37]. These perceptions vary systematically across cultural contexts, creating different political pressures on regulatory systems. European public concerns about biotechnology emphasize issues of naturalness, corporate control, and distributional justice that have significantly influenced regulatory frameworks. American public discourse has focused more on safety and religious concerns while generally maintaining greater acceptance of technological intervention. Chinese public perceptions emphasize economic benefits and food security implications while expressing less concern about naturalness boundaries. These varying public perceptions create different political environments for regulators, constraining available policy options and shaping regulatory priorities across jurisdictions.

5.3.3 Political Economy and Regulatory Capture Variations

The political economy of regulation represents a third critical dimension of risk culture that shapes biotechnology governance approaches. Nations maintain distinctive relationships between government, industry, and civil society that influence regulatory decision-making processes and outcomes. Carpenter and Moss identify varying susceptibility to regulatory capture across governance systems, where industry interests exert disproportionate influence over regulatory decisions [38]. American regulatory systems demonstrate particular vulnerability to "corrosive capture" where incremental weakening of oversight occurs through information asymmetries and revolving door relationships between industry and regulators. European Union institutions maintain greater separation between regulators and regulated entities but face challenges from fragmented oversight and differential implementation across member states. Authoritarian systems like China's demonstrate distinctive capture dynamics through direct state ownership of major biotechnology enterprises and coordinated leadership across regulatory and industry bodies [39].

These varying political economy arrangements create different incentive structures for both regulators and regulated entities. Vogel identifies how the organizational goals of regulatory institutions shape their approach to uncertainty and risk assessment [40]. European food safety and environmental agencies established in the aftermath of regulatory failures prioritize precaution and thoroughness to rebuild public trust. American agencies with longer institutional histories and closer industry relationships emphasize efficiency and innovation facilitation alongside safety considerations. Singaporean regulators operate within a coherent development framework that coordinates regulatory approaches with national economic priorities. These institutional incentives penetrate regulatory cultures, shaping how front-line regulators interpret ambiguous evidence and exercise discretion when confronting scientific uncertainties inherent in synthetic biology applications.

Industry structure similarly varies across jurisdictions, creating different patterns of interest group influence on regulatory development. Bernauer and Meins demonstrate how the distribution of industry capabilities shapes regulatory preferences and political mobilization [41]. European biotechnology regulations emerged in contexts where domestic industries maintained limited competitive advantages, reducing political opposition to stringent oversight. American regulatory frameworks developed alongside domestic industry leadership, creating stronger political constituencies for facilitative approaches. This pattern extends to synthetic biology, where jurisdictions with established corporate research capacities typically advocate regulatory frameworks compatible with commercialization pathways while those with nascent industries may implement more restrictive approaches. These political economy dynamics interact with institutional structures to create distinctive regulatory cultures that interpret identical technological uncertainties through different lenses.

5.3.4 Implications for Synthetic Biology Governance Under Uncertainty

The convergence of synthetic biology with artificial intelligence creates unprecedented governance challenges that expose and potentially amplify differences in national risk cultures. Synthetic biology presents what Funtowicz and Ravetz term "post-normal science" conditions, where facts remain uncertain, values are contested, stakes are high, and decisions are urgent [42]. Under such conditions, regulatory frameworks that evolved to address conventional risks with established assessment methodologies face fundamental challenges. The integration of machine learning algorithms into biological design processes introduces additional opacity that further strains risk assessment capabilities. National risk cultures will likely respond differently to these heightened uncertainties based on their institutional characteristics, trust relationships, and political economy arrangements. Precautionary systems may impose significant restrictions on autonomous design systems until comprehensive assessment methodologies develop, while proactionary frameworks may permit development with limited oversight until concrete harms emerge.

These divergent approaches to uncertainty management generate different consequences for both risk mitigation and innovation development. Marchant and Wallach identify fundamental tensions between precautionary approaches that may prevent potential harms but impose innovation costs and permissive approaches that facilitate development but accept potential unintended consequences [43]. Neither approach escapes the fundamental condition of uncertainty inherent in emerging technologies, but each distributes potential error costs differently. Precautionary approaches risk over-regulation that prevents beneficial applications, while adaptive or proactive approaches risk under-regulation that permits harmful ones. The diversity of national approaches to synthetic biology creates a natural experiment in risk governance, with different jurisdictions implementing varying regulatory strategies that generate evidence about both governance efficacy and innovation impacts.

This regulatory diversity creates distinct technological trajectories that may ultimately prove beneficial for synthetic biology development. Permissive jurisdictions may generate evidence about technological possibilities and limitations while precautionary systems develop comprehensive assessment methodologies and identify potential risk pathways [44]. The resulting regulatory learning potentially benefits all jurisdictions through observation of outcomes across different governance approaches. However, this diversity also creates coordination challenges for transboundary risk management, market fragmentation for developers, and potential competitive distortions in international markets. The tension between beneficial experimentation and harmful inconsistency represents a central challenge for international governance of synthetic biology under persistent differences in national risk cultures.

The emerging governance landscape for synthetic biology will likely reflect these enduring differences in risk culture rather than converging toward a unified international framework. Jasanoff argues that regulatory divergence in biotechnology reflects not temporary informational gaps but fundamental differences in how societies organize technological decision-making within democratic systems [45]. These differences remain remarkably durable despite globalization pressures and harmonization efforts, suggesting that synthetic biology governance will continue to reflect distinct national approaches. However, selective convergence may emerge around specific governance mechanisms or risk assessment methodologies that demonstrate effectiveness across cultural contexts. The resulting governance landscape will likely feature persistent variation in risk management philosophies alongside pragmatic coordination mechanisms that facilitate international research collaboration and commercial development while accommodating enduring differences in national risk cultures.

5.4 Implications of Risk Culture Divergence for Biotechnological Advancement

The divergence in risk cultures across nations creates profound implications for biotechnology development and the emergence of sustainable bioeconomies. These variations in how societies perceive, evaluate, and govern technological uncertainty shape the trajectory of synthetic biology advancement with consequences extending far beyond regulatory details to fundamental questions of national security, economic competitiveness, and global influence. As biotechnology increasingly functions as a strategic domain, understanding these risk culture differences becomes essential for effective policymaking and international engagement.

Regulatory divergence stemming from these differing risk cultures creates immediate challenges for cross-border technology development and commercialization. The European Union's precautionary approach, China's state-driven model, Singapore's adaptive governance, and the U.S. product-based framework represent not merely technical variations, but manifestations of deeply embedded cultural perspectives on appropriate technological governance. Organizations developing synthetic biology applications must navigate these different systems simultaneously, creating complex regulatory strategies that address multiple, sometimes contradictory requirements across markets. This navigation significantly increases compliance costs and extends development timelines for globally distributed products, creating competitive disadvantages for some applications while favoring others based not on technical merit but regulatory adaptation.

These regulatory differences increasingly function as competitive tools through what might be termed "lawfare" tactics in international biotechnology competition. Nations embed their risk governance preferences in proposed international standards, certification requirements, and market access regulations, creating structural advantages for domestic firms already adapted to these frameworks. The planned Malta workshop focusing on resilient biomanufacturing standards represents a direct response to these challenges, aiming to establish protocols that can facilitate market access while ensuring appropriate risk management. These standardization efforts represent contested terrain where risk culture differences manifest as competing specifications with significant economic implications. Nations that successfully embed their approaches in international standards gain substantial advantages through aligned compliance requirements that favor their domestic industries while potentially creating barriers for competitors operating under different regulatory paradigms.

The dual-use potential inherent in synthetic biology capabilities further amplifies these risk governance challenges across cultural contexts. Dual-use research of concern (DURC) refers to the phenomenon that a technology developed for beneficial purposes for society could also be used to create harms to that same or other societies. Technologies enabling transformative applications in medicine, materials, and environmental remediation simultaneously create capabilities for harmful applications if misused—a reality that different risk cultures interpret and manage through distinctly different approaches. Nations with precautionary orientations implement extensive oversight mechanisms that may impede legitimate research, while those with more permissive frameworks potentially allow capabilities to develop without adequate safeguards. These variations create inconsistent security environments that complicate international collaboration while potentially enabling technology transfer to jurisdictions with inadequate controls. Finding governance approaches that address legitimate security concerns without unnecessarily constraining innovation represents perhaps the central challenge for maintaining leadership in an increasingly fragmented global regulatory landscape.

The limited international standardization resulting from these divergent risk perceptions directly impacts market development and supply chain resilience for emerging biotechnology products. Without globally recognized standards for evaluating synthetic biology applications, manufacturers face regulatory uncertainty that impedes investment and delays commercialization of potentially transformative technologies. Establishing internationally recognized standards aligned with specific national approaches would expand market access while creating technical requirements that leverage existing strengths. Such standards would position firms advantageously within global value chains while simultaneously addressing legitimate safety and quality concerns across different risk cultures. The nation that successfully leads this standardization process potentially secures substantial first-mover advantages through frameworks that align with their domestic capabilities while creating adaptation requirements for competitors operating under different systems.

Recognizing these interconnected challenges, the establishment of comprehensive national bioeconomy strategies represents an essential response that integrates technological development, regulatory frameworks, and international engagement. Such strategies must explicitly address risk culture variations not as temporary obstacles to be eliminated but as enduring features of the international landscape that require sophisticated governance approaches. By emphasizing domains where national values and capabilities align with global needs—sustainable production, health security, and environmental applications—these strategic visions create opportunities for leadership that transcend narrow commercial interests. The resulting policy frameworks provide direction for both public and private sector investments while positioning nations advantageously within the competitive global bioeconomy landscape shaped by these divergent risk perceptions.

The advancement of standardization for biomanufactured products represents a critical mechanism within these broader strategies for addressing market fragmentation resulting from risk culture differences. Standardized protocols for production, testing, and verification enable cross-border commerce while maintaining appropriate safety and quality controls acceptable across different risk governance systems. Leadership in developing these standards requires strategic engagement through international organizations and bilateral partnerships focused on specific technical domains where consensus remains possible despite broader governance differences. These standardization initiatives should prioritize areas where technological leadership aligns with substantial market opportunities, creating reinforcing advantages for continued innovation while establishing governance frameworks that accommodate different risk perceptions within workable technical parameters.

Strategic biotechnology partnerships with nations sharing similar security concerns and technological capabilities offer particularly promising avenues for addressing risk culture challenges through coordinated approaches to shared priorities. Collaborative platforms create opportunities for establishing common standards within specific application domains that respect different risk cultures while enabling progress toward shared objectives. These focused collaborations potentially overcome broader governance differences by demonstrating effective risk management within defined technological spaces, establishing models for balancing appropriate safeguards with innovation advancement. Such partnerships strengthen influence in shaping biotechnology development trajectories while building governance frameworks acceptable across multiple jurisdictions despite persisting differences in underlying risk perceptions.

These international collaborations directly support enhanced supply chain resilience—another critical dimension affected by risk culture variations across production jurisdictions. Nations maintaining different standards for containment, verification, and quality control create inconsistent operating environments that complicate integrated production systems for biomanufactured products. Distributed bioindustrial manufacturing capabilities address these challenges by establishing production networks across jurisdictions sharing similar risk governance approaches. These distributed systems enhance resilience against both market disruptions and regulatory changes while positioning participating nations advantageously within evolving global value chains. By linking supply chain development with governance standardization, this approach creates mutually reinforcing technical and regulatory capabilities essential for navigating an increasingly fragmented international regulatory landscape.

The convergence of synthetic biology with artificial intelligence further amplifies these governance challenges by introducing new dimensions of complexity and opacity that strain traditional risk assessment methodologies across all cultural contexts. Machine learning approaches to biological design create unprecedented capabilities that different risk cultures will inevitably interpret and regulate through their distinctive governance lenses, potentially widening existing regulatory divergences.

Nations that develop governance frameworks accommodating this convergence while maintaining appropriate oversight will likely establish leadership positions in next-generation biotechnology platforms. In practice, it is possible that some nations would not consider AI as a method that deserved special attention in the regulatory framework for biotechnology at all, while others may specifically limit or prohibit use of AI in particular contexts of biotechnology development. This governance challenge represents both risk and opportunity, potentially enabling leadership through frameworks that balance innovation facilitation with risk management considerations appropriate to these transformative technologies while remaining adaptable to different cultural contexts.

The distribution of benefits from synthetic biology advances will ultimately reflect how effectively nations navigate these interconnected risk culture challenges while developing their bioeconomy capabilities. Countries that establish both innovative capacity and appropriate governance frameworks position themselves advantageously within emerging value chains by enabling domestic development while facilitating international market access. The optimal approach requires balancing technological leadership with addressing legitimate risk concerns in ways compatible with both domestic values and international market requirements. This balanced strategy demands coordinated policy spanning research funding, regulatory oversight, and international engagement that collectively addresses the fundamental risk culture variations shaping biotechnology governance worldwide. Nations mastering this complex balance will secure disproportionate benefits through both direct innovation returns and structural advantages within global markets increasingly defined by divergent perceptions of technological risk.

These implications highlight why risk culture understanding must form a central element of biotechnology strategy rather than representing a peripheral consideration. The development of a sustainable, resilient bioeconomy requires governance approaches aligned with both technological capabilities and cultural values regarding appropriate risk management. By addressing these challenges through comprehensive approaches spanning technology development, standardization, and international engagement, nations can establish enduring leadership in biotechnology domains increasingly critical to national security, economic competitiveness, and global influence in the decades ahead. The nation that most effectively navigates this complex landscape of divergent risk cultures will likely define the trajectory of the global bioeconomy—shaping not only technological development but also the governance frameworks that determine how these powerful capabilities transform societies worldwide.

5.5 Roadblocks and Strategic Pathways for U.S. Biotechnology Modernization

The United States faces significant interconnected challenges in its pursuit of biotechnology modernization and bioeconomy development. These roadblocks stem not just from technical or financial limitations, but from deeper structures within America's distinctive risk culture and governance approaches. Understanding these barriers and their relationships reveals potential strategic pathways that could transform challenges into opportunities for leadership in the coming decade.

Regulatory fragmentation represents a potential barrier to biotechnology advancement in the American context. With oversight distributed across FDA, EPA, USDA, and other agencies, synthetic biology applications that inherently cross traditional boundaries face a complex maze of diverse requirements and uncertain jurisdictions. This fragmentation is no accident, but rather a reflection of America's preference for distributed authority and checks on centralized power—values deeply embedded in the nation's risk governance traditions. The consequences extend far beyond mere procedural inconvenience, as this uncertainty drives investment toward applications with clearer oversight pathways regardless of strategic importance, extends development timelines, and significantly increases compliance costs. Novel applications lacking regulatory precedent face particular difficulties, often stalling in development despite promising capabilities and substantial market potential.

This regulatory complexity directly feeds infrastructure limitations that further constrain the translation of laboratory innovations into commercial realities. Despite world-leading research capabilities, America lacks sufficient pilot-scale and demonstration-scale biomanufacturing facilities to bridge the critical gap between proof-of-concept and industrial production. This infrastructure deficit stems partly from historical investment patterns that prioritized breakthrough science over implementation capabilities—another manifestation of American risk culture that traditionally values disruptive innovation over incremental improvement. The resulting dependency on foreign manufacturing for many biologically-derived products creates vulnerabilities painfully evident during recent supply chain disruptions, affecting critical materials across healthcare, defense, and industrial applications. Without addressing this infrastructure gap, even the most promising innovations remain stranded in laboratories, their economic and strategic potential unrealized despite substantial public research investments.

The infrastructure challenge cannot be separated from workforce limitations that constrain biotechnology scaling across the value chain. The interdisciplinary nature of synthetic biology demands expertise spanning molecular biology, engineering, computer science, and other specialized domains—creating distinctive educational requirements poorly served by traditional siloed programs. This skills shortage extends beyond scientific positions to include bioprocess engineers, regulatory specialists, and technical operations personnel essential for establishing and maintaining production systems. American technical education has historically emphasized theoretical knowledge over practical implementation skills, creating a persistent gap between research capabilities and manufacturing expertise. These workforce limitations mean that even substantial investments in physical infrastructure would deliver suboptimal returns without corresponding investments in human capital development focused specifically on biomanufacturing capabilities.

Against this backdrop of implementation challenges, the dual-use potential inherent in synthetic biology capabilities introduces security considerations that further complicate modernization efforts. Many advanced applications simultaneously enable both transformative benefits and potential misuse scenarios that raise legitimate concerns in an increasingly complex geopolitical environment. Traditional biosecurity approaches focused on restricting access to specific pathogens prove inadequate for addressing synthetic biology capabilities distributed across research institutions, companies, and international collaborations. This security dimension reflects a fundamental tension within American risk culture between openness that facilitates innovation and controls necessary for preventing harmful applications. Finding governance approaches that address security concerns without imposing excessive restrictions that impede beneficial development represents a central challenge for sustainable advancement, particularly given inconsistent security standards across international jurisdictions.

These security considerations connect directly to challenges stemming from international market fragmentation resulting from divergent regulatory approaches across nations with different risk cultures. The absence of globally recognized standards for evaluating synthetic biology products creates significant market access barriers, particularly for innovative applications lacking established precedents. Different jurisdictions impose conflicting requirements for demonstrating safety, environmental impact, or containment measures that substantially increase compliance costs for products targeting global markets. American firms face particular challenges navigating these inconsistent requirements while operating within a domestic governance system that emphasizes both innovation facilitation and appropriate safeguards. This international fragmentation effectively constrains global scaling opportunities despite substantial technological capabilities, limiting both economic returns and strategic influence for U.S. biotechnology advancements.

These challenges in scaling and market access intersect with capital availability limitations that create critical financing gaps in the development pipeline. While American venture capital readily supports early-stage research and development, the extended timelines and substantial capital requirements for scaling biomanufacturing capabilities exceed traditional investment models expecting rapid returns. This mismatch proves particularly problematic for applications requiring significant infrastructure development before generating substantial revenue—precisely the types of capabilities most essential for bioeconomy development and national resilience. These capital constraints reflect risk perception patterns within financial markets that inadequately value long-term strategic advantages associated with domestic biomanufacturing capabilities. Applications with significant national security or resilience implications face particular disadvantages in attracting private capital despite their strategic value, creating a persistent gap between innovation potential and deployment reality.

Completing this interconnected web of challenges, sustainability assessment frameworks represent a critical but underdeveloped dimension of biotechnology governance. While many synthetic biology applications promise significant environmental advantages over traditional manufacturing approaches, these benefits require verification through standardized methodologies that accurately measure full lifecycle impacts. Without transparent frameworks for evaluating resource utilization, energy consumption, waste generation, and environmental footprint, biotechnology products face skepticism or barriers despite potential sustainability improvements. This challenge affects both domestic acceptance and international market access, particularly in jurisdictions with stringent sustainability requirements or different assessment protocols. Developing credible, standardized sustainability metrics represents both a necessity for market development and an opportunity for differentiation in increasingly environmentally conscious markets.

These interconnected challenges, while substantial, reveal promising strategic pathways that could significantly advance U.S. biotechnology modernization in the coming decade. Rather than requiring fundamental restructuring of American institutions or values, these pathways leverage existing strengths while addressing specific limitations through targeted interventions. A coordinated biotechnology regulatory framework that maintains distributed agency authorities while implementing harmonized assessment approaches would significantly reduce current barriers without contradicting American governance preferences. Such a framework could standardize risk assessment methodologies across agencies, establish coordinated review timelines for multi-jurisdictional applications, and provide clear determination procedures for novel biological systems—maintaining necessary safeguards while improving predictability essential for investment and development.

This regulatory coordination would complement strategic investments in distributed biomanufacturing infrastructure designed for flexibility and resilience. Next-generation bioreactor systems incorporating automation, continuous processing capabilities, and modular designs would address critical scaling barriers while enhancing domestic production capabilities for critical materials. Establishing a network of such facilities across geographic regions with complementary specializations would create redundant production capabilities while facilitating technology transfer from research institutions to commercial applications. This distributed approach aligns with American preferences for resilience through diversity rather than centralized control, creating infrastructure that enhances national security and economic development while respecting cultural values regarding appropriate governance structures.

The effectiveness of this infrastructure depends critically on parallel workforce development initiatives spanning educational institutions, industry partnerships, and government programs. Integrating theoretical foundations with practical applications through hands-on training, industry apprenticeships, and continuing education would address immediate skills gaps while building long-term capacity for bioeconomy expansion. Specialized training in biological manufacturing processes, quality control protocols, and regulatory compliance would fill critical workforce needs currently constraining industrial scaling. Establishing industry-recognized credentials that validate specific skill sets would create clear career pathways while ensuring workforce capabilities align with evolving industry requirements—an essential foundation for translating America's substantial research advantages into manufacturing capabilities that drive economic growth and enhance national resilience.

Addressing security considerations within this development framework requires integrated biosecurity protocols that balance innovation facilitation with appropriate safeguards against potential misuse. Rather than imposing blanket restrictions that impede beneficial applications, these protocols should emphasize "designed-in" safety features that mitigate specific risk factors during early development stages. A rapid-response interagency capability for assessing emerging synthetic biology developments could provide timely guidance on risk management approaches appropriate to specific applications. Integrating advanced monitoring systems capable of identifying anomalous research activities or concerning patterns would enhance security awareness while maintaining the openness essential for continued scientific advancement—a balance particularly important within American research traditions that value both innovation and responsible governance.

These domestic governance improvements would support international engagement strategies focused on establishing standardized metrics and certification protocols for biomanufactured products. Developing transparent, science-based standards addressing production methods, performance characteristics, and safety parameters would facilitate market access across different regulatory environments. By leading international standardization efforts while building bilateral agreements with key trading partners, the United States could establish assessment frameworks that accommodate different risk perspectives within workable technical parameters. This approach would enhance the global competitiveness of American biotechnology products while addressing a critical barrier to international scaling—leveraging existing strengths in technical development while creating governance frameworks acceptable across different cultural contexts.

Financing these scaling efforts requires strategic public–private partnerships that share development costs during the critical transition from laboratory to commercial production. Focusing these partnerships on applications with both commercial potential and strategic significance would bridge critical financing gaps, while validating technologies for subsequent private investment. Performance-based funding models incorporating milestone-driven financing would ensure accountability, while maintaining innovation flexibility essential for addressing emerging challenges. This approach leverages American strengths in entrepreneurial development and public–private collaboration in biotechnology scaling—creating pathways for translating research advantages into manufacturing capabilities essential for bioeconomy leadership.

Uniting these strategies, standardized frameworks for assessing environmental and economic sustainability would verify the benefits of biomanufacturing, while addressing both domestic and international market requirements. Evaluating feedstock sourcing, production efficiency, waste valorization, and lifecycle impacts through transparent methodologies would allow comparative assessment against traditional manufacturing approaches. Incentivizing private sector adoption through procurement preferences, certification advantages, and market access benefits would align economic incentives with sustainability objectives. This approach would potentially differentiate American products in international markets increasingly sensitive to environmental considerations, creating competitive advantages while ensuring that biotechnology advancement contributes to broader sustainability goals.

These strategic pathways collectively create an integrated approach to biotechnology modernization aligned with American values, institutional structures, and competitive advantages. By addressing regulatory fragmentation, infrastructure limitations, workforce constraints, risk assessment, security considerations, market barriers, capital availability, and sustainability assessment through coordinated initiatives, the United States can establish foundations for a sustainable bioeconomy that enhances national security, economic resilience, and global influence. This approach recognizes that effective biotechnology modernization requires not merely technical advancement, but also governance innovations that navigate the complex landscape of risk perceptions, institutional structures, and cultural values that shape domestic development and international competitiveness in this increasingly critical domain.

References

1. Mizutani M, Glass JI, Fukatsu T, Suzuki Y, Kakizawa S (2025) Robust and highly efficient transformation method for a minimal mycoplasma cell. J Bacteriol 207(3):e0041524
2. Titus AJ, van Opstal E, Rozo M (2020) Biotechnology in defense of economic and national security. Health Secur 18(4):310–312
3. O'Brien JT, Nelson C (2020) Assessing the risks posed by the convergence of artificial intelligence and biotechnology. Health Secur 18(3):219–227
4. Trump BD (2017) Synthetic biology regulation and governance: lessons from TAPIC for the United States, European Union, and Singapore. Health Policy 121(11):1139–1146
5. Douglas M, Wildavsky A (1983) Risk and culture: an essay on the selection of technological and environmental dangers. University of California Press
6. Elowitz MB, Leibler S (2000) A synthetic oscillatory network of transcriptional regulators. Nature 403(6767):335–338

7. Cameron DE, Bashor CJ, Collins JJ (2014) A brief history of synthetic biology. Nat Rev Microbiol 12(5):381–390
8. Gibson DG, Glass JI, Lartigue C, Noskov VN, Chuang RY, Algire MA et al (2010) Creation of a bacterial cell controlled by a chemically synthesized genome. Science 329(5987):52–56
9. Groff-Vindman CS, Trump BD, Cummings CL, Smith M, Titus AJ, Oye K, Prado V, Turmus E, Linkov I (2025) The convergence of AI and synthetic biology: the looming deluge. npj Biomed Innovations 2(1):20
10. Paddon CJ, Keasling JD (2014) Semi-synthetic artemisinin: a model for the use of synthetic biology in pharmaceutical development. Nat Rev Microbiol 12(5):355–367
11. Wells E, Trump BD, Finkel AM, Linkov I (2020) A solution-focused comparative risk assessment of conventional and emerging synthetic biology technologies for fuel ethanol. In: Synthetic biology 2020: frontiers in risk analysis and governance, pp 223–255
12. Roell MS, Zurbriggen MD (2020) The impact of synthetic biology for future agriculture and nutrition. Curr Opin Biotechnol 61:102–109
13. Liu AP, Appel EA, Ashby PD, Baker BM, Franco E, Gu L, Chaudhuri O et al (2022) The living interface between synthetic biology and biomaterial design. Nat Mater 21(4):390–397
14. Rylott EL, Bruce NC (2020) How synthetic biology can help bioremediation. Curr Opin Chem Biol 58:86–95
15. Hodgson A, Maxon ME, Alper J (2022) The US bioeconomy: charting a course for a resilient and competitive future. Ind Biotechnol 18(3):115–136
16. Holland C, Shapira P (2024) Building the bioeconomy: a targeted assessment approach to identifying biobased technologies, challenges and opportunities. Eng Biol 8(1):1–15
17. Mao N, Aggarwal N, Poh CL, Cho BK, Kondo A, Liu C, Chang MW et al (2021) Future trends in synthetic biology in Asia. Adv Genet 2(1):e10038
18. Beck U (2006) Risk society revisited: theory, politics and research programmes. In: The sociology of risk and gambling reader. Routledge, pp 68–90
19. Jasanoff S (2015) Future imperfect: science, technology, and the imaginations of modernity. In: Dreamscapes of modernity: sociotechnical imaginaries and the fabrication of power, pp 1–33
20. Kelemen RD (2011) Eurolegalism: the transformation of law and regulation in the European Union. Harvard University Press
21. Allhoff F (2020) Risk, precaution, and emerging technologies. In: Emerging technologies, pp 273–300
22. More M (2013) The proactionary principle: optimizing technological outcomes. In: The transhumanist reader, pp 258–267
23. Douglas M (1970) Natural symbols: explorations in cosmology. Barrie & Rockliff, Cresset Press
24. Silbergeld EK, Graham JP (2008) The Cuyahoga is still burning. Environ Health Perspect 116(4):A150
25. National Research Council (1983) Risk assessment in the Federal Government: managing the process. Washington DC
26. National Research Council (2009) Science and decisions: advancing risk assessment. Washington DC
27. Jasanoff S (2005) Technologies of humility: citizen participation in governing science. In: Governing science. VS Verlag für Sozialwissenschaften, pp 370–389
28. Vogel D, Lynch D (2001) The regulation of GMOs in Europe and the United States: a case-study of contemporary European regulatory politics. Council on Foreign Relations, New York/Washington
29. Freeman HB (2002) Trade epidemic: the impact of the mad cow crisis on EU–US relations. BC Int'l Comp L Rev 25:343
30. Patterson LA, Josling T (2012) Regulating biotechnology: comparing EU and US approaches. In: Environmental policy in the EU. Routledge, pp 201–218
31. Parthasarathy S (2019) Patent politics: life forms, markets, and the public interest in the United States and Europe. University of Chicago Press

32. Lofstedt RE (2006) How can we make food risk communication better: where are we and where are we going? J Risk Res 9(8):869–890
33. Löfstedt R, Way D (2016) Transparency and trust in the European pharmaceutical sector: outcomes from an experimental study. J Risk Res 19(9):1082–1103
34. Executive Office of the President (1986) Office of science and technology policy: coordinated framework for regulation of biotechnology, 51 FR 23302, at 23302–23303
35. Renn O, Roco M (2006) White paper on nanotechnology risk governance. International Risk Governance Council, Geneva
36. Lee HL, Vavitsas K (2021) Driving innovation in biotechnology with private-public partnerships: a Singapore perspective. Biotechnol Notes 2:59–62
37. Slovic P (1987) Perception of risk. Science 236(4799):280–285
38. Carpenter D, Moss DA (eds) (2013) Preventing regulatory capture: special interest influence and how to limit it. Cambridge University Press
39. Trump B, Cummings C, Kuzma J, Linkov I (2020) Synthetic biology 2020: frontiers in risk analysis and governance. Springer, New York
40. Vogel D (2012) The politics of precaution: regulating health, safety, and environmental risks in Europe and the United States. In: The politics of precaution. Princeton University Press
41. Bernauer T, Meins E (2003) Technological revolution meets policy and the market: explaining cross-national differences in agricultural biotechnology regulation. Eur J Polit Res 42(5):643–683
42. Funtowicz SO, Ravetz JR (1993) Science for the post-normal age. Futures 25(7):739–755
43. Marchant GE, Wallach W (2015) Coordinating technology governance. Iss Sci Technol 31(4)
44. Greer SL, Trump B (2019) Regulation and regime: the comparative politics of adaptive regulation in synthetic biology. Policy Sci 52(4):505–524
45. Jasanoff S (2018) Just transitions: a humble approach to global energy futures. Energy Res Soc Sci 35:11–14 or Jasanoff A (2018) The biological mind: how brain, body, and environment collaborate to make us who we are. Basic Books

Chapter 6
Governance Frameworks and Scenario Evaluation Exercises on AI and SB Ethical Dilemmas

Juan C. Cruz, Ruth Mampuys, German Reyes, and Sarah Dryhurst

Abstract This chapter meticulously examines the profound ethical dilemmas arising from the rapid convergence of Artificial Intelligence (AI) and Synthetic Biology (SB), advocating for innovative and adaptive governance frameworks. Through a critical analysis of existing international, regional, and national governance structures, and by employing illustrative scenario evaluations in healthcare (e.g., AI-driven gene editing), agriculture (AI-enhanced crop development), environmental conservation (AI-designed bioremediation organisms), and public perception, the chapter highlights how this technological fusion intensifies concerns. These include data privacy, algorithmic opacity, dual-use applications, intellectual property, equitable access, and ecological integrity, often outpacing traditional regulatory oversight. A core argument is the necessity for governance to be anticipatory, culturally sensitive, and participatory, addressing the unique challenges posed by AI's predictive power combined with SB's transformative potential. The chapter underscores recurring ethical themes—transparency, equity, autonomy, environmental impact, and public trust—and stresses the risk of ethical imperialism if diverse global perspectives are not integrated. It concludes by emphasizing the need for interdisciplinary collaboration, robust stakeholder engagement, and flexible governance mechanisms to ensure that AI-SB innovations are developed responsibly, aligning technological advancements with diverse societal values and ethical principles.

J. C. Cruz (✉) · G. Reyes
Department of Biomedical Engineering, Universidad de los Andes, Bogotá, Colombia
e-mail: jc.cruz@uniandes.edu.co

G. Reyes
e-mail: g.reyes@uniandes.edu.co

R. Mampuys
The Netherlands Commission on Genetic Modification (COGEM), Bilthoven, Utrecht, The Netherlands
e-mail: mampuys@wrr.nl

S. Dryhurst
Institute for Risk and Disaster Reduction, University College London, London, UK
e-mail: s.dryhurst@ucl.ac.uk

C. L. Cummings et al. (eds.), *Biotechnology and AI: Technological Convergence and Information Hazards*, NATO Science for Peace and Security Series A: Chemistry and Biology, https://doi.org/10.1007/978-3-032-05246-9_6

Keywords AI-synthetic biology governance · Ethical dilemmas · Scenario evaluation · Cultural considerations · Adaptive regulation

6.1 Introduction

The rapid incorporation of Artificial Intelligence (AI) into Synthetic Biology (SB) poses a complex landscape of ethical challenges that require the timely formulation of innovative approaches for governance and regulation. With the advent of this new technological age, there is a noticeable sense of both excitement and concern about the capacity of AI and SB to significantly shape our environment. These technologies offer potential resolutions to some of humanity's most urgent challenges, ranging from climate change to food instability and incurable diseases. However, they also raise significant questions about the limits of human interference in natural systems and the fair allocation of technological advantages across society. In the case of standalone SB, regulatory pathways are relatively established. By contrast, AI–SB introduces dynamic, learning-based modifications, prompting real-time adjustments that may bypass traditional checkpoints.

Herein, the term AI primarily refers to machine learning and data-driven predictive tools that can identify patterns and generate insights from large datasets at unprecedented speed and scale. SB is used here to describe the rational engineering of biological components—such as genes or metabolic pathways—often toward designing new functions in living organisms. Genetic Modification (GMO) can be one aspect of SB, though GMOs typically involve transferring or altering specific genes, whereas SB can extend to creating entirely new synthetic pathways and genetic circuits. Where these fields intersect—an AI–SB convergence—AI algorithms significantly accelerate and optimize the design, testing, and refinement of synthetic organisms. Throughout this chapter, we highlight the ethical, cultural, and governance challenges arising from this convergence, which combines the predictive power of AI with the transformative potential of SB.

In this chapter, we begin by establishing the scope of AI and SB convergence, highlighting how these rapidly advancing fields hold promise for addressing global challenges while also raising new and urgent ethical questions. We then examine existing governance structures—at the international, regional, and national levels—to see how these frameworks grapple with the unique demands of AI–SB integration. Building on this groundwork, we offer a series of illustrative scenarios in healthcare, agriculture, environmental conservation, and public perception to show how ethical dilemmas play out in real or near-real-world contexts. From there, we broaden our lens to consider the significance of cultural and regional differences, arguing that any effective governance approach must reflect a sensitivity to diverse ethical perspectives and priorities. Finally, we conclude by synthesizing the key insights and suggesting strategies for ensuring that AI–SB technologies develop responsibly, balancing innovative potential with ethical and biosafety imperatives.

6.1.1 Defining the AI-SB Convergence

The integration of AI with SB is significantly shaping several fields, each presenting its own distinct ethical dilemmas. In precision health, AI-driven diagnostics and personalized therapies offer the potential for more precise and customized healthcare delivery. This progress, however, gives rise to issues around data privacy, algorithmic bias, and fair access to these technologies [1]. Robust controls are necessary to preserve individual privacy and prevent misuse of the vast quantities of sensitive health data required and generated by these systems [2]. Moreover, incorporating AI into healthcare decision-making processes raises inquiries regarding the significance of human judgment in medical practice and the possibility of overdependence on algorithmic suggestions [3].

In agriculture, AI and SB are transforming farming methods, improving productivity, and strengthening food security. However, these technological breakthroughs are not devoid of ethical dilemmas either. A broader evaluation of genetically modified organisms (GMOs)—encompassing both risk assessments and socio-economic considerations—is necessary due to their possible influence on small-scale farmers, the potential risk of monocultures, and the long-term ecological consequences [4]. These issues are longstanding in GM agriculture, but the convergence with AI can intensify them by accelerating trait development and complicating regulatory oversight. Although formal risk assessment processes exist in many countries, socio-economic evaluations are often not mandatory, highlighting the need for voluntary or policy-driven frameworks to address distributive impacts, rural livelihoods, and cultural preservation. The inherent conflict between technological advancements and the conservation of traditional agricultural methods underscores the necessity for comprehensive discourse and policy formulation. Additionally, the use of AI in agriculture gives rise to apprehensions over the ownership and regulatory authority over data, particularly concerning the independence of farmers and the possibility of corporate monopolization of agricultural expertise [5].

Implementing AI and SB technologies for biodiversity protection poses distinct ethical dilemmas. The prospect of preventing extinctions and manipulating genes in wild populations raises significant questions regarding our influence on ecosystems and the possible unforeseen outcomes of such actions [6]. These interventions blur the boundaries between natural and engineered domains, complicating our conventional ideas of preservation and environmental management. When AI accelerates or refines genetic interventions in wild populations, it adds layers of complexity: algorithms can propose novel gene edits at a pace and scale that challenge traditional conservation paradigms. The notion of using SB interventions instead of longstanding conservation efforts thus compels us to reassess our moral obligations towards the environment and future generations [7].

6.1.2 Key Ethical Concerns

The task of developing ethical frameworks to tackle these various problems is further complicated by the fast rate of technological advancement. Established regulatory systems typically face difficulties keeping up with the rapid progress in AI and SB, resulting in governance deficiencies that may cause unanticipated negative consequences. Therefore, it is imperative to create flexible and responsive frameworks that can adapt and evolve in real-time alongside the technologies they control [8]. In this domain, the notion of "anticipatory governance" assumes major importance, necessitating policymakers and ethicists to anticipate prospective problems and formulate adaptable principles capable of accommodating emerging challenges [9].

One key obstacle in creating governance frameworks is achieving a proper equilibrium between promoting innovation and guaranteeing ethical adherence. Excessively stringent constraints pose the danger of impeding growth and possibly denying society advantageous innovations. In contrast, inadequate supervision may result in ethical incompetence and potential harm. An optimal framework should possess the necessary adaptability to support swift innovation while upholding strong ethical principles [10]. This delicate balance is especially vital in the domain of SB, where the possibility of transformative breakthroughs must be carefully considered in relation to the hazards of generating new organisms that may have unexpected effects on ecosystems and human well-being [11].

Effective anticipatory governance necessitates proactive rather than responsive approaches. Conventional methods of ethical evaluation frequently address problems after they occur, but the swift progress and possible consequences of AI and SB require anticipatory approaches. This entails foreseeing potential ethical dilemmas and establishing precautions before issues arise [12]. Developing comprehensive foresight methods becomes essential in this context. Techniques such as scenario planning, horizon scanning, Delphi studies, and trend analysis enable the exploration of possible futures and their ethical implications [13]. Additionally, technology assessment approaches like Constructive Technology Assessment (CTA) and Real-Time Technology Assessment (RTTA) can evaluate the potential impacts of emerging technologies on society, helping to anticipate and address ethical concerns proactively.

The global scope of SB and AI research adds another layer of intricacy to the formulation of ethical frameworks. Cultural disparities in values and norms result in divergent perspectives on what is deemed valuable and ethical across different societies. To address this variation, it is essential to create frameworks that are adaptable, culturally sensitive, and inclusive, enabling the application of ethical norms worldwide while honoring local beliefs [14]. Although Jasanoff focused primarily on the transnational implications of biotechnology, her insights regarding the cross-border impact of emerging technologies also illuminate how AI—especially when converged with SB—can produce effects that transcend national and cultural boundaries [15]. Consequently, the formulation of ethical principles for AI–SB is challenged by the wide range of cultural viewpoints and societal values, reinforcing the need for governance that is both globally minded and locally responsive.

Tackling these complex issues necessitates broad, interdisciplinary cooperation. Collaboration among ethicists, technologists, policymakers, and stakeholders from various disciplines is necessary to create inclusive frameworks that consider heterogeneous viewpoints [16]. Adopting a collaborative approach is key to developing comprehensive guidelines for navigating the intricate ethical terrain of AI and SB convergence. In this setting, incorporating social sciences and humanities into STEM disciplines becomes essential, as it enables a comprehensive recognition of the societal consequences of modern innovations [17].

The design of adaptable ethical frameworks is not only a technical or regulatory obstacle but also a societal one. Sustained public involvement and discussion are necessary to ensure that the ethical standards governing emerging technologies align with society's values and concerns. This interaction should take an intersectional approach, considering various viewpoints within the intersections of gender, race, disability, and other dimensions to establish genuinely inclusive frameworks [18]. Participatory modes of technology governance, such as citizen juries and consensus conferences, are crucial instruments in ensuring that the advancement of these technologies meets society's values and expectations [8].

6.1.3 Governances and Policy Challenges

The convergence of AI and SB is not just a scientific milestone; it's a transformative force reshaping the very fabric of life, because it merges rapid, data-driven insights with the power to modify living systems in ways that can surpass traditional biological engineering. These questions—posed since the early days of recombinant DNA technology—remain deeply relevant today, yet become more pressing when AI accelerates and automates processes beyond human comprehension. How far should we go in altering natural systems when AI-driven gene editing can design and execute modifications at breakneck speeds? Who decides which AI-driven technologies are pursued and who benefits from them, particularly when automated bioengineering may bypass traditional oversight? The urgency of these questions is underscored by the rapid pace of AI–SB development, which often outstrips the evolution of ethical guidelines and regulatory frameworks.

Although biosafety risks and data privacy concerns have long accompanied genetic modification efforts, these issues grow more complex when AI algorithms manage or optimize biological design. By hastening each iteration and analyzing enormous genomic datasets, AI amplifies existing challenges around environmental containment, patient data protection, and equitable access. At the same time, AI–SB convergence introduces concerns that rarely surfaced under traditional SB or GMO processes, such as minimal human oversight in automated gene edits, algorithmic opacity that complicates risk assessment, and the rapid emergence of novel phenotypes before adequate review can take place. As a result, the moral and regulatory questions we once treated as familiar now demand updated approaches that match AI's dynamic influence on synthetic biology.

In this rapidly evolving landscape, the stakes are high. The decisions we make today will shape the future of humanity and the planet. Therefore, it is imperative to engage in proactive ethical deliberation, interdisciplinary collaboration, and inclusive policymaking to navigate the ethical dilemmas presented by the convergence of AI and SB (Fig. 6.1).

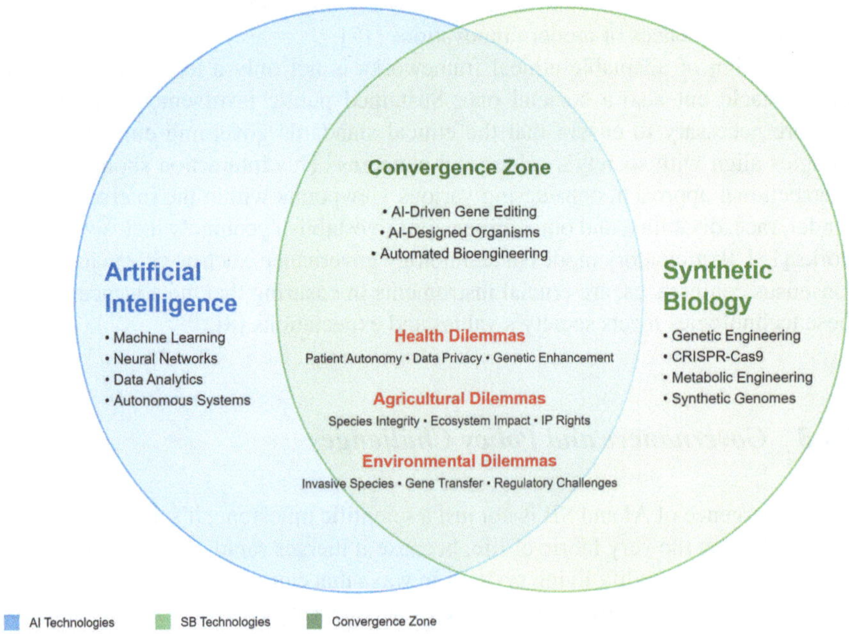

Fig. 6.1 Integration landscape of AI and synthetic biology: a conceptual framework illustrating the convergence of artificial intelligence (AI) and synthetic biology (SB), highlighting their distinct capabilities and shared applications. The intersecting domains reveal transformative convergence applications like AI-driven gene editing and automated bioengineering. The framework identifies critical ethical dilemmas emerging from this technological fusion across three key sectors: healthcare (patient autonomy, data privacy, genetic enhancement), agriculture (species integrity, ecosystem impact, intellectual property rights), and environmental conservation (invasive species risks, horizontal gene transfer, regulatory challenges). This visualization underscores the complex interplay between technological advancement and ethical considerations in the emerging bio-digital age

6.2 Current Ethical Frameworks and Governance Approaches

Ensuring public and environmental safety is not merely a technical or procedural matter—it is fundamentally an ethical imperative. While terms like "biosafety" and "biosecurity" may suggest narrow regulatory or scientific protocols, they speak directly to principles of non-maleficence and stewardship. In the context of AI–SB convergence, containing potential risks—whether unintentional ecological harm or deliberate dual-use applications—demands that scientists, policymakers, and society at large acknowledge their moral responsibility to protect both current and future generations. This perspective situates biosafety and biosecurity within the broader ethical framework, underscoring how safeguarding human and environmental well-being must be a core priority in any governance system for emerging biotechnologies.

Given the rapid evolution of these technologies, it is imperative that ethical frameworks maintain their flexibility and responsiveness to emerging concerns. The intersection of AI and SB gives rise to distinct ethical concerns, notably the possibility for AI to independently craft and enhance biological systems. This phenomenon prompts the exploration of novel inquiries on agency, accountability, and the limits of human interference in the natural world. The application of machine learning algorithms in gene editing technologies such as CRISPR-Cas9 raises intricate ethical concerns about the level of human supervision and the possibility of unforeseen outcomes in genetic alterations [19]. In the future, ethical frameworks will need to tackle new issues, including the long-term consequences of human genome editing, the possibility of autonomous AI systems in SB research, and the fair allocation of gains from these technologies. Furthermore, there is an increasing acknowledgment of the necessity to take into account the environmental and ecological consequences of new technologies, which corresponds to wider societal issues regarding sustainability and environmental management [20].

In this section, we primarily examine the ethical principles and guidelines that inform the governance of AI–SB convergence, including those developed by international bodies, regional regulators, and national oversight committees. Although our main emphasis is on ethical aspects—such as transparency, consent, equity, and cultural sensitivity—we also acknowledge the biosafety and dual-use (security) frameworks that often overlap with ethical concerns. For instance, environmental risk assessments and controls designed to prevent misuse can be viewed as extensions of the ethical obligation to protect public health, maintain ecological balance, and uphold social responsibility. Consequently, while we refer to biosafety and security measures where relevant, our overarching perspective remains grounded in the ethical dimension of these governance structures.

6.2.1 International Ethical Guidelines

In recent decades, regulators have established mechanisms to address biosafety and ethical concerns tied to genetic modification, including guidelines for contained trials, data protection, and risk assessments. However, when AI tools enter the mix—enabling in silico modeling of complex genomic pathways or real-time editing—these mechanisms often prove too narrow or slow.

Established codes and committees now face the added challenge of overseeing algorithmic decision-making, where machine learning can identify previously unexamined gene interactions and rapidly implement them in biological systems. Traditional approval pipelines can struggle to address questions about black-box AI logic, continuous iterative design, or the pace at which new traits can be introduced into crops, organisms, or human therapies.

Core ethical concepts relevant to both AI and SB have been articulated by international bioethics committees. This effort has been spearheaded by the UNESCO International Bioethics Committee (IBC), whose Universal Declaration on Bioethics and Human Rights is now universally acknowledged as a key guiding statement on global bioethics. This declaration underscores the vital significance of human dignity, human rights, and basic freedoms within the framework of developing technologies. The statement emphasizes the need to optimize advantages while reducing possible negative consequences, a concept especially important considering the potentially dual-use capabilities of AI and SB [21]. Building upon UNESCO's efforts, the Council for International Organizations of Medical Sciences (CIOMS) provides critical guidance for regulating SB and AI in health-related contexts. The CIOMS guidelines prioritize principles such as informed consent, respect for vulnerable populations, and fair distribution of research benefits and burdens. Furthermore, these frameworks underscore that biosafety and biosecurity measures are driven by the ethical responsibility to protect public health and the environment. By applying the precautionary principle, they acknowledge that the duty to prevent harm extends to anticipating potential misuse or unintended outcomes of AI–SB research. These resources are essential for researchers and policymakers grappling with the ethical implications of employing AI and SB in medical research and practice [22].

International organizations play a pivotal role in fostering global deliberation on bioethics and the ethics of AI, not merely as standard-setters but as facilitators of ethical capacity building both locally and globally. UNESCO exemplifies this approach through the development of national bioethics committees and the implementation of programs for teaching ethics. These committees act as nodes within a global network of ethical deliberation, providing opportunities for the exchange of ideas and best practices while respecting cultural diversity [23]. UNESCO's Ethics Education Program, initiated in 2004, has been instrumental in designing standard teaching curricula on ethics and organizing regional expert meetings to explore and validate such programs. This initiative facilitates the dissemination of a common understanding of ethical principles and the incorporation of local ethical perspectives into the global discourse. However, in some cultural contexts, consenting to

AI–SB interventions goes beyond individual choice, reflecting communal norms or family-based decision-making. This variation underscores the challenge of designing truly universal ethical guidelines.

A notable example of an international body advancing ethical harmonization in a complex scientific field is the Global Alliance for Genomics and Health (GA4GH). Its work extends beyond setting standards; it actively builds a global community of practice in ethical genomic data sharing. The organization's Framework for Responsible Sharing of Genomic and Health-Related Data is based on human rights principles, providing a universally applicable foundation that allows for cultural adaptation. GA4GH also develops practical tools and policies to operationalize ethical principles, such as consent policies and data access agreements.

By merging high-level ethical frameworks with practical implementation guidance, international organizations demonstrate how to translate general ethical principles into tangible practices governing emerging technologies. These bodies play essential roles in advising decision-makers on emerging technologies through the diversity of ethical perspectives. However, their influence may be limited by factors such as composition, mandate, and working methods. For example, the International Bioethics Committee of UNESCO comprises 36 independent experts chosen for their competence and geographical representation. While this diversity brings various cultural and disciplinary views to ethical issues, the challenge lies in synthesizing these views into consistent guidelines.

This variation in ethical norms is particularly evident in the international governance of CRISPR-Cas9 technology. While the United Kingdom has established a regulatory pathway for specific germline editing applications through the Human Fertilisation and Embryology Authority, China initially operated under less restrictive oversight, leading to the controversial case of CRISPR-edited babies in 2018. Meanwhile, Germany prohibits any embryo research that doesn't directly benefit the embryo itself, effectively precluding many germline editing applications. These divergent national approaches to a single technology illustrate how cultural and societal values fundamentally shape the ethical frameworks governing AI-SB, particularly regarding consent models, acceptable risk thresholds, and the balance between individual and societal benefits.

6.2.2 Regional and National Frameworks

The European Union (EU) has emerged as a prominent global authority in formulating comprehensive regulatory and ethical frameworks for both AI and SB. The EU's approach is characterized by a strong emphasis on human rights, safety, and transparency. For instance, Directive 2001/18/EC primarily focuses on environmental risk assessments and procedural requirements for the deliberate release of GMOs, although it intersects with broader ethical questions around biotech governance. Similarly, the General Data Protection Regulation (GDPR) highlights the importance of safeguarding personal data and individual rights in an increasingly data-driven landscape, a pivotal consideration for AI-based Synthetic Biology as it relies heavily on large-scale data collection and analysis [24]. The EU High-Level Expert Group on AI contributed further with its "Ethics Guidelines for Trustworthy AI," providing a thorough structure that emphasizes crucial criteria such as human agency, technical robustness, privacy, transparency, and accountability [25].

At the national level, countries have formulated their own ethical principles and regulatory structures. In the United States, the National Institutes of Health (NIH) provides explicit ethical guidelines for genetic research and SB, prioritizing risk assessment, ethical oversight, and responsible dissemination of research findings [26]. China has made notable progress by implementing new standards that emphasize public engagement and ethical governance in SB research and applications. These initiatives demonstrate a commitment to responsible innovation that considers ethical concerns and the welfare of the general population [27]. Tang et al. highlight differences in emphasis and implementation approaches between China and other nations, reflecting diverse cultural and regulatory environments [28].

Unlike the EU, the U.S. adopts a more risk-based and market-driven regime for regulating AI and SB, emphasizing the potential of these technologies for economic growth while maintaining appropriate safeguards. Formal regulation in the U.S. is combined with self-regulation and is implemented largely through the oversight of agencies such as the U.S. Department of Agriculture and the Environmental Protection Agency. The U.S. approach systematically evaluates potential risks against anticipated benefits, fostering innovation while maintaining targeted safety protocols, which has facilitated more rapid adoption of new technologies compared to the EU framework.

In Asia, particularly in China, the rapid advancement of AI-driven synthetic biology research has generated multifaceted ethical debates extending beyond laboratory concerns. The extensive collection and analysis of genetic data have sparked critical discussions regarding gene privacy, with questions about consent models that align with cultural values emphasizing collective over individual benefits. Additionally, the strategic importance of these technologies has elevated them to national security considerations, with implications for international collaboration and data-sharing protocols. Public transparency mechanisms have also been contested, as stakeholders navigate the balance between intellectual property protection, national competitiveness, and the public's right to information about technologies that may fundamentally alter biological systems [29]. Asian cultural values, especially collectivism versus individual rights, influence regulatory approaches. This has made public health and agricultural sector regulators relatively accepting and comfortable with SB.

China has been proactive in dealing with GMOs and SB, allowing for significant increases in agricultural productivity and public health interventions. However, this emphasis on collective gains can sometimes lead to overlooking individual rights and broader ethical concerns. Millett and Alexanian point out that while adaptive governance in Asia effectively promotes technological advancement and economic development, it may face challenges in ensuring individual rights and broader ethical considerations are adequately addressed [30]. As these technologies continue to advance, Asian countries may need to update their governance frameworks to incorporate new ethical challenges.

Regulations may diverge not only because of legal philosophies but also due to cultural perspectives on autonomy, privacy, and acceptable risk. For instance, what is deemed ethically permissible in one region can conflict with the spiritual or societal values in another.

6.2.3 Gaps in Existing Frameworks

Despite comprehensive efforts, significant gaps remain in addressing critical issues such as intellectual property, equitable access, and the potential for technological leapfrogging. However, another key challenge arises from the inherent dual-use nature of AI and SB technologies, which current frameworks often struggle to address effectively. While acknowledging the potential for misuse, existing control mechanisms can be insufficient to prevent it. From an ethical standpoint, this shortfall reflects a duty to prevent harm, wherein biosafety and biosecurity protocols must evolve alongside technological developments. Treating these measures merely as technical fixes overlooks the moral imperative to proactively safeguard society and ecosystems from emerging threats. Undheim argues for stronger governance systems to prevent abuse and ensure biosafety, including more stringent oversight of research and development activities that may pose biosecurity risks [31]. Yet in the fast-moving AI–SB landscape, purely procedural approaches may lag behind. Continuous investment in adaptive governance, international collaboration, and transparent risk assessment processes is crucial for mitigating dual-use concerns, especially as AI accelerates the speed and scale of synthetic biology research beyond the reach of traditional oversight mechanisms.

These dual-use concerns are exemplified by scenarios such as AI systems designed to identify novel pathogen variations that could be repurposed to engineer more virulent or transmissible infectious agents by circumventing existing detection systems. Similarly, synthetic biology applications for environmental bioremediation could potentially be adapted for military purposes, such as engineered microorganisms capable of degrading specific materials critical to infrastructure or defense systems. The algorithmic acceleration of these processes—where AI can rapidly iterate through thousands of potential genetic designs without human intervention—significantly intensifies biosecurity concerns and underscores the urgent need for more comprehensive oversight mechanisms with appropriate technical expertise and international coordination.

Taken together, these issues—ranging from benefit distribution to dual-use risks—underscore how current governance frameworks can fall short when tested by the rapid progress in AI–SB. Such gaps highlight the need for both global coordination and ethically grounded adaptability, ensuring that regulations keep pace with evolving technologies and societal values.

The task of ethical frameworks is to strike a balance between universal application and sensitivity to unique cultural situations. Regarding matters such as privacy, data ownership, or genetic modification, what is deemed ethical in one community may be perceived differently in another. Heavey contends that this requires an adaptable strategy that can accept and incorporate many viewpoints while upholding fundamental ethical principles [32]. Rather than identifying a gap, Heavey highlights that cultural adaptability is a core requirement of ethical frameworks, ensuring they remain relevant and equitable across diverse societal contexts. Tang et al. emphasize these cultural disparities in their comparative analysis of ethical strategies applied to the regulation of SB in China and other nations [28]. The authors observe that although safety and responsible innovation are often discussed topics, there are notable differences in the focus and methods of implementation that mirror diverse cultural and legal contexts. Furthermore, existing frameworks often assume a universal notion of informed consent. Yet, in many societies, communal or spiritual beliefs significantly shape how AI–SB innovations are evaluated, creating ethical tensions where these beliefs clash with Western-derived norms. There is considerable variation in attitudes towards personal data privacy in AI applications across different countries. Although several countries give priority to the rights of individual data, others may place greater emphasis on communal advantages than personal privacy. The presence of cultural disparities can give rise to disputes in multinational partnerships and hamper the formulation of universally relevant ethical norms.

However, although acknowledging the significance of public involvement, several existing frameworks—such as the regulatory strategies described by Tang et al. and the communication approaches highlighted by Kronberger et al.—are inadequate for successfully including the public in ethical deliberations and decision-making procedures [28, 33]. This absence of active involvement might result in a deficit of trust and impede the societal adoption of novel technology. In their 2009 study on conveying SB, Kronberger et al. emphasize this problem by pointing out that as knowledge disseminates from the laboratory through the media to the general population, it progressively emphasizes practical uses rather than ethical concerns [33]. In the field of SB, public opinion generally centers around possible hazards or exaggerated uses rather than the ethical dilemmas that scientists and politicians actively confront. Insufficient and ineffective communication can lead to uninformed public discussions and possibly erroneous policy choices, such as rushed or lopsided regulations that either stifle beneficial research or overlook legitimate public concerns.

The rapid rate of technical progress in AI and SB requires continuous, open discussion among ethicists, scientists, politicians, and the public to ensure that ethical governance remains up-to-date with innovation. A crucial aspect of this collaborative strategy is not just the ability to tackle present ethical issues but also the capacity to anticipate and prepare for future advancements in these rapidly changing domains. To successfully construct comprehensive and effective ethical frameworks, it is crucial to engage in interdisciplinary collaboration and global cooperation as we negotiate these intricate ethical landscapes. While ensuring the protection of human rights, dignity, and the wider interests of society and the environment, these frameworks must promote responsible innovation (i.e., development that respects fundamental rights to privacy, equality, and environmental stewardship). The incorporation of technological proficiency with ethical deliberations will be essential in formulating governance frameworks that can efficiently tackle the distinct obstacles presented by the interaction of AI and SB (Fig. 6.2).

In addition to the gaps identified, developing nations may benefit from 'technological leapfrogging,' where they bypass intermediary tech stages by adopting advanced AI–SB tools directly. This phenomenon has already been observed in sectors like telecommunications, where many regions in Africa adopted mobile banking services without going through traditional banking infrastructure. Similarly, there may be opportunities for developing nations to leapfrog intermediate stages of healthcare and agricultural development by adopting sophisticated AI-based SB technologies directly. For instance, AI-driven precision agriculture could allow these countries to achieve significant advancements in food security, while AI-enhanced diagnostics could provide improved healthcare without the need for extensive traditional healthcare infrastructure. However, the success of technological leapfrogging depends

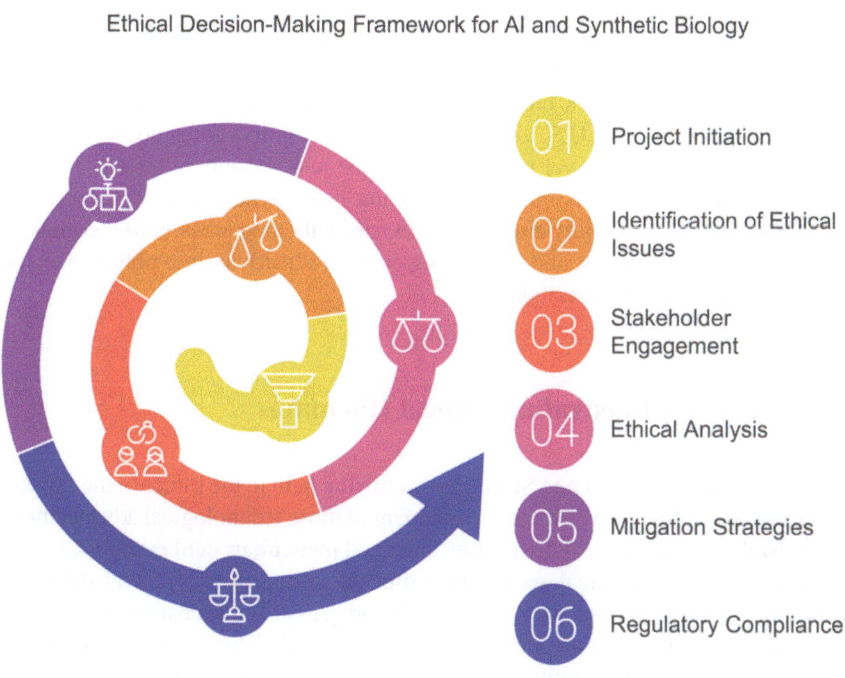

Fig. 6.2 Ethical decision-making framework for AI and synthetic biology: this spiral diagram outlines a six-step ethical decision-making process designed to ensure responsible project development in AI and synthetic biology. It begins with project initiation (step 1), followed by the identification of ethical issues (step 2) to recognize potential risks. Stakeholder engagement (step 3) involves key voices in deliberation, while ethical analysis (step 4) critically examines impacts. Mitigation strategies (step 5) aim to address identified concerns, and regulatory compliance (step 6) ensures adherence to ethical standards and legal requirements. This iterative approach promotes a structured, ethical progression through each phase of a project

on deliberate investments in capacity building, technology transfer, and fostering international partnerships to facilitate access to these advanced tools [34]. Such efforts could mitigate the risks of increased inequalities that are associated with these technologies and ensure a more equitable distribution of their benefits.

In this section, we explored international, regional, and national frameworks guiding the ethical governance of AI–SB convergence. Key issues include uncertainties around dual-use applications, data ownership, and equitable benefit-sharing, all of which current governance systems address only partially. Some existing measures—like UNESCO's bioethics declarations, EU regulations on genetically modified organisms, and U.S. agency oversight—provide a foundation for ethical and safety considerations, yet remain fragmented and variably enforced. To move forward, potential solutions include developing more agile regulatory approaches capable of addressing rapidly emerging risks, strengthening cross-border collaboration to share best practices, and embedding cultural sensitivity into policy to ensure that diverse values are respected.

In this section, we explored international, regional, and national frameworks guiding the ethical governance of AI–SB convergence. Key issues include uncertainties around dual-use applications, data ownership, and equitable benefit-sharing, all of which current governance systems address only partially. Some existing measures—like UNESCO's bioethics declarations, EU regulations on genetically modified organisms, and U.S. agency oversight—provide a foundation for ethical and safety considerations, yet remain fragmented and variably enforced. To move forward, potential solutions include developing more agile regulatory approaches capable of addressing rapidly emerging risks, strengthening cross-border collaboration to share best practices, and embedding cultural sensitivity into policy to ensure that diverse values are respected.

6.3 Scenarios Illustrating Ethical Dilemmas

Rapid advancements in AI and SB offer a promising path to the future in the fields of health and agriculture. Nevertheless, the advent of those technological advancements brings forth complex ethical quandaries that need meticulous deliberation. Outlined below are four hypothetical scenarios crafted to exemplify the ethical dilemmas emerging at this important stage of scientific progress. The scenarios presented in this section are fictional yet grounded in real-world trends. They are designed to highlight illustrative ethical dilemmas that may arise from the convergence of AI and SB, rather than to provide an exhaustive forecast of how these fields will evolve. Consequently, readers should note that not all real-world regulatory processes—such as the multiple permit and review stages typically required in many regions—are depicted in detail. This choice allows us to focus on how AI–SB might raise novel or intensified ethical questions, without being constrained by every procedural safeguard that could exist in practice. With these considerations in mind, we now present four hypothetical scenarios that illustrate the pressing ethical dilemmas emerging from AI–SB convergence.

These scenarios touch on biosafety and biosecurity to some extent, but their primary aim is to illustrate the ethical challenges associated with AI–SB convergence—challenges that can be intensified or reshaped when AI tools drive rapid or complex developments in SB. Throughout each scenario, we occasionally reference incomplete regulatory processes (e.g., field trial permits, post-market monitoring) to underscore that in real-world applications, multiple oversight mechanisms are typically in place. By selectively omitting certain procedural details, we keep the spotlight on core ethical dilemmas like privacy, equity, and autonomy, which remain central concerns even under robust governance frameworks.

The scenarios demonstrate varying focuses in their ethical considerations. The precision health scenario uniquely explores issues related to autonomy in genetic interventions, particularly regarding children, while the environmental conservation scenario delves into the regulatory challenges and biosafety concerns associated with adaptive synthetic organisms. The agricultural scenario focuses extensively on public perception, emphasizing the importance of cultural, societal, and consumer acceptance in the responsible deployment of AI and SB technologies. These differences reflect the multifaceted nature of ethical dilemmas in the convergence of AI and SB, requiring tailored governance and ethical frameworks to address each specific context.

6.3.1 Scenario 1: Precision Health

Conventional synthetic biology already grappled with questions about patient autonomy and consent, particularly when modifying genetic material to treat or prevent disease. Data privacy was likewise a priority, as genetic information can be highly sensitive. Now, AI's capability to propose and refine genetic solutions in near-real time intensifies those concerns, especially if physicians or patients cannot fully understand—or challenge—the AI's rationales. The following scenario underscores how these longstanding ethical debates become more urgent. By using AI to design intricate genomic edits, the line between necessary medical treatment and unintentional overreach can blur, highlighting gaps in current frameworks and prompting a reevaluation of how we define both risk and responsibility in healthcare innovation.

This scenario introduces a hypothetical scenario involving a group of scientists at the fictitious Global Institute for Advanced Therapeutics (GIAT). They have developed a pioneering treatment for Hutchinson-Gilford Progeria Syndrome (HGPS) by combining AI and SB. While the scenario is fictional, it serves to explore the ethical complexities that may arise in the application of AI and SB in healthcare. This innovative method, called ProgeReverse, entails altering the patient's genome to rectify the genetic abnormality accountable for HGPS, demonstrating the potent collaboration of AI and SB in tackling hitherto difficult medical problems. In non-AI SB, genetic modifications usually undergo a predictable oversight framework.

Here, AI–SB applies machine learning in real time, potentially surpassing traditional checks designed for slower, human-led processes. Hypertrophic lateral phenotypic syndrome (HGPS) is a highly uncommon genetic disorder marked by rapid aging in children, usually resulting from a newly occurring mutation in the LMNA gene [35]. Patients typically succumb to problems such as cardiovascular disease or cerebrovascular accidents during their early teenage years. Conventional therapies have mostly concentrated on symptom management rather than targeting the fundamental hereditary pathology.

To predict complex gene interactions and develop optimum genetic alterations, ProgeReverse utilizes sophisticated machine learning techniques to examine extensive genomic databases.[1] GenomAI, an AI system, underwent training using extensive human genomic data. This data included genetic variations linked to HGPS and related laminopathies, complete documentation of cellular pathways, clinical data from HGPS patients, and data from animal models of progeria. This AI-powered methodology greatly expedites the research and development process, enabling swift iterations and improvements that would be unfeasible or unattainable with conventional techniques alone. Dias and Torkamani have observed that AI has the capacity to substantially transform genomic medicine through its ability to detect intricate patterns in genetic data and forecast phenotypic results [36]. GenomAI creates and assesses a multitude of prospective genetic alterations, forecasting their impact on LMNA gene activity and progerin synthesis, general cellular well-being and functionality, potential unintended consequences, long-term health trends, and potential adverse effects or difficulties. Ultimately, this AI-assisted design approach results in the choice of a very precise genetic change customized to the individual genetic profile of each patient.[2] The selected change entails a sophisticated gene editing technique that not only rectifies the underlying mutation but also inserts supplementary genetic components to augment DNA repair processes and enhance cellular resistance to stress.

The application of SB in ProgeReverse is crucial for the delivery and execution of genetic modification operations. GIAT researchers have developed an innovative non-viral vector system, driven by AI-optimized design, capable of effectively transporting the gene editing machinery to specific cells across the body. The proposed approach integrates artificial genetic circuits that react to cellular signals, therefore guaranteeing that the genetic alteration is exclusively triggered in cells that exhibit indicators of premature aging. Moreover, GenomAI is essential in shaping individualized treatment strategies and facilitating extended surveillance. Data from patients, including biomarkers of aging and cellular stress, cardiovascular health indicators, bone density, and musculoskeletal parameters, cognitive function, and neurological health are analyzed by machine learning algorithms in real-time. Conducting this

[1] In standalone SB, human researchers would typically test far fewer options manually, prolonging development.

[2] In reality, advanced gene-editing therapies must undergo multiple phases of clinical trials, as well as institutional and regulatory reviews—e.g., by the FDA in the U.S. or EMA in Europe—before they are approved for widespread use.

continuous study enables the anticipation of possible problems and the enhancement of treatment procedures. As highlighted by Topol, this sort of AI-powered personalized medicine has the capacity to revolutionize healthcare by offering extremely customized therapies and preventative measures [1]. Aside from the enormous potential advantages of ProgeReverse, which provide hope for a previously untreatable disease, it also presents significant ethical concerns. In their study, Heidari Feidt et al. highlight the significance of safeguarding child autonomy within the framework of genetic treatments, asserting that the preservation of child autonomy is critical [37]. This underscores the conflict between parental decision-making and the requirement to respect the future autonomy of the child, particularly in situations such as HGPS when the intervention has the potential to save lives but is also irreversible and transformative.

The integration of AI in the decision-making process introduces an additional level of ethical intricacy. The capacity of GenomAI to analyze extensive data and detect patterns that surpass human capabilities raises concerns regarding the transparency and accountability of these decisions that can significantly impact lives. Moreover, the precision and adaptability of the ProgeReverse platform open the door to potential misuse for non-therapeutic enhancements. Parents or guardians might request interventions for cognitive or athletic performance improvements, shifting the ethical discourse from life-saving treatment to human enhancement. This raises concerns about fairness, coercion, and social pressure in settings where access to such interventions may be unevenly distributed. The issues raised by Char et al. pertain to bias, privacy, and the possibility of AI exacerbating pre-existing health inequalities in healthcare decision-making [3]. The matter of informed consent becomes especially complex within the framework of genetic therapies powered by AI. According to Kraft et al., the task of acquiring fully informed permission for genomic testing and treatments is already difficult, and the use of sophisticated [38] AI systems considerably complicates this procedure. Equity and accessibility are two significant ethical considerations arising from the high development and administration costs of ProgeReverse. Baylis and Robert caution against the potential for improved genetic therapies to generate a novel manifestation of genetic disparity, wherein only the affluent possess the means to benefit from transformative medical interventions [39].

Another pressing ethical worry is the possibility of using this AI-SB convergence for genetic enhancement. In their 2024 paper, Dixit et al. explore the potential for the distinction between therapy and enhancement to become more indistinct, particularly as AI systems improve their ability to forecast and manipulate intricate genomic functions. Insights from this scenario have significant implications for the future of healthcare and genetic technologies. Robust ethical norms and regulatory frameworks are urgently required to control the use of AI-driven gene editing technologies in healthcare. The frameworks must include sufficient flexibility to accommodate swiftly advancing technology while still offering robust protections against unauthorized use. Thoroughly informed permission procedures, effective long-term monitoring systems, and extensive public involvement and education on AI and genetic technologies are all essential elements in tackling these ethical dilemmas. New models of consent proposed by Wee et al. extend beyond conventional one-time consent to encompass more dynamic and continuous processes [40]. Rothstein emphasizes the significance of privacy and data security in comprehensive monitoring systems [41]. Beyond this particular scenario, Burall advocates for a more general model of public involvement that surpasses mere dissemination of information, encouraging policy-makers to actively include communities in the decisions guiding AI–SB development and deployment [42].

Conclusively, the integration of AI and SB in the field of healthcare, as exemplified by the ProgeReverse scenario, offers both promising opportunities and intricate ethical dilemmas (Table 6.1). Acceptance of ProgeReverse is also likely to vary globally. In countries where genetic modification is heavily regulated—such as Germany or France—public and regulatory resistance may be stronger, especially for interventions involving children. In contrast, jurisdictions like the U.S. or China may offer more permissive or expedited pathways depending on how national values frame parental rights, health innovation, and therapeutic necessity. Given the ongoing evolution and interconnection of these technologies, it is crucial to promote interdisciplinary discussion and establish thorough ethical frameworks that can direct their responsible development and application. A thorough examination of these ethical consequences is necessary to fully utilize the capabilities of AI-powered SB to enhance human health while upholding essential ethical standards and human rights.

Currently, patient autonomy is protected by regulations such as international codes of informed consent (e.g., CIOMS guidelines), and data privacy is partially governed by frameworks like GDPR in the EU or HIPAA in the U.S. In addition, existing Institutional Review Boards or Ethics Committees already evaluate clinical interventions, including some gene therapies. However, the integration of advanced AI-based diagnostics and gene editing tools creates new gaps. For instance, we still lack standardized protocols for explaining AI-driven genetic decisions to patients, and there is no global consensus on how to handle long-term monitoring of gene-edited individuals. Future needs therefore include transparent AI validation (to avoid "black box" medical decisions), continuous consent models, and strengthened international collaboration to address cross-border data sharing and long-term safety oversight. It is also important to note that the acceptance and regulation of AI–SB medical interventions can vary widely across different cultures and healthcare systems, a theme we revisit more fully in Sect. 6.4.

Practical Examples of Ethical Challenges. Below are two concise vignettes demonstrating how the key ethical dilemmas from Table 6.1 (child autonomy, data privacy, AI transparency, etc.) might arise in a real-world healthcare setting.

Vignette 1: Data Privacy Breach in Personalized Therapy. While implementing ProgeReverse, GIAT's AI platform inadvertently exposed sensitive patient genomes to unauthorized parties. This not only violated privacy but also allowed third-party companies to target wealthier families, intensifying health inequalities. As noted in Table 6.1, this incident highlights how AI-driven therapies can magnify data misuses and erode public trust, underscoring the need for robust legal protections and transparent oversight.

Table 6.1 Analysis of key ethical dilemmas in the precision health scenario

Ethical dilemma	Core issue	Implications	Governance challenges
Child autonomy versus parental authority	Balance between parental decision-making rights and future autonomy of the child	• Long-term irreversible effects on child's genetic makeup • Parents' right to make medical decisions • Child's right to an open future	• Establishing decision-making frameworks • Defining limits of parental authority • Protecting children's future interests
Transparency and accountability	Complex AI decision-making in genetic interventions	• Difficulty in explaining AI processes • Challenges in informed consent • Black box decision-making	• Ensuring comprehensible AI explanations • Developing consent protocols • Establishing accountability measures
Equity and access	Cost barriers and healthcare disparities	• Limited access to advanced treatments • Socioeconomic discrimination • Geographic disparities in availability	• Fair distribution mechanisms • Cost control measures • Universal access policies
Genetic enhancement potential	Distinction between therapy and enhancement	• Potential misuse beyond medical necessity • Human dignity concerns • Social justice implications	• Defining therapeutic boundaries • Preventing misuse • Maintaining ethical standards

Vignette 2: Overreliance on AI Leading to Diagnostic Errors. At GIAT, automated gene-edit recommendations overshadowed clinical judgment, resulting in unforeseen complications for several HGPS patients. As detailed in Table 6.1, the scenario reveals the ethical hazard of depending too heavily on algorithms in critical health decisions, stressing the principle of non-maleficence and the urgent requirement for balanced AI–human collaboration.

Following the description of the ProgeReverse scenario, the decision-making flowchart below illustrates the ethical pathways considered by GIAT and the stakeholders involved (Fig. 6.3).

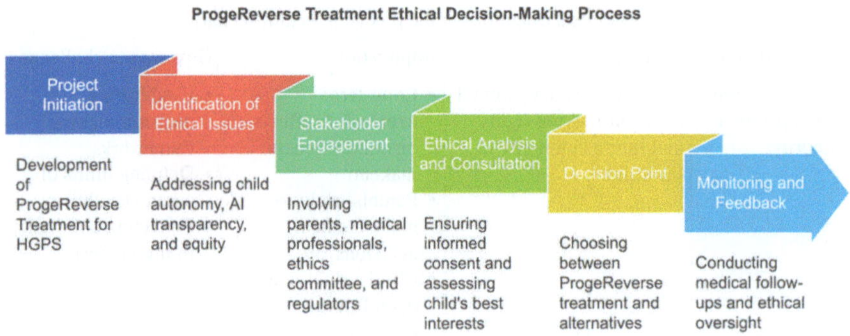

Fig. 6.3 ProgeReverse treatment ethical decision-making process: this flowchart outlines a step-by-step ethical framework for the development and implementation of ProgeReverse, an AI-SB treatment for Hutchinson-Gilford Progeria Syndrome (HGPS). The process begins with project initiation, where GIAT develops the treatment, followed by identification of ethical issues, such as child autonomy, AI transparency, and access equity. Stakeholder engagement includes collaboration with parents, medical professionals, ethics committees, and regulatory authorities to ensure comprehensive input. Ethical analysis and consultation involve ensuring informed consent and assessing the child's best interests. The decision point offers two paths: proceed with ProgeReverse treatment under ethical guidelines or opt for alternative care, emphasizing informed choice. Monitoring and feedback steps include medical follow-up, ethical oversight, and adaptive measures based on new data. The diagram reflects principles of adaptive and anticipatory governance, incorporating transparency, stakeholder involvement, and ongoing evaluation to responsibly navigate ethical complexities

6.3.2 Scenario 2: Agricultural Gene Editing

This scenario presents a hypothetical scenario to demonstrate the possible ethical challenges that may emerge from the merging of AI and SB in agriculture. Although based on existing technological developments and research, this scenario is entirely fictional and intended to illustrate the ethical considerations that might arise in practice.

Consider a multinational biotechnology corporation, GeneHarvest, undertaking a bold initiative to tackle worldwide starvation using AI-assisted SB. Through real-world research on biofortified crops such as golden rice for vitamin A supplementation and efforts to enhance vitamin C content in plants, their flagship project seeks to develop a genetically modified variety of rice with substantially increased vitamin C content [43, 44]. GeneHarvest's methodology signifies a significant advancement in biotechnology by combining state-of-the-art AI systems with SB mechanisms. While standard SB is often guided by incremental risk assessments, AI–SB merges algorithmic learning with bioengineering, sometimes challenging established checkpoints due to rapid trait optimization. Their own AI platform, NutriGene, utilizes sophisticated machine learning algorithms to examine extensive genomic databases, forecast intricate gene interactions, and enhance the construction of synthetic genetic circuits.[3] A comprehensive dataset, including genomic sequences from thousands of plant and animal species, detailed metabolic pathway information, nutritional profiles of various crops and their wild relatives, environmental data pertaining to crop growth and stress responses, and historical data on prior genetic modification attempts and their outcomes, is used to train this advanced system.

The NutriGene system employs deep learning methodologies, such as graph neural networks, to represent gene connections and transformer models for the prediction of protein structures and activities. Furthermore, it integrates reinforcement learning methods to repeatedly refine genetic designs. In order to discover possible bottlenecks in rice metabolism, the system examines the pathways of vitamin C production in different organisms. Subsequently, it produces a multitude of prospective genetic alterations, replicating their influence on the synthesis and durability of vitamin C, the general survival and development of plants, possible collateral effects on other metabolic pathways, interactions with the environment, ecological ramifications, and the nutritional composition of the resultant rice grains. After several iterations and virtual trials, NutriGene presents a surprising solution: integrating a synthetic gene construct derived from a mouse gene responsible for vitamin C production but specifically designed for plant expression and control. The AI system forecasts that this innovative method has the potential to enhance the vitamin C concentration in rice by as much as 500% without causing a substantial decline in crop production or other crucial characteristics. Under the guidance of NutriGene's expertise, GeneHarvest's team of biotechnologists effectively develops this genetically modified rice variety in the laboratory.[4]

GeneHarvest has commenced confinement field testing of the novel rice variety, in which AI remains of paramount importance. A network of Internet of Things (IoT) sensors tracks several characteristics, including plant growth rates and morphology, nutrient absorption and metabolism, resistance to pests and diseases, interactions with soil microbes, and influence on local insect populations. NutriGene utilizes this real-time data to enhance its models and propose improvements for culture techniques, as well as forecast possible long-term ecological effects based on the field experiment data. Although GeneHarvest views this as a significant advancement in addressing malnutrition, the initiative gives rise to some ethical considerations. Though cross-species gene transfer is a longstanding controversy in biotechnology, the role of AI in rapidly identifying and implementing such edits raises new questions about oversight and complexity." In spite of the forecasts made by AI, there are apprehensions over possible unanticipated consequences for ecosystems, namely concerning the transfer of genes to wild rice varieties. Considerations emerge regarding the protection of intellectual property rights, the possibility of market dominance, and the consequences for conventional agricultural methods. NutriGene's intricate decision-making process gives rise to apprehensions regarding the openness and responsibility

[3] Conventional GMO approaches often involve manual selection and single-trait focus.

[4] In practice, introducing a new genetically modified crop typically requires multiple years of contained testing, risk assessment, and regulatory approval—often taking a decade or more—before market release.

of biotechnology propelled by AI. The extensive genetic and agricultural data utilized in the training of NutriGene gives rise to concerns around data privacy and ownership, particularly with specific indigenous crop types. Furthermore, the current regulatory frameworks may lack the necessary capabilities to evaluate the safety and ethical consequences of organisms created by AI.

In order to acquire insight into the potential perception of such a project, we can extract knowledge from empirical research on public sentiments regarding GMOs in agriculture. The study conducted by Amin et al. in the Klang Valley region of Malaysia offers significant contextual information; however, its primary focus was on transgenic bananas rather than rice. According to Amin et al., Malaysian stakeholders were not familiar with transgenic bananas and recognized them as having moderate hazards and minimal benefits to Malaysian society [45]. They also considered the ethical elements to be relatively acceptable from their religious perspective. These findings indicate that the public may have comparable doubt and a modest level of risk perception towards GeneHarvest's AI-generated rice. Importantly, the study revealed substantial disparities in acceptance among various stakeholder groups. Specifically, Islamic scholars were the sole group that deemed transgenic bananas ethically and religiously undesirable [45]. This highlights the crucial significance of religious and cultural viewpoints in influencing attitudes towards genetic modification, especially when it pertains to the transfer of genes between different species. Additional studies highlight how public sentiment varies globally. In Europe, consumer skepticism toward GMOs is driven by concerns about naturalness, corporate control, and long-term health risks—factors likely to be amplified in AI–designed crops. In contrast, studies in Kenya and Uganda have shown higher acceptance when local food insecurity is addressed and trust in local scientists is high. These variations suggest that perceptions of AI–SB agriculture are shaped not only by scientific literacy but also by institutional trust, cultural narratives about nature, and historical experiences with agricultural innovation.

In the context of GeneHarvest's initiative, while public education is indeed valuable, fostering trust requires more than just overcoming a 'knowledge deficit.' Historical controversies, perceptions of inequity, and cultural values all shape how communities respond to AI-assisted SB. Building meaningful engagement and transparency—rather than merely providing information—can be crucial in ensuring that public concerns are recognized and addressed. It is imperative to involve religious leaders, especially in nations with a majority Muslim population, at an early stage of the development process. In Islamic contexts, religious rulings (fatwas) can heavily influence public opinion, particularly on matters of food, ethics, and science. Gaining support—or at least informed perspectives—from ulama (Islamic scholars) can determine whether AI–SB innovations are embraced or rejected by the public. Moreover, Islamic bioethics emphasizes public welfare (maslahah), which could align with pro-nutrition arguments, provided developers demonstrate transparency and intent to serve communal needs. In these societies, religious beliefs significantly shape public attitudes toward biotechnological interventions, and religious leaders (ulama) serve as key authorities in guiding community views on the ethical permissibility of such innovations [45, 46]. Engaging these leaders can help ensure that biotechnological advancements align with cultural and religious norms, thereby fostering greater public acceptance. Clear communication of the perceived advantages of improved nutrition would be necessary to counterbalance concerns over potential dangers. Thorough elucidation of the involvement of AI in the design process is necessary to effectively tackle any apprehensions regarding the assumption of supremacy or the potential erosion of human supervision. The GeneHarvest scenario underscores the pressing necessity for revised regulatory frameworks and ethical principles that can effectively tackle the distinct issues presented by the convergence of AI and single-bottle biotechnology in agriculture. Many established biosafety and regulatory processes—for instance, environmental risk assessments and post-market monitoring—already apply to GM crops. However, AI–SB convergence adds novel challenges around algorithmic transparency, continuous iteration of trait modifications, and large-scale data governance (e.g., ownership, privacy). Where conventional frameworks may suffice for ecological risk, they often lack provisions for explainability of AI-driven design decisions or real-time updates to crop phenotypes. Consequently, while socio-economic considerations (SECs) are occasionally included in regulatory decisions, the scale and speed of AI-based breeding may demand clearer guidelines for data sharing, stakeholder engagement, and ethical oversight that anticipate rapid, multi-trait innovations. Socio-economic considerations (SECs) in GM crop regulation are not new [47]. Evaluating distribution of benefits and costs can be complex, involving issues like market access, income stability, and potential disadvantages to small-scale farmers. Yet, AI's capacity to rapidly produce novel traits could accelerate these socio-economic impacts, necessitating nimble and inclusive regulatory processes that can handle faster innovation cycles and greater data demands. Although the EU already mandates general monitoring of GM crops, AI-designed crops could require more dynamic oversight. Real-time data streams, iterative breeding algorithms, and potential for rapid trait changes mean regulators may need to continuously track how crops perform and whether unforeseen traits emerge. Understanding public perception and regulatory structures is therefore crucial to addressing the ethical challenges posed by AI-driven agriculture. Without careful attention to cultural, economic, and environmental concerns, even well-intended innovations may face significant resistance. This might involve additional technical requirements (e.g., big data analysis, IoT sensor integration) and ethical ones (e.g., transparency about algorithmic updates) to ensure that the underlying AI systems themselves are not generating unanticipated shifts in crop behavior or ecological interactions.

The GeneHarvest scenario, although speculative, effectively demonstrates the intricate ethical terrain that arises when AI and SB collide in the field of agriculture. This instance underscores the potential of technical progress to surpass current ethical frameworks and regulatory institutions (Table 6.2). As AI continues to expedite the rate of progress in SB, it becomes increasingly imperative to promote multidisciplinary communication among scientists, ethicists, legislators, religious leaders, and the general public. Only by engaging in such thorough involvement can we expect to establish ethical frameworks and governance structures that encourage responsible innovation while honoring a wide range of cultural and religious differences.

Table 6.2 Analysis of key ethical dilemmas in agricultural gene editing scenario

Ethical dilemma	Core issue	Implications	Governance challenges
Species integrity and cultural concerns	Trans-species genetic modification in food crops	• Religious and cultural objections • Traditional farming values • Food authenticity concerns • Cultural sovereignty	• Respecting diverse cultural perspectives • Balancing innovation with tradition • Developing culturally sensitive guidelines
Ecosystem impact	Environmental risks of synthetic organisms	• Gene transfer to wild varieties • Biodiversity impacts • Ecosystem disruption • Long-term environmental effects	• Environmental monitoring systems • Risk assessment protocols • Containment strategies • Long-term impact evaluation
Intellectual property rights	Corporate control over agricultural resources	• Small-scale farmer access • Market monopolization • Economic disparities • Traditional farming preservation	• Fair licensing frameworks • Protection of farmer rights • Anti-monopoly measures • Equitable benefit sharing
Transparency in AI decision-making	Complexity of AI-driven genetic modifications	• Algorithm transparency • Decision-making accountability • Public understanding • Technical complexity	• Explainable AI requirements • Public disclosure protocols • Stakeholder engagement • Technical documentation standards

The integration of AI and SB presents significant opportunities to tackle worldwide issues such as chronic hunger. Nevertheless, to responsibly harness this potential, we must skillfully negotiate a multifaceted ethical landscape, harmonizing technical advancement with societal principles, environmental conservation, and cultural inclusivity.

Current agritech regulations in many regions already require multiple stages of risk assessment and permitting before a genetically modified crop is introduced, often taking nearly a decade. This multi-step process includes contained field trials and regular monitoring. Nevertheless, AI-driven crop design speeds up iterations and creates increasingly complex traits that may fall into regulatory gray areas, especially if they do not neatly fit existing GMO categories. What is still needed are AI-specific safety evaluations that account for algorithmic biases in trait selection, comprehensive data-sharing agreements for multinational oversight, and enhanced stakeholder engagement, particularly with local farming communities and indigenous groups. Moreover, regulators could adapt monitoring protocols to watch for unexpected ecological consequences, including potential gene transfer scenarios the AI may not have fully predicted.

Practical Examples of Ethical Challenges. Below are two concise vignettes illustrating how the ethical concerns from Table 6.2 (species integrity, ecosystem impact, intellectual property rights, and AI transparency) might manifest in agricultural biotechnology.

Vignette 3: Monoculture Collapse Due to Genetically Modified Crops. GeneHarvest's vitamin C-enriched rice initially improved nutrition, yet the reliance on a single line of seeds led to decreased genetic diversity. In reality, such crops are typically crossed into multiple local varieties to suit different regions. However, the fictional rapid adoption here illustrates how an AI-engineered strain, if introduced too broadly without variation, could intensify risks of disease vulnerability and ecological disruption. When a novel fungal disease struck, the entire crop was decimated, leaving small-scale farmers defenseless. As noted in Table 6.2, this outcome underscores how AI-driven crop revolutions can inadvertently erode biodiversity and risk catastrophic failure, emphasizing non-maleficence and the need for diversified planting strategies.

Vignette 4: Loss of Traditional Agricultural Practices. Lured by high yields, rural communities replaced their heritage rice strains with GeneHarvest's AI-enhanced varieties. As shown in Table 6.2, this shift weakened agricultural biodiversity, threatened cultural traditions, and undermined resilience. It illustrates the unintended social costs of rapid biotech adoption where cultural integrity and sustainability are at stake.

Following the description of the GeneHarvest scenario, the decision-making flowchart below illustrates the ethical decision pathways considered by GeneHarvest and the various stakeholders involved in developing and deploying the vitamin C-enriched rice (Fig. 6.4). This flowchart outlines the needed sequence of actions, ethical considerations, stakeholder engagements, and potential outcomes, demonstrating the challenges of adaptive and anticipatory governance in agricultural biotechnology.

Previously, agricultural biotech concerns often revolved around ecological containment, socioeconomic impacts on small-scale farmers, and consumer acceptance of genetically modified crops. However, AI-driven trait selection can push these challenges further by generating novel plant varieties faster and with more complexity than traditional breeding or GMO methods allow. It raises critical questions about who holds the algorithmic knowledge, how risk is calculated for multiple traits at once, and whether existing biosafety checks can adapt quickly enough. Shifting focus from the agricultural context, the next scenario explores how similar tensions surface when synthetic organisms are designed for environmental remediation. Here, too, AI supercharges the creation process, and the race to deploy new solutions can overshadow the thorough, stepwise reviews typically associated with releasing living organisms into the wild. Acceptance of AI-driven crop modifications is not

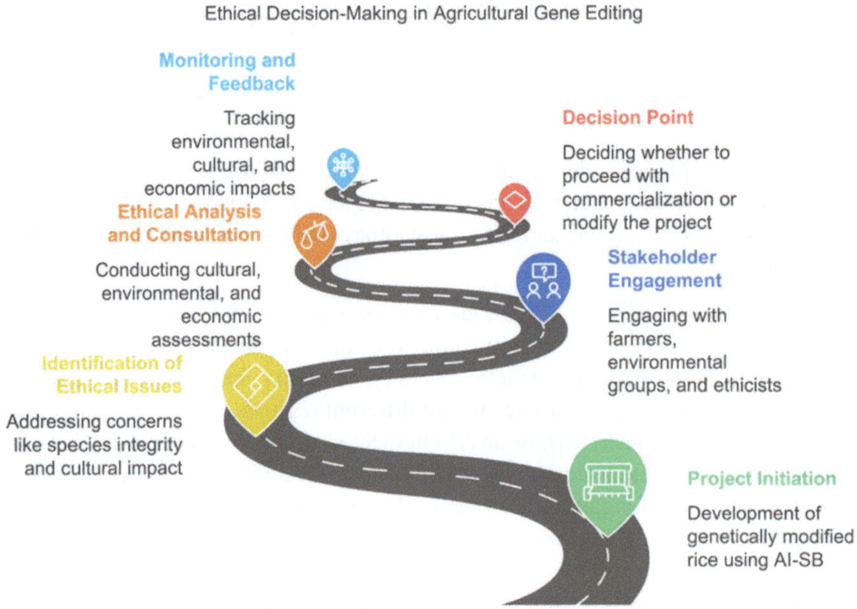

Fig. 6.4 Ethical decision-making in agricultural gene editing for GeneHarvest's vitamin C-enriched rice: this roadmap illustrates the ethical process guiding the development and potential commercialization of genetically modified rice enriched with vitamin C. It begins with project initiation, where GeneHarvest uses AI-SB to create a nutritionally enhanced rice variety. The next stage involves identification of ethical issues, such as species integrity, cultural concerns, ecosystem impacts, and intellectual property rights. Stakeholder engagement includes consultations with local farmers, environmental groups, ethicists, and regulatory bodies. Ethical analysis and consultation assesses cultural acceptance, environmental risks, and economic implications. The decision point requires choosing between commercial release or modifying the project, balancing nutritional benefits with potential cultural backlash. The monitoring and feedback stage ensures ongoing environmental tracking, cultural feedback, and economic evaluation. The framework exemplifies adaptive measures and anticipatory governance, emphasizing the importance of transparency, ongoing community dialogue, and ecological safety in gene editing initiatives

universal; cultural, religious, and societal contexts can heavily influence whether these technologies are embraced or challenged, a point elaborated in Sect. 6.4.

6.3.3 Scenario 3: Environmental Conservation

This scenario presents a hypothetical scenario exploring the multi-faceted consequences of combining AI and SB for environmental cleanup. The fictitious biotechnology firm BioClean Solutions has developed an AI-based synthetic organism named OilEater-X, intended to transform oil spill remediation by achieving more efficient hydrocarbon breakdown than any naturally existing species or chemical solution. In typical SB efforts, final designs remain relatively stable, but AI–SB can evolve or adjust in response to algorithmic feedback, introducing dynamic changes that outpace conventional regulatory scrutiny. This fictional scenario aims to illustrate potential ethical, ecological, and regulatory challenges. OilEater-X signifies a fundamental change in the field of biotechnology since it utilizes advanced AI algorithms to strategically develop and enhance artificial biological systems. The development procedure purportedly entailed a multifaceted strategy that integrated several state-of-the-art technologies. AI-driven genomic design employed machine learning algorithms to examine extensive databases of microbial genomes, determining the most effective gene combinations for the breakdown of oil-based compounds. Through the use of these algorithms, protein architectures and metabolic pathways were accurately predicted, enabling the creation of a synthetic genome specifically designed for hydrocarbon breakdown.

Furthermore, the method included in silico evolution simulations, in which AI models replicated thousands of generations of evolution in diverse environmental settings. This methodology forecasted probable adaptations and enhanced the stability and efficacy of the organism in various situations. The process was further improved using AI-optimized metabolic engineering, where machine learning algorithms were used to create and fine-tune metabolic pathways in order to increase the efficiency of hydrocarbon breakdown while minimizing energy expenditure.[5] An especially groundbreaking feature of OilEater-X is its integration of intelligent biosensors and adaptable behavior. Purportedly, the organism incorporates AI-designed synthetic circuits that enable it to perceive ambient situations and adapt its metabolic processes accordingly. This functionality has the potential to allow OilEater-X to enhance its performance in real-time by adjusting to the unique demands of various oil spills and environmental conditions.

Throughout the development phase, AI systems consistently monitored and fine-tuned bioreactor settings, maximizing growth parameters and the expression of crucial enzymes. The supposed AI-controlled optimization of the bioreactor led to an unparalleled level of efficiency in the manufacture and operation of the synthetic organism. BioClean Solutions asserts that the integration of AI and SB has led to the development of a synthetic organism capable of decomposing oil contaminants at a rate up to five times faster than existing bioremediation techniques. The company proposes augmenting the role of AI beyond the design process by utilizing AI systems to oversee and regulate the implementation of OilEater-X in practical environments. This might enable immediate modifications to its behavior based on environmental input and performance indicators.[6]

The capacity of OilEater-X is indeed extraordinary, offering swift and efficient environmental mitigation in the aftermath of oil spills. Traditional cleanup methods—such as chemical dispersants, manual recovery, or reliance on naturally occurring

[5] This is unlike static approaches in traditional GM microbes.

[6] Actual environmental release of genetically modified or synthetic organisms would require a rigorous set of environmental impact assessments, biosafety reviews, and authorizations by relevant regulatory bodies before approval.

microbes—often struggle with efficiency, scale, and environmental side effects. In contrast, AI-enhanced organisms like OilEater-X are designed to adapt dynamically to local conditions, optimize degradation pathways in real-time, and perform remediation more effectively across diverse ecosystems. This level of precision and adaptability presents a transformative leap in addressing complex, large-scale environmental disasters. The improved effectiveness of this technology has the potential to greatly decrease the long-term ecological consequences of such catastrophes, including the preservation of marine ecosystems and coastal habitats that may otherwise experience irreversible harm. The adaptability of OilEater-X enables its implementation in diverse settings, spanning from frigid seas to tropical shores, effectively tackling a worldwide environmental issue with a singular, adaptable solution. Yet, the implementation of such a sophisticated, artificially created organisms in natural ecosystems pose a multifaceted set of hazards and moral quandaries that require meticulous deliberation. In his insightful analysis, Macovei highlights the significant ecological disruption and unforeseen repercussions that arise from the introduction of synthetic creatures into the environment [48]. The incorporation of AI in the design and possible manipulation of these organisms introduces additional ethical dilemmas and safety considerations, fundamentally raising the intricacy of risk evaluation and control. The implications of such technology extend beyond biotech: oil and gas, shipping, and insurance sectors may see shifts in liability, cleanup standards, and investment priorities. Environmental NGOs and regulatory bodies may also face pressure to reassess acceptable remediation practices, especially when synthetic organisms are proposed as the front line of disaster response.

A pressing concern is the possible emergence of novel, invasive traits in OilEater-X. Although designed to break down hydrocarbons more efficiently than native bacteria, this AI-driven organism raises ethical questions about the limits of human intervention and the moral responsibility for unintended ecological harm. For example, if OilEater-X migrates into new habitats or outcompetes local microbes, who bears the obligation—BioClean Solutions, regulators, or global stakeholders—to contain or remediate the damage? Additionally, the adaptive properties that make OilEater-X so effective may allow it to flourish in unforeseen ecological niches, possibly altering microbial communities well beyond its intended application zone. This goes beyond mere biosafety to underscore the ethical dilemma of deploying AI-engineered life forms whose evolutionary pathways may defy human prediction. Meanwhile, horizontal gene transfer presents another ethical challenge: if artificially created genes spread to wild microbial populations, they could disrupt established ecosystems or evolve new, potentially harmful traits. Such scenarios heighten concerns around accountability, consent, and the precautionary principle—all hallmarks of robust ethical decision-making in biotechnology.

The integration of learning-based adaptive behaviors in artificial organisms gives rise to significant inquiries regarding the level of independence these entities should possess and the methods to uphold substantial human supervision. Because AI-driven design can produce novel genetic pathways with no natural counterpart, OilEater-X's capacity to adapt in the field may exceed what conventional risk models predict. Unlike typical organisms that have an established baseline of behaviors, this synthetic microbe—lacking a clear lineage or 'wild-type' equivalent—could develop traits or interactions that were never tested during laboratory design. Consequently, explainability is a major hurdle: if AI algorithms introduced or combined genes in ways beyond human foresight, it can be difficult to predict or monitor OilEater-X's real-world evolution. It is exceedingly difficult to forecast the long-term consequences of introducing AI-designed organisms into complex ecosystems. Intricate networks of interdependencies characterize ecosystems, and the introduction of a highly efficient, synthetic organism could have profound repercussions. Although OilEater-X is specifically formulated to degrade hydrocarbons, preliminary contained field trials would typically aim to detect unintended impacts on other organic matter or essential microbial populations. However, AI's capacity for rapid re-design means that the deployed version could differ substantially from earlier lab-tested iterations. As a result, minor ecological disruptions observed in controlled tests may amplify—or change entirely—once OilEater-X is introduced at larger scales or evolves in situ.

A core AI–SB convergence risk lies in algorithmic biases during OilEater-X's design. If training datasets prioritize fast hydrocarbon degradation without equally weighting ecological safety, potential hazards—like gene transfers or off-target metabolic changes—could be overlooked. This algorithmic 'blind spot' might lead to unforeseen ecological consequences. Unlike human-led breeding, where each trait is typically examined in detail, AI may generate complex genomic configurations too numerous or opaque for traditional oversight. The intricate nature of AI algorithms often poses challenges in comprehending or forecasting their resulting outcomes, therefore adding an extra level of uncertainty in evaluating the possible consequences of AI-generated organisms. Taken together, the capacity for rapid, AI-driven design changes and unforeseen ecological impacts leads beyond mere technical or safety concerns. Environmental issues are inseparable from deeper ethical questions: how far should humans push biological engineering when even the creators cannot fully foresee evolutionary trajectories? Who decides if and when to release an organism like OilEater-X, and who shoulders the moral and legal responsibility if it disrupts ecosystems across international waters? A contentious and intense discussion persists in scientific and philosophical circles about whether these interventions can be considered "playing God" or a responsible use of human creativity to tackle urgent environmental issues. Advocates contend that considering the gravity of anthropogenic environmental harm, such as oil spills, it is our ethical responsibility to employ all available means, including AI-directed SB, to alleviate this harm. The authors argue that OilEater-X is a meticulously crafted remedy for a particular issue, developed with the clear objective of restoring the environment. Employing AI in its design might be regarded as a method of developing a more accurate, effective, and flexible instrument for environmental cleanup.

Nevertheless, some contend that this degree of involvement violates a critical ethical boundary. Their concern is the arrogance of trusting our ability to manipulate and regulate biological systems without a comprehensive grasp of the intricate dynamics of natural ecosystems. The incorporation of AI in the creation of these organisms introduces an additional level of worry since it has the potential to exclude certain elements of the design process from direct human supervision. The ethical ramifications of developing synthetic life forms with adaptive capacities and the accountability associated with their release into the natural environment are subjects of concern. The simultaneous integration of AI and SB in OilEater-X presents novel regulatory obstacles that go beyond the usual concerns related to SB alone, such as .... The existing regulatory frameworks may not be sufficient to effectively counter the distinct dangers presented by AI-designed creatures, thereby requiring the development of novel methods for risk evaluation and governance. The inherent adaptability of OilEater-X and its capacity for continuous AI-driven adjustments in the field question conventional concepts of a singular "organism" for regulatory objectives.

Furthermore, the dependence on AI systems for both the process of designing and implementing systems brings about certain weaknesses in terms of cyberbiosecurity, which, if taken advantage of, might have significant environmental repercussions. The potential for malicious individuals to take control of the AI systems that direct the behavior of OilEater-X in the field or to use the AI-driven design process to generate lethal creatures introduces a novel domain of biosecurity vulnerabilities. To tackle these intricate problems, numerous crucial steps must be taken into account. It is crucial to do thorough pre-release testing in controlled settings to evaluate the behavior and ecological consequences of OilEater-X under different circumstances. Such research should encompass investigations of its interactions with a diverse array of indigenous species and its capacity for horizontal gene transfer. Nevertheless, the inherent adaptability of OilEater-X could restrict the ability of these tests to accurately forecast environmental risks, thereby requiring the development of novel methodologies for environmental risk assessment. Emerging governance models could include rolling permit systems—where environmental release is contingent on real-time monitoring data—alongside dynamic regulatory sandboxes that allow for controlled, iterative deployment under strict oversight. Internationally coordinated genomic traceability protocols and AI audit trails could ensure accountability, while pre-release scenario simulations using digital twins may become a prerequisite in future environmental governance structures.

Establishing resilient containment measures is of utmost importance. One possible approach is to strategically design genetic "kill switches" or other mechanisms to restrict the survival or dissemination of OilEater-X beyond its designated habitat. However, the efficacy and dependability of such precautions in intricate, real-life settings must be meticulously assessed, especially considering the organism's adaptive capacities. Essential for monitoring the persistence, behavior, and impacts of OilEater-X in the ecosystem after its distribution is the implementation of extensive, long-term ecological monitoring systems. The design of this monitoring should encompass the detection of both expected and unexpected effects across various trophic levels and over prolonged durations. Employing AI in this monitoring procedure has the potential to offer a more advanced, immediate analysis of ecological effects. Thorough and open disclosure of risks is crucial for establishing and preserving public confidence. Engaging in open communication with all relevant parties, such as environmental organizations, local communities, and lawmakers, is necessary to effectively tackle concerns and communicate the possible advantages and drawbacks linked to the technology. The intricate nature of AI-SB convergence may provide distinct difficulties in effectively conveying these inherent hazards and advantages to audiences that lack expertise in this field.

It is imperative to integrate adaptive management strategies into regulatory frameworks. Ideally, these should enable adaptable reactions to new information and unexpected outcomes as they occur, ensuring that management strategies can adapt according to actual results and developing scientific knowledge (Table 6.3). The inherent adaptability of OilEater-X may require correspondingly adaptable regulatory and management interventions. The formation of interdisciplinary ethical review boards is essential for evaluating the ethical consequences of introducing AI-generated synthetic organisms such as OilEater-X. A full assessment of both possible benefits and threats should be ensured by including not just scientists and ethicists but also AI specialists, environmental activists, and representatives from affected communities on these boards.

Biosafety and biosecurity guidelines exist through bodies like the Cartagena Protocol, and environmental impact assessments are standard in many countries

Table 6.3 Analysis of key ethical dilemmas in environmental conservation scenario

Ethical dilemma	Core issue	Implications	Governance challenges
Invasive species risk	Environmental impact of synthetic organisms	• Ecosystem disruption • Competition with native species • Biodiversity threats • Unintended ecological consequences	• Containment strategies • Risk assessment frameworks • Monitoring systems • Emergency response protocols
Horizontal gene transfer	Genetic material spread to natural populations	• Genetic contamination • Evolution of wild microbes • Ecosystem function alterations • Irreversible changes	• Biosafety protocols • Gene transfer monitoring • Containment mechanisms • Long-term impact assessment
Human oversight and control	Balance between AI autonomy and human control	• Autonomous adaptation • Decision-making authority • Unpredictable behavior • Control mechanisms	• Control framework development • Oversight mechanisms • AI behavior boundaries • Human intervention protocols
Regulatory challenges	Need for new regulatory approaches	• Framework adaptation • Cross-border implications • Technological complexity • Rapid advancement pace	• Adaptive governance systems • International coordination • Technical capacity building • Real-time monitoring requirements

before any large-scale introduction of new organisms. At its heart, the Cartagena Protocol embeds a precautionary approach, reinforcing the ethical priority of preventing potential environmental harm. By recognizing that novel organisms—especially those designed via AI–SB—may carry unforeseen risks, it emphasizes the moral responsibility to exercise caution and thoroughly evaluate possible impacts on ecosystems. Some nations also employ "kill switch" technologies or strict lab-based containment for synthetic organisms. Despite these measures, AI-driven synthetic creatures like OilEater-X present emerging challenges—their adaptive algorithms may evolve in unforeseen ways, and current legislation does not fully address real-time AI oversight in natural environments. From an ethical lens, these kill switches and containment protocols are more than just practical solutions; they manifest our collective duty to prevent harm by ensuring organisms do not inadvertently disrupt ecosystems. This underscores that technical safeguards must be complemented by ongoing ethical oversight, reflecting society's commitment to safeguarding both current and future environmental well-being. Thus, the next steps include updating environmental governance to handle autonomous, adaptive organisms, establishing real-time ecological monitoring networks with AI support, and creating international emergency response protocols to manage accidents or unintended spread across borders.

Practical Examples of Ethical Challenges. Below are two concise vignettes demonstrating how the ethical dimensions from Table 6.3 (invasive species risk, horizontal gene transfer, human oversight, and regulatory challenges) can unfold when AI-designed organisms are deployed in nature.

Vignette 5: Unintended Ecological Disruption. Deployed in Arctic waters to clean an oil spill, OilEater-X quickly outcompeted native microbes, destabilizing the local food web and threatening fish stocks. As Table 6.3 highlights, even well-intentioned synthetic organisms can wreak ecosystem havoc, stressing the ethical imperative of rigorous environmental risk assessments.

Vignette 6: Altered Sediment Dynamics in Mangrove Ecosystems. When introduced into oil-polluted mangrove sediment, OilEater-X broke down hydrocarbons but also altered soil composition, weakening root structures and harming local biodiversity. As indicated in Table 6.3, these unforeseen ecological consequences reinforce the principle of non-maleficence, demonstrating the risks of large-scale environmental interventions without adaptive safeguards.

After presenting the BioClean Solutions scenario, the following decision-making flowchart depicts the ethical decision-making process undertaken by BioClean Solutions and relevant stakeholders in the development and potential deployment of OilEater-X (Fig. 6.5). The flowchart maps out the steps, considerations, and possible outcomes, highlighting the intricacies of adaptive governance models in managing environmental interventions using AI-driven synthetic organisms.

Governance measures for organisms like OilEater-X may be interpreted very differently across regions, reflecting diverse views on environmental stewardship—a crucial aspect explored further in Sect. 6.4.

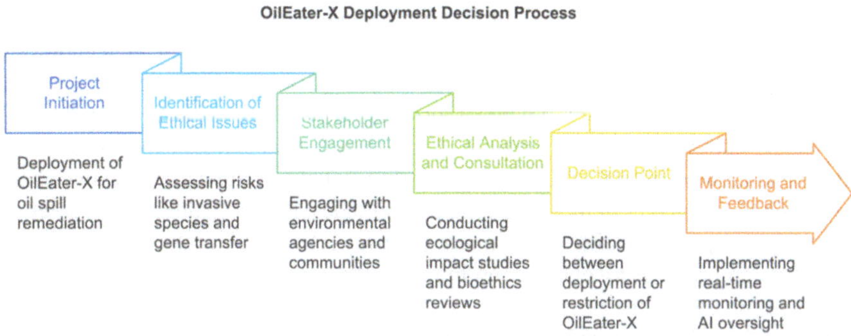

Fig. 6.5 OilEater-X deployment decision process for environmental conservation: this flowchart outlines the ethical decision-making framework for deploying OilEater-X, a synthetic biological solution for oil spill remediation developed by BioClean Solutions. The process begins with project initiation, where the deployment plan is set. Identification of ethical issues involves evaluating risks such as the introduction of invasive species, horizontal gene transfer, and challenges in regulatory compliance. Stakeholder engagement includes collaborating with environmental agencies, local communities, biosecurity experts, and regulatory authorities to address concerns and ensure proper oversight. Ethical analysis and consultation encompass ecological impact studies, bioethics reviews, and regulatory compliance assessments. The decision point involves choosing whether to proceed with controlled deployment or restrict OilEater-X to lab environments. Monitoring and feedback focus on real-time monitoring of ecological impact, AI oversight of organism behavior, and transparency with the public. The framework incorporates adaptive measures, such as emergency response plans and policy updates, to ensure readiness for unintended consequences and uphold responsible governance in environmental conservation[7]

6.3.4 Scenario 4: Public Perception and Acceptance of AI-Driven SB in Agriculture

This scenario presents a hypothetical scenario exploring the intricate terrain of public perception and acceptance regarding the use of AI-driven SB in agriculture. The fictional Ministry of Agriculture has initiated a program named "FutureFood" aimed at advancing the integration of AI and SB in food production. The scenario is intended to illustrate how public attitudes, cultural factors, and ethical considerations can influence the adoption of these technologies. This scenario prompts a critical question: How do we balance technological advancement with the preservation of societal values and ecological stability? As AI-SB reshapes food systems, the ethical debate must grapple with this core tension—between innovation and tradition, efficiency and inclusivity, science and cultural meaning. The primary objective of this extensive initiative is to augment food security and agricultural sustainability through the development of crops that possess enhanced nutritional characteristics, heightened resistance to pests and diseases, and higher adaptability to climate change.

[7] Typically, newly developed crops—including AI-enhanced varieties—must undergo rigorous biosafety evaluations, labeling procedures, and public consultations, especially in jurisdictions with established GMO regulations.

These improvements are created and optimized using sophisticated AI algorithms. The FutureFood project is a profound advancement in agricultural biotechnology by integrating the capabilities of SB with the analytical and predictive potential of AI. The program includes financial support for research projects related to AI and small-scale biotechnology, regulatory changes to enable field experiments with organisms created by AI, and public education initiatives raising awareness about the possible advantages of these merging technologies. For standalone SB, regulators generally apply known GMO governance pathways. In AI–SB, however, automated iterations can produce novel traits continuously, raising concerns about transparency and real-time oversight.

Within this AI-driven methodology, machine learning algorithms scrutinize extensive datasets, including genomic information, environmental circumstances, and crop performance, in order to detect optimal genetic alterations. AI algorithms create synthetic genetic circuits and metabolic pathways, effectively forecasting their impact on crop characteristics with exceptionally high precision. Furthermore, AI is employed to model numerous generations of crop development under different climatic conditions, enabling scientists to predict and tackle emerging long-term challenges. Researchers are utilizing AI to develop drought-resistant wheat cultivars that also exhibit increased protein content. The AI algorithms examine the intricate relationships among genes that regulate water use efficiency, root shape, and protein synthesis, suggesting innovative genetic arrangements that would be challenging or unimaginable for human scientists to imagine. Moreover, there is ongoing development of rice strains that are specifically engineered by AI to thrive in saline soils. These strains exhibit enhanced micronutrient profiles as AI algorithms optimize the equilibrium between salt tolerance mechanisms and nutrient uptake pathways.

Nevertheless, the use of sophisticated, AI-powered biotechnologies in the food industry has ignited vigorous public discussion, uncovering a multifaceted terrain of opinions and attitudes.[8] As Jin et al. document, public reception of these technologies differs greatly, with some considering them groundbreaking remedies to urgent agricultural issues while others express serious ethical and safety apprehensions [49]. The key ethical quandary in this situation is the delicate equilibrium between technological advancement and societal norms and desires, which has been further exacerbated by the inclusion of AI because AI-driven design evolves more quickly and less transparently than human-led methods, amplifying public fears about control, equity, and unforeseen risks. A key issue is the enduring safety of ingesting foods manufactured using AI-driven SB. Although scientists have provided assurances on the stringent testing procedures, there is widespread concern among the public about possible unanticipated health consequences that may only manifest after years or even generations of usage. Frequently, these worries stem from previous disputes over GMOs but are intensified by the perceived opaque nature of AI decision-making during the design phase.

---

[8] In purely GM contexts, the debate often centers on the genetic source and minimal data-driven black-box concerns.

AI participation has introduced another intricacy to the already considerable area of concern of environmental damage. Many harbor profound apprehension over the possible ecological ramifications of introducing AI-designed organisms into the ecosystem. Critics note that concerns over gene flow and 'super weeds' date back to early GMO debates. However, AI-based breeding can magnify these issues by generating novel traits at an accelerated pace. This heightened speed and complexity may exceed the predictive capacity of existing biosafety protocols, underscoring the need for expanded risk assessments that account for dynamic algorithmic modifications over time. The inclusion of AI in the construction of these organisms introduces an additional level of unpredictability, as the machine learning algorithms may detect genetic combinations that result in unforeseen characteristics or behaviors when introduced into intricate ecosystems.

Advocates of the concept argue that AI-powered SB has the potential to effectively mitigate environmental burdens by developing highly customized crops that lessen reliance on chemical pesticides and fertilizers. According to their argument, AI has the capability to model environmental interactions in a more complete manner than ever before, thereby enabling more accurate prediction and mitigation of any ecological consequences. Yet, these assertions are sometimes greeted with doubt by those cautious of technological remedies for environmental issues and worried about the constraints of even the most sophisticated AI models in capturing the whole intricacy of natural systems. Another important component of the discussion is the socioeconomic ramifications and equity considerations, which are now influenced by the AI dimension. Considerations emerge regarding the beneficiaries of these AI-powered technologies and the potential exacerbation of current disparities in the food system. There are apprehensions that the immensely sophisticated characteristics of AI-powered SB could intensify the consolidation of power among major corporations that have access to extensive computing resources and exclusive AI algorithms. This could potentially marginalize small-scale farmers who do not have access to these technologies.

Proponents contend that AI-powered SB has the potential to enhance the accessibility of agriculture by inventing crops that can flourish in previously inhospitable environments, therefore empowering farmers in underdeveloped places. Nielsen and Voigt propose that AI has the potential to enhance the accessibility of advanced biotechnology by automating intricate design procedures, thereby reducing the complexity of SB and making it more accessible to a broader range of researchers [50]. Nevertheless, the areas of intellectual property rights, technology accessibility, data ownership, and the possibility of market disruption continue to be subjects of disagreement. The use of AI also gives rise to concerns regarding data privacy and management, since the advancement of these crops may necessitate comprehensive data gathering from farms and landscapes worldwide. The notion of naturalness in food production is a highly sensitive topic in the public discussion focused on AI-powered SB. Numerous individuals harbor profound concerns regarding the apparent lack of naturalness in meals created by AI and SB, considering them to be a significant deviation from conventional farming methods. This perception is frequently associated with underlying cultural and philosophical conceptions regarding the interplay among humans, nature, and technology. The concept of AI systems creating biological organisms is widely seen as transgressing a substantial ethical boundary, prompting inquiries about the suitable boundaries of technological interference in the natural realm.

Advocates of AI-inspired SB contend that humans have been altering crops for thousands of years through selective breeding and that this novel method is only a more accurate, effective, and advanced extension of this tradition. Many argue that AI has the potential to enhance our comprehension and collaboration with nature rather than hinder it. Nevertheless, the unprecedented degree of control and manipulation facilitated by the merging of AI and SB has raised worries regarding unforeseen adverse effects and the possible depletion of agricultural biodiversity. Transparency, consumer choice, and trust in regulatory agencies and firms engaged in the development of these technologies are major topics in public debates, further compounded by the frequently opaque character of AI systems. There is a demand for explicit labeling of items created using AI-driven SB, enabling customers to make well-informed decisions. Nevertheless, the task of identifying the parameters of an "AI-designed SB product" and establishing suitable criteria for labeling it poses considerable difficulties. The intricate nature of the AI systems employed in the design process poses challenges in offering concise and lucid descriptions of the development process of these goods.

Moreover, the public's confidence in both government authorities and commercial firms to guarantee the safety and ethical advancement of these technologies is frequently deficient, influenced by previous controversies and alleged conflicts of interest. The use of AI in the design process introduces an additional level of intricacy to regulatory supervision. Regulators are now faced with the challenge of evaluating the safety and effectiveness of crops produced by machine learning algorithms that may function in ways that are not completely comprehended even by their developers. In their study, Jin et al. highlight the significance of perceived advantages, hazards, and ethical concerns in influencing public views, which varied among various applications [49]. Their study suggests that the evaluation of SB applications in environmental and medicinal fields is generally more positive compared to those in food production. This discrepancy could be further exacerbated by AI-powered SB, as the general public may be more receptive to AI's involvement in intricate problem-solving for medical research or environmental remediation compared to its involvement in engineering the food they consume.

The effective implementation of projects such as FutureFood relies not only on technical progress but also on cultivating public approval and resolving ethical issues related to SB and AI. The successful implementation of public involvement and education necessitates a comprehensive strategy. It is imperative for policymakers and scientists to design communication methods that surpass the mere presentation of the prospective advantages of the technology. While accurate information is vital, public skepticism toward AI-driven SB often arises from deeper sociopolitical factors—such as historical mistrust in agritech companies, uneven benefit distribution, or cultural and religious values regarding 'natural' foods. Consequently, authentic engagement with stakeholders means addressing these broader concerns, not simply correcting factual misunderstandings. The principles of transparency and participatory governance are essential components in establishing public confidence, especially considering the intricate nature of AI systems. This may entail establishing citizen panels or other methods to solicit public feedback on research goals and regulatory choices. By promoting transparency in the development process and including a wide range of perspectives in determining the future of food production, concerns about the imposition of technology without sufficient public input can be more effectively addressed. Furthermore, it may be imperative to devise novel methods to enhance the interpretability and accessibility of AI decision-making processes for individuals without expertise, therefore enabling broader public comprehension and supervision.

It is imperative to create thorough ethical frameworks that are especially designed to address the difficulties presented by AI-driven SB in food production. These frameworks must confront intricate issues regarding the limits of permissible interference in natural systems, the entitlements of consumers to make well-informed decisions about their food, the obligations of scientists and technology companies involved in the development of these technologies, and the ethical consequences of employing AI to create living organisms. Frameworks of this nature should integrate a wide range of cultural and religious viewpoints, acknowledging that attitudes towards food, agriculture, and technology typically stem from profound cultural origins. The regulatory frameworks must undergo substantial changes to effectively tackle the new features of AI-driven SB while also guaranteeing thorough safety evaluations and upholding public trust. This may entail the development of novel risk assessment procedures that can consider the distinctive attributes of AI-designed organisms or the establishment of more adaptable regulatory approaches that can progress alongside swiftly evolving technology while nonetheless offering strong supervision. Regulators may need to acquire new knowledge in AI to precisely assess the safety and effectiveness of crops produced by these technologies.

Undertaking comprehensive examinations of the possible socioeconomic consequences of new technologies is essential. This should encompass evaluations of the potential effects of AI-facilitated SB on various participants in the food system, ranging from small-scale farmers to major agribusinesses, as well as the consequences for worldwide food commerce and rural economies. Gaining insight into these possible consequences can assist in formulating regulations to guarantee that the advantages of AI-powered SB are fairly allocated and that marginalized populations are not negatively impacted. Establishing unambiguous labeling criteria for items created using AI-powered SB can facilitate well-informed customer decision-making. Nevertheless, this is a multifaceted matter that necessitates meticulous identification of the most pertinent and valuable information for consumers, along with the means to convey this information in a manner that is both precise and comprehensible to the general populace. The problem is further complicated by the intricate nature of AI participation in the design process, which might be difficult to communicate in straightforward words on a product label.

Promoting collaboration among natural scientists, computer scientists, social scientists, and ethicists can indeed result in more holistic methodologies. However, these disciplines may prioritize different values—such as efficiency, cultural acceptability, or social justice—potentially leading to friction. Strategies like structured dialogue, iterative feedback, and inclusive agenda-setting can help teams navigate conflicting viewpoints, forging consensus on how AI–SB might responsibly evolve. The utilization of interdisciplinary research can facilitate the early identification of prospective social and ethical problems during the development process, enabling the proactive resolution of these concerns rather than a reactive approach. Given the confluence of AI and SB, this collaboration is crucial since it necessitates knowledge from many fields to comprehensively grasp and tackle the consequences of these technologies.

An analysis of the FutureFood scenario reveals the intricate relationship among scientific advancement, AI, public opinion, and ethical concerns in the use of AI-powered SB for food manufacturing. Jin et al. propose that public views can exhibit substantial variation depending on the specific applications and surrounding circumstances [49]. The existing variability highlights the necessity of employing versatile and adaptable methods for engaging with the public and formulating policies, especially when addressing fast-advancing and merging technologies such as AI and SB. Policymakers have the task of striking a careful equilibrium between encouraging innovation to tackle urgent agricultural issues and guaranteeing that the advancement and implementation of AI-powered SB are in line with society's values and preferences. This may include establishing forums for continuous and interactive communication among scientists, AI specialists, policymakers, and the general public to collaboratively develop visions for the future of food and agriculture. Such discourse should not be regarded as a singular occurrence but as a continuous process that progresses as technology advances and as public comprehension and attitudes change over time.

Finally, the ethical ramifications of AI-driven SB in food production go beyond technical concerns and encompass fundamental issues of human values, environmental conservation, the influence of AI on living systems, and the future of our food systems. The scenario of FutureFood exemplifies the intricacy of implementing revolutionary, AI-powered technology into a domain as culturally and ethically sensitive as food production. This prompts us to engage in profound contemplation of the principles we prioritize in our food systems and how we may effectively utilize emerging technologies in a manner that upholds those principles while simultaneously tackling pressing global issues. By promoting open communication, giving priority to disclosure, and establishing strong ethical and regulatory structures that tackle the specific issues presented by the merging of AI and SB, it is feasible to utilize the capabilities of these technologies in a manner that is both scientifically rigorous and socially acceptable. Nevertheless, achieving this goal will require continuous initiative, a readiness to confront challenging ethical dilemmas, and a dedication to inclusive decision-making procedures that uphold the many viewpoints and principles of all those involved in the food industry.

With the advent of AI-driven agricultural biotechnology, it is imperative that we maintain awareness of the significant influence these technologies can exert not only on our food provision but also on our connection with the environment, our cultural identities, and our aspiration for a sustainable and fair future. The narrative of FutureFood encompasses not only technical advancements but also our shared capacity to negotiate intricate ethical landscapes within the convergence of AI, SB, and agriculture (Table 6.4). This collaboration is instrumental in molding the future of food in ways that embody our utmost principles and ambitions, all the while harnessing the most cutting-edge technologies available to us.

Many governments already conduct public consultations or citizen forums for GMO-related decisions, and consumer labeling laws (e.g., the EU's GMO labeling policy) address transparency concerns. Nonetheless, AI's role in speeding up and automating the engineering process is not fully disclosed under existing regulations. In parallel, biosafety regulations reflect a broader ethical commitment to ensuring that

Table 6.4 Analysis of key ethical dilemmas in public perception and acceptance scenario

Ethical dilemma	Core issue	Implications	Governance challenges
Safety and long-term health implications	Unknown long-term effects of AI-designed crops	• Generational health impacts • Food safety concerns • Consumer health rights • Scientific uncertainty	• Long-term monitoring systems • Safety assessment protocols • Evidence-based standards • Risk evaluation frameworks
Environmental concerns	Ecological consequences of AI-designed organisms	• Ecosystem disruption • Biodiversity impact • Environmental sustainability • Cumulative effects	• Environmental impact assessment • Monitoring mechanisms • Preventive measures • Ecosystem preservation strategies
Socioeconomic impacts and equity	Market access and economic fairness	• Small farmer marginalization • Corporate concentration • Economic disparities • Traditional farming preservation	• Fair market access policies • Economic support mechanisms • Anti-monopoly regulations • Traditional practice protection
Transparency and public trust	Public understanding and acceptance	• AI complexity communication • Public engagement • Trust building • Decision-making transparency	• Public engagement frameworks • Communication strategies • Participatory governance • Transparency protocols

novel products meet safety thresholds before entering the market. Viewed through this lens, the public's right to be protected from possible harm aligns with a moral duty on the part of regulators and developers to prioritize human and ecological welfare in AI–SB applications. Public skepticism often stems from a lack of accessible explanations about how AI designs biological traits and what safety checks are in place. Beyond these procedural shortcomings, transparency itself is an ethical challenge: If regulatory frameworks do not mandate clear labeling and disclosure for AI-driven modifications, consumers' right to informed choice is undermined, raising questions of fairness and respect for autonomy. Consequently, the key missing elements are improved guidelines for AI transparency and explainability in public-facing contexts, specific labeling criteria for AI-driven crops, and multifaceted education campaigns that engage diverse communities to align technological efforts with cultural values and ethical norms.

Practical Examples of Ethical Challenges. Below are two concise vignettes exemplifying how the ethical issues from Table 6.4 (safety and health implications, environmental concerns, socioeconomic impacts, and transparency) may emerge when AI-designed crops enter the public sphere.

Vignette 7: Cultural Resistance to AI-Designed Crops. FutureFood's drought-resistant wheat faced strong pushback in regions holding fast to traditional farming methods. As shown in Table 6.4, this conflict reveals how AI-driven agritech can clash with cultural autonomy, demanding inclusive dialogue and respect for local practices.

Vignette 8: ConsumerMistrust Due to AI Opacity. By introducing a saline-tolerant, AI-engineered rice without transparent explanations, FutureFood inadvertently fueled public skepticism about hidden health or ecological risks. Table 6.4 underscores the critical role of openness in fostering consumer acceptance, highlighting the call for explicit labeling and disclosure of AI's role in designing novel crops.

Subsequent to the description of the FutureFood program scenario, the decision-making flowchart below illustrates the ethical decision pathways navigated by the Ministry of Agriculture and stakeholders in addressing public perception and acceptance of AI-driven synthetic biology in agriculture (Fig. 6.6). This flowchart showcases the sequence of actions, public engagement strategies, ethical considerations, and potential outcomes, reflecting the challenges and complexities of anticipatory governance in aligning technological innovation with societal values.

On one hand, public skepticism about food safety, labeling, and technology concentration was already familiar territory in GMO debates. On the other, integrating AI into crop design adds layers of concern about data-driven decision-making, algorithmic bias, and lack of transparency. In many regions, existing regulations never anticipated the possibility of AI playing a primary role in breeding seeds or engineering agricultural traits. Thus, the very same cultural and ethical questions that once focused on genetic modification must now grapple with machine-learning algorithms that optimize, predict, and update genetic lines. As this scenario shows, a convergence of AI and SB intensifies old controversies while simultaneously presenting entirely

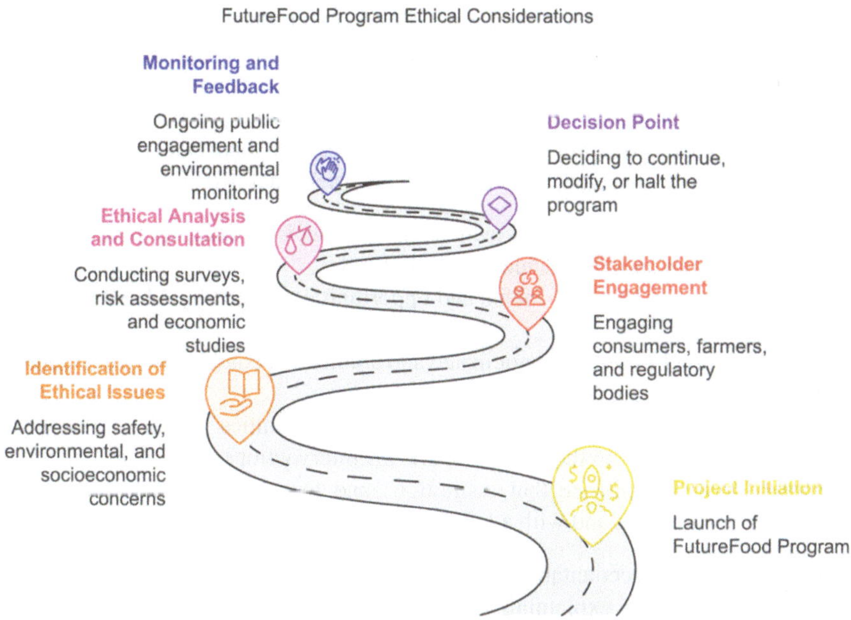

Fig. 6.6 FutureFood program ethical considerations: this roadmap illustrates the ethical decision-making process for the FutureFood program, focusing on public perception and acceptance of AI-designed crops. The journey begins with project initiation, followed by the identification of ethical issues such as safety, environmental impacts, and socioeconomic concerns related to small-scale farmers. Stakeholder engagement plays a crucial role, involving consumers, farmers, environmental groups, regulatory bodies, and educators to ensure a comprehensive understanding of public perceptions and practical implications. ethical analysis and consultation include conducting surveys, risk assessments, and economic impact studies to gauge the potential effects. At the decision point, stakeholders must decide whether to continue, modify, or halt the program, balancing public feedback with program goals. Monitoring and feedback involve ongoing public engagement, environmental monitoring, and economic assessments to ensure transparency and responsiveness. Adaptive governance principles are reflected in policy adjustments and technology modifications, emphasizing the integration of stakeholder input to achieve a balanced and ethical approach

new issues—ones that may require more inclusive dialogue, innovative labeling standards, and adaptive governance strategies. While these consumer and public trust issues are global in scope, cultural and regional nuances heavily affect attitudes toward AI-designed crops—a matter we address in Sect. 6.4 when examining cross-cultural ethical perspectives.

6.3.5 Lessons Learned

The four scenarios presented illustrate a diverse range of ethical dilemmas emerging from the convergence of AI and SB. Across all scenarios, common ethical challenges include the need for greater transparency in AI-driven decision-making, ensuring equitable access to technological advancements, and addressing the societal and environmental impact of introducing synthetic organisms. As shown in Table 6.5, these themes manifest differently across the scenarios examining precision health, environmental applications, and agricultural innovations, yet share fundamental ethical concerns. To guide interpretation, each theme is briefly defined as follows:

- Transparency: The extent to which AI–SB decision-making processes are explainable and understandable to stakeholders.
- Equity: Fair access to the benefits and opportunities created by AI–SB technologies.
- Environmental Impact: The anticipated and unintended consequences of synthetic interventions on ecosystems.
- Autonomy: The ability of individuals or communities to make informed, uncoerced decisions in the context of AI–SB interventions.
- Public Trust: Confidence that institutions and technologies are acting in the public's best interest and with adequate oversight.

Transparency and accountability are recurring themes, highlighting the challenges of AI's complexity in explaining and justifying its decisions, which is particularly crucial when these technologies directly impact human health, agriculture, and ecosystems. A key similarity across the scenarios is the concern for equitable access, whether it pertains to healthcare treatments or advanced biotechnologies in agriculture and environmental remediation. These concerns underscore the risk that these advancements could exacerbate existing disparities if not implemented thoughtfully.

Table 6.5 Key ethical themes across scenarios

Theme	Health applications scenario	Agricultural gene editing scenario	Environmental conservation scenario	Public perception scenario
Transparency	AI decision-making in genetic interventions	AI design processes in genetic modification	Algorithm transparency in ecosystem management	Public disclosure of AI-SB processes
Equity	Access to personalized therapies	Access to AI-enhanced biotechnology	Distribution of environmental benefits	Small-scale farmer access to technology
Environmental impact	Biosafety considerations	Gene transfer and biodiversity risks	Ecosystem disruption risks	Long-term ecological effects
Autonomy	Patient/child rights	Corporate versus farmer independence	Synthetic organism control	Consumer choice and cultural values
Public trust	Medical data privacy	Religious and cultural acceptance	Environmental safety assurance	Societal adoption of AI-SB foods

As detailed in Table 6.6, these challenges require specific governance responses across technical, social, environmental, and economic dimensions.

The four scenarios—ranging from personalized healthcare to environmental conservation—demonstrate how AI-driven SB intensifies existing ethical questions regarding privacy, autonomy, biosafety, and equity. Key issues revealed by these hypothetical cases include the tension between rapid innovation and the need for robust safeguards, the challenge of maintaining human oversight in algorithmic decision-making, and the reality that cultural values significantly affect perceptions of risk. Existing tools such as risk assessments, consent protocols, and environmental monitoring frameworks offer partial solutions, but gaps persist, especially when scaled to global contexts. Looking ahead, refining these tools to account for AI's accelerating capabilities and promoting transparent stakeholder engagement will be critical for ensuring responsible innovation. This task requires governance frameworks that are both responsive and precautionary. One approach is to integrate ethical impact assessments at key decision points in the development cycle—structured tools that anticipate societal implications and evaluate trade-offs. In high-risk contexts, emergency stop mechanisms could be mandated, allowing regulators to pause or revoke deployment when ethical thresholds are breached or unexpected harms emerge. These tools aim to balance rapid advancement with careful ethical reflection. Across all scenarios, gaps in current governance frameworks—be they unclear labeling requirements, insufficient containment strategies, or inadequate public engagement—are ultimately ethical challenges that go beyond mere procedural fixes. Determining accountability for cross-border ecological harm, balancing AI-driven optimization with public trust, and ensuring equitable access to breakthroughs all hinge on broader moral considerations.

Indeed, these dilemmas cannot be viewed in isolation from the cultural and regional contexts in which AI–SB technologies unfold. While the examples above highlight universal tensions—such as privacy, equity, and public trust—their impact varies significantly across different societies. Section 6.4 therefore shifts the lens to

Table 6.6 Governance implications and recommended responses

Challenge category	Governance implications	Recommended responses
Technical	Complexity of AI-SB systems and need for oversight	Enhanced monitoring systems and standardized assessment protocols
Social	Public acceptance and cultural sensitivity	Stakeholder engagement frameworks and inclusive decision-making processes
Environmental	Long-term ecosystem impacts and biosafety	Comprehensive risk assessment and containment strategies
Economic	Access disparities and benefit distribution	Fair benefit-sharing mechanisms and support for disadvantaged communities

show how diverse ethical perspectives—shaped by cultural, religious, and societal values—demand adaptive, context-sensitive governance strategies that can respond effectively to local needs and priorities.

6.4 Global and Cultural Ethical Considerations

6.4.1 Diversity of Ethical Perspectives

Building on the ethical dilemmas identified in the scenarios, this section examines how regional values, traditions, and regulatory philosophies reshape the challenges of AI–SB governance. By recognizing cultural diversity, we can better address the complexities that arise when universal principles meet local realities. The rationale behind global inclusiveness in the governance of AI (AI) and SB (SB) stems from the realization that these technologies have effects that transcend national borders. Recognizing that inclusive policymaking needs to be sensitive to and include diverse societies and cultures in decision-making, ÓhÉigeartaigh et al. noted that different ethical worldviews and objectives must be acknowledged [51]. This paradigm ensures that governing systems embrace a broader spectrum of global values rather than being dominated by a specific cultural perspective.

Cultural sensitivity in governance becomes particularly significant when overseeing technologies that have the power to alter our understanding of life and intelligence. Bensuade observes that some cultures base their ethical concerns on specific genetic modifications or the introduction of AI into biological processes, often rooted in religious or traditional beliefs [52]. For example, aboriginal tribes that believe life is sacred and interconnected may have concerns about tampering with genes and genetic matter in general [53]. Various cultural and religious traditions hold distinct values regarding nature, technology, and human intervention. For instance, some Eastern philosophies emphasize harmony with nature and a holistic worldview that may seem at odds with Western techno-scientific ideologies [52]. Meanwhile, other religious groups center on family or procreation as paramount goals, sometimes differentiating less between natural and assisted methods. As noted in studies of monotheistic religions, attitudes toward reproductive techniques (e.g., IVF or embryo transfer) can range from near-total acceptance (as in certain Jewish and Protestant denominations) to more restrictive views (e.g., the Vatican's stance on assisted reproduction), highlighting how theological frameworks shape bioethical positions [54]. Such intra- and inter-religious variations suggest that no single culture or faith uniformly accepts or rejects emerging biotechnologies. These differences can lead to practical tensions in international governance. For instance, the Western emphasis on individual consent—central to U.S. bioethics and EU GDPR—can conflict with collectivist norms in some Asian or Indigenous communities, where family or tribal elders may be the primary decision-makers. This creates challenges for establishing universally applicable consent protocols for AI–SB interventions, especially in cross-border research collaborations or multinational field trials.

One risk is ethical imperialism, where external actors impose moral norms on local communities [53]. Historically, biotech programs introduced in regions with strong cultural traditions or non-Western worldviews have clashed when developers assumed universal ethical standards. AI–SB could repeat these patterns if frameworks fail to adapt to local conceptions of stewardship, property rights, or spiritual beliefs surrounding the modification of life. This perspective warns against assuming universal ethics; practices considered permissible in one society might conflict with another's norms. However, it is equally important to acknowledge points of convergence—for example, widespread concern for human well-being, ecological balance, or the sanctity of life—where collaborative consensus may be found. Governance strategies that outright assume the absence of common ground may miss opportunities for shared principles, while failing to respect deep differences can spark conflict. For example, algorithmic bias often reflects the values of the society where the AI is developed, potentially marginalizing groups in different cultural settings [55]. When Western ethical norms dominate AI development, they may clash with non-Western values regarding privacy, autonomy, and collective well-being [56]. Furthermore, the implementation of global AI guidelines without local adaptation can stifle innovation and acceptance in regions with different ethical priorities [10]. For instance, while data privacy is paramount in some cultures, others may prioritize communal benefits over individual rights, affecting how AI technologies should be governed.

Notably, Western ethical frameworks often place a high value on human liberty and informed consent in medical research. However, in many non-Western societies, decision-making is bound by greater mutual engagement, and individual informed consent may be an alien or even inappropriate concept [51]. "Cultural colonialism," "moral imperialism," and "ethical imperialism" are often used interchangeably to describe scenarios in which one cultural or ethical system is imposed on another. In biotechnology, this can manifest when Western-dominated scientific norms or corporate strategies overlook or marginalize local traditions and values [53]. For instance, a one-size-fits-all approach to AI–SB governance might disregard a community's religious convictions regarding gene manipulation or ecological stewardship, leading to resistance or mistrust. To mitigate such risks, governance frameworks must be flexible and inclusive, reflecting the diversity of moral perspectives in both policy formation and decision-making processes. One potential remedy is establishing research centers in multiple countries, staffed by culturally diverse professionals who can ensure that local worldviews are integrated from the outset [52].

A plurality of ethical viewpoints enhances the resilience of governance arrangements by enriching the volume and quality of applied ethical reasoning. Heavey suggests that multiple perspectives can elucidate ethical dilemmas that might not be detected within the limitations of a single cultural perspective [32]. Such a multidimensional approach is particularly relevant to the complex worlds of AI and SB, where the implications of research and development are often not immediately clear.

For example, Western moral and ethical perspectives may be more concerned about potential costs to personal privacy and autonomy as AI is developed, while Asian perspectives might focus on potential social benefits and collective well-being [51]. Conversely, indigenous opinions tend to emphasize fears of persistent environmental impacts of artificial organisms [53]. By considering these diverse views, governance systems are better positioned to design comprehensive risk analysis frameworks and ethical guidelines.

Rather than presupposing that every cultural perspective diverges irreconcilably, effective governance recognizes that exposure to diverse ethical traditions can strengthen oversight mechanisms [29]. By weaving together convergent principles (e.g., the importance of safety, the desire to support community well-being) and openly addressing divergences (e.g., conceptions of nature, permissible interventions), policymakers can develop architectures with advanced and adaptable ethical principles. This approach aligns with the insight that public participation and transparency foster institutional legitimacy [57].

Wareham and Nardini propose a deliberative sensibility, using focus groups and other participatory methods to co-create acceptable ethical standards for AI–SB technologies [58]. Beyond simple "awareness campaigns," genuine involvement may include inviting local stakeholders—small-scale farmers, faith leaders, consumer advocacy groups, or non-Western ethicists—to collaboratively shape research priorities and regulatory guidelines. In this process, acknowledging similarities in overarching concerns (such as health, sustainability, or equity) can bridge cultural gaps, while respectful negotiation of differences avoids flattening nuanced worldviews under a dominant ethical framework.

Understanding and integrating diverse ethical perspectives are paramount in the governance of AI and SB. These technologies have global impacts that transcend national borders, making it essential to acknowledge and respect different ethical worldviews and objectives. Cultural sensitivity involves more than just understanding other viewpoints; it requires active engagement with diverse communities to comprehend their concerns and values. This approach helps prevent ethical imperialism and ensures that governance structures are responsive to a wide range of cultural contexts (Fig. 6.3).

6.4.2 Regional Perspectives

Regional approaches to governing AI and SB vary significantly based on cultural values, political structures, and economic priorities. While some regions emphasize individual rights and privacy, others prioritize collective welfare or environmental stewardship. These variations shape how ethical concerns—such as data protection, bioterrorism risks, and biosafety—are addressed. The following subsections illustrate how cultural norms and regional governance models influence the adoption of AI–SB, highlighting both overarching frameworks and specific case studies.

Western Approaches. Most Western countries focus on balancing individual rights, data privacy, and public safety. Concerns about bioterrorism or unethical genetic modifications have shaped robust regulatory efforts, particularly in healthcare and environmental applications [59]. Western nations typically favor a precautionary approach, seeking comprehensive risk assessments before approving new technologies [58]. In the United States, SB oversight involves multiple agencies, including the Environmental Protection Agency (EPA) and the Food and Drug Administration (FDA). Although this broad governance structure sometimes prolongs the approval process [29], U.S. regulations generally aim to foster innovation through more permissive frameworks. Critics argue that gaps persist in long-term ecological monitoring and ethical oversight, prompting calls for continuous regulatory adaptation. In contrast, the European Union (EU) has adopted stringent regulations guided by the precautionary principle. For instance, Directive 2001/18/EC mandates risk assessments, labeling, and post-market monitoring for the deliberate release of genetically modified organisms (GMOs), including those derived from SB [60]. This risk-averse strategy underscores public safety and consumer protection, although it can slow technological advancement [61]. Certain EU countries—like the United Kingdom[9]—have integrated Ethical, Legal, and Social Implications (ELSI) studies into funding mechanisms [62]. However, uneven implementation across Europe can pose challenges to a unified SB governance model.

North America: Balancing Innovation and Oversight. Compared to the EU, the U.S. regime is more innovation-driven, combining formal regulation with a degree of self-regulation across agencies such as the U.S. Department of Agriculture and the EPA. This approach encourages rapid technological progress and economic growth, often leading to faster adoption of new AI and SB applications [63]. However, gaps in coverage may emerge—particularly in long-term ecological monitoring and ethical oversight—because regulations tend to focus on immediate safety assurances [64]. Public trust can be undermined when regulations appear too permissive, leading to calls for more robust ethical and environmental safeguards [29]. Balmer et al. observe that transitioning to more cautious international standards, such as those of the EU, is challenging given the U.S.'s strong emphasis on market-driven growth and rapid innovation [17]. This tension reverberates beyond national borders, shaping international collaboration and regulatory alignment.

Asia: Rapid Technological Adoption andAdaptive Governance. Many Asian cultures place greater value on social welfare and community well-being compared to individual rights [51]. This collectivist orientation fosters increased openness toward new technologies if they promise societal benefits, such as enhanced agricultural output or public health interventions. Nations like China and Singapore have swiftly adapted governance frameworks to maintain global competitiveness and address local needs [29]. China's significant investments in AI and SB, coupled with a

[9] Although the UK formally exited the European Union in 2020 (Brexit), the cited study and associated policies predate that event, reflecting the UK's then-current engagement with EU frameworks.

governance strategy emphasizing fast and practical implementation, have spurred robust innovation yet drawn criticism regarding ethical oversights [51]. Singapore's Bioethics Advisory Committee (BAC) provides guidelines on emerging biomedical research and technology, striking a balance between innovation and ethical responsibility [65]. Through such national bodies, Singapore attempts to reconcile economic ambitions with culturally grounded ethical standards. However, this emphasis on collective gains can sometimes lead to overlooking individual rights and broader ethical concerns. Millett and Alexanian point out that while adaptive governance in Asia effectively promotes technological advancement and economic development, it may face challenges in ensuring individual rights and broader ethical considerations are adequately addressed [30].

Indigenous Perspectives. Indigenous communities often prioritize environmental harmony and holistic ways of life, highlighting ethical considerations that revolve around maintaining ecosystems and cultural practices [53, 66]. Resistance toward certain AI–SB interventions, such as genetically modified organisms (GMOs), may stem from potential disruptions to biodiversity and traditional livelihoods. Community consent and representation are critical, guided by "free, prior, and informed consent" (FPIC) outlined in the United Nations Declaration on the Rights of Indigenous Peoples [53, 67]. Operationalizing FPIC in AI–SB governance is challenging given the global nature of these technologies. Still, indigenous perspectives—such as Traditional Ecological Knowledge (TEK)—could enrich risk assessments and promote more sustainable outcomes [53]. Incorporating indigenous representatives in biosafety committees or integrating TEK into environmental review processes exemplifies practical steps toward pluralistic governance.

Deplazes-Zemp et al. argue that more attention should be paid to preserving natural ecosystems and carefully assessing the ethical aspects of introducing technological interventions that disturb traditional lifestyles [53]. Such societies often require technological interventions to align with traditional knowledge and cultures. Resistance may arise against technologies that could harm natural ecosystems or cultural practices. For instance, some indigenous communities are concerned that the introduction of GMOs may negatively impact ecosystems, posing threats to biodiversity and traditional farming methods.

Latin America: Hybrid Models and Agricultural Priorities. Latin American countries display diverse approaches to AI and SB governance, significantly influenced by both U.S. and EU models. This diversity reflects different economic priorities and levels of regulatory capacity across the region. Countries like Brazil and Argentina have developed specialized regulatory systems for genome-edited plants, similar to the U.S. model [68]. The agricultural economies in Latin America are central to their regulatory policies from a SB perspective. Most countries strive to expand robust agricultural export targets while satisfying biosafety concerns, ensuring that technological advancements do not compromise public and environmental health. Argentina's regulatory framework pioneers dealing with cases according to the principles of SB and GMOs, using a case-by-case assessment to balance innovation with safety. However, aligning these heterogeneous approaches with international standards proves challenging. According to Laird and Wynberg, while many Latin American regulatory approaches have supported economic growth and technology adoption, ensuring comprehensive biosafety measures that meet global standards remains difficult [69]. The alignment process is further complicated by the diversity in regulatory capacity and public perceptions within the region. Emerging economies in the region often face regulatory capacity gaps, limiting effective governance despite a strong reliance on agricultural exports and public concern about ecological integrity.

Balancing Global Standards with LocalAdaptations. Governance of AI and SB across these regions exemplifies the delicate interplay between innovation, safety, and ethical considerations on a global scale. Regions prioritizing rapid innovation risk compromising extensive safety measures, as seen with the U.S. and parts of Asia. Conversely, regions like the EU that adopt a precautionary approach may hamper technological advancement due to stringent regulations. These regional differences have significant consequences for international collaboration and the development of a global governance framework for AI and SB. The challenge lies in striking a balance between promoting innovation and ensuring adequate protection and ethical behavior across multiple cultural and economic settings. Although previous sections highlighted how ethical values vary widely across cultures, it remains possible to align on core principles—such as environmental stewardship, human dignity, and equitable access—at a global level. These high-level guidelines serve as common ground for international collaboration on AI–SB. Local adaptations, informed by cultural, religious, or historical contexts, ensure that these guidelines remain relevant and acceptable to distinct communities. Harmonizing across diverse governance approaches is essential when addressing transnational concerns like food security and climate change, which benefit from shared technological development. By maintaining global benchmarks—for instance, around biosafety or data-sharing protocols—while allowing tailored solutions (e.g., region-specific consent processes), governance frameworks can both honor local ethical stances and preserve global safety standards (Table 6.7).

Table 6.7 emphasizes that coordinated efforts can yield broad yet flexible ethical principles, simultaneously fostering global coherence and respecting regional differences. For instance, a universal principle like "informed consent" may be implemented differently in individualist societies versus communal cultures, but the core remains consistent: ensuring that affected communities understand and agree to AI–SB interventions. This dual approach—universal in principle yet adaptable in practice—helps reconcile the seemingly contradictory need for an integrated international structure with the reality of diverse local values.

Table 6.7 Key insights from regional ethical impacts of AI and synthetic biology

Key insight	Description	Implications for governance
Cultural values shape ethical priorities	• Local beliefs and traditions influence how communities perceive technology, accept risks, and define ethical conduct • Cultural norms affect attitudes toward innovation, privacy, environmental stewardship, and technology's role in society	• Governance frameworks must be culturally sensitive, integrating local values to ensure policies are accepted and effective • Ethical guidelines should be adaptable to align with regional beliefs and societal norms, avoiding ethical imperialism
Regulatory environments reflect societal expectations	• Legal frameworks and regulatory approaches express regional ethical priorities • They determine control mechanisms for technologies and the extent to which ethical concerns are formalized into law • Variations reflect differences in balancing innovation with ethics	• Policymakers need to understand and incorporate societal expectations into regulations, aligning policies with public values • Harmonizing international standards requires acknowledging differences and finding common ground without imposing external values
Public trust and engagement are crucial	• Acceptance of innovations relies heavily on public trust built through transparency, communication, and involvement • Ignoring public sentiment can lead to resistance, ethical controversies, and rejection of technologies	• Engaging stakeholders through consultations, education, and participatory governance enhances legitimacy and acceptance • Policies should prioritize transparency and responsiveness to public concerns, building trust and facilitating adoption
One-size-fits-all approaches are ineffective	• Applying uniform ethical frameworks globally without considering regional contexts can result in ethical imperialism and neglect of local needs • Diverse conditions necessitate tailored governance strategies addressing specific dynamics and perspectives	• Governance should be flexible and context-specific, allowing adaptations that respect local customs and priorities • Collaborative policy development ensures relevance and effectiveness across regions, accommodating diverse perspectives

(continued)

Table 6.7 (continued)

Key insight	Description	Implications for governance
Interconnectedness requires global collaboration	• Despite regional differences, the global nature of impacts demands international cooperation • Issues like data sharing, cross-border environmental effects, and global markets necessitate collaborative efforts • Sharing best practices is essential	• Establishing international platforms for dialogue and collaboration helps address cross-border challenges • Coordinated efforts can develop universally applicable yet adaptable ethical standards, fostering global harmony while respecting regional differences

6.4.3 Bridging the Gap: Inclusive and Adaptive Governance

Effectively integrating diverse ethical perspectives into the governance of AI and Synthetic Biology (AI–SB) requires multifaceted strategies that extend beyond high-level international frameworks. While such frameworks offer foundational norms, ensuring legitimacy and responsiveness demands approaches grounded in specific cultural contexts and local realities. The following strategies are essential for operationalizing ethical pluralism in AI–SB governance:

Engagement and Public Participation. One key strategy is the incorporation of participatory approaches in policy development and decision-making processes. Engaging stakeholders—including scientists, policymakers, industry representatives, civil society organizations, and affected communities—in meaningful dialogue ensures that a wide range of perspectives is considered. Techniques such as public consultations, citizen assemblies, and deliberative workshops facilitate inclusive discussions, allowing for the expression of diverse ethical viewpoints and fostering mutual understanding [8]. For example, the UK's "GM Nation?" public debate and Norway's consensus conferences have served as participatory models that informed national biotechnology policies through structured citizen input, shaping both regulatory and societal positions on gene technologies.

Integration of Local and Indigenous Knowledge. Local and indigenous knowledge systems offer valuable ethical insights that can enrich governance frameworks. Recognizing and respecting traditional values and practices can help bridge the gap between global ethical principles and local realities. For instance, involving indigenous communities in the governance of biotechnologies that impact their lands and resources acknowledges their rights and integrates their ethical considerations into broader decision-making [66]. This inclusion respects cultural sovereignty and ensures that technological advancements do not undermine local traditions and ecosystems. One applied model is the inclusion of Free, Prior, and Informed Consent (FPIC) requirements for biotech trials in countries like Canada and Peru, ensuring community-driven governance over indigenous lands and ecological systems.

Capacity Building and Education. Education and capacity building at the grassroots level are essential for empowering communities to participate effectively in ethical deliberations. Providing accessible information about AI and Synthetic Biology, along with their potential benefits and risks, enables individuals to make informed contributions to policy discussions. Tailoring educational programs to local languages and cultural contexts enhances engagement and ensures that diverse voices are heard [70]. This approach also helps demystify complex technologies, reducing fear and misconceptions. An example of this is UNESCO's Ethics Teacher Training Program, which equips educators in over 60 countries to deliver culturally adapted bioethics content, supporting widespread ethical literacy and governance engagement.

Collaborative Networks and Regional Forums. Collaborative networks and partnerships at regional and national levels can facilitate the exchange of ethical perspectives and best practices. Establishing forums where stakeholders from different sectors and regions can share experiences and discuss ethical challenges promotes cross-cultural learning and helps identify common values. These networks can support the development of context-specific guidelines that align with both global standards and local ethical frameworks [16]. For example, regional ethics committees or councils can address issues pertinent to specific cultural or environmental contexts. The African Union's Science, Technology, and Innovation Strategy and regional bioethics hubs in Asia provide such examples, linking local priorities to broader ethical dialogues through regional committees and policy harmonization efforts.

Adaptive Governance Mechanisms. Adaptive governance mechanisms that are flexible and responsive to emerging ethical concerns are crucial. Implementing iterative policy processes that incorporate feedback from diverse stakeholders allows governance structures to evolve alongside technological advancements. This adaptability ensures that ethical frameworks remain relevant and effective in addressing new challenges posed by the convergence of AI and Synthetic Biology [71]. Such mechanisms might include regular reviews of policies, open channels for stakeholder input, and the ability to adjust regulations promptly in response to new information.

Ethical Impact Assessments and Cultural Sensitivity. Ethical impact assessments that include cultural considerations can be integrated into the development and deployment of technologies. By evaluating the potential ethical implications of AI and Synthetic Biology projects through a culturally sensitive lens, policymakers and developers can identify and mitigate adverse effects on specific communities. This proactive approach enhances ethical oversight and promotes socially responsible innovation [72]. Assessments should consider factors such as community values, environmental sustainability, and social justice implications. For example, iGEM's Safety and Security Program provides a youth-led sandbox for adaptive biosafety norms, updated annually based on project risks and global input from hundreds of synthetic biology teams.

Interdisciplinary Research and Cross-Sectoral Collaboration. Fostering interdisciplinary research that brings together experts from various fields—including ethics, social sciences, law, and engineering—can lead to a more comprehensive understanding of ethical issues. Interdisciplinary collaborations encourage the examination of ethical dilemmas from multiple angles, uncovering hidden assumptions and biases that may be overlooked within a single discipline [73]. Such collaborations can result in more holistic governance frameworks that account for technical feasibility, ethical soundness, legal compliance, and social acceptance. In the Netherlands, for example, the Rathenau Institute includes cultural impact metrics in technology assessments, integrating values-based foresight into national innovation policy. The European Group on Ethics in Science and New Technologies (EGE) is one example of an interdisciplinary advisory body that incorporates legal, philosophical, and scientific input into EU guidance on emerging technologies.

Data-Driven Insights and Technological Tools. Moreover, leveraging technology itself can aid in integrating diverse ethical views. For instance, AI-driven tools can analyze large-scale public opinion data to identify prevailing ethical concerns across different cultures and regions [74]. These insights can inform policymakers about the values and priorities of various communities, guiding the development of more representative governance strategies. Pilot platforms using natural language processing to analyze citizen sentiment—such as those developed in civic tech initiatives in Taiwan—are already being used to inform government policy on AI governance and ethics.

Transparency and Accountability. Transparency in communication and decision-making processes builds trust among stakeholders. Providing clear and accessible information about how ethical considerations are integrated into governance fosters accountability and allows for public scrutiny. This openness is particularly important in areas where mistrust or skepticism may exist due to past experiences or cultural differences [75]. Transparent reporting on the outcomes of ethical assessments, regulatory decisions, and the rationale behind them can enhance legitimacy and acceptance. The NIH ClinicalTrials.gov registry and its international analogs exemplify how mandated transparency can promote public trust, especially in fields involving genomic data and biomedical AI.

Bottom-Up Governance and Community Empowerment. Lastly, embracing a bottom-up approach to governance can empower communities and ensure that ethical frameworks are not imposed but co-created. Engaging local leaders, community organizations, and citizens in the governance process ensures that policies reflect the lived experiences and values of those directly affected by technological developments [76]. This participatory governance model promotes inclusivity and democratizes decision-making. Brazil's community-driven science-policy platforms, such as the Biodiversity Knowledge Integration Network (Rede Biocon), demonstrate how rural and Indigenous communities can shape national bioeconomy policies in ways that honor cultural and environmental ethics.

In conclusion, integrating diverse ethical views into the governance of AI and Synthetic Biology requires strategies that are participatory, context-sensitive, and adaptable. By embracing local knowledge, fostering dialogue, promoting interdisciplinary collaboration, and ensuring transparency, governance frameworks can effectively reflect the plurality of ethical perspectives. These approaches not only enhance the ethical robustness of policies but also build public trust and facilitate the responsible advancement of these transformative technologies (Fig. 6.7).

This section underscored how ethical frameworks must adapt to diverse regional values and contexts, with cultural norms shaping everything from regulatory priorities to public acceptance of AI–SB technologies. Key issues include the risk of ethical imperialism when a single worldview dominates, and the challenge of

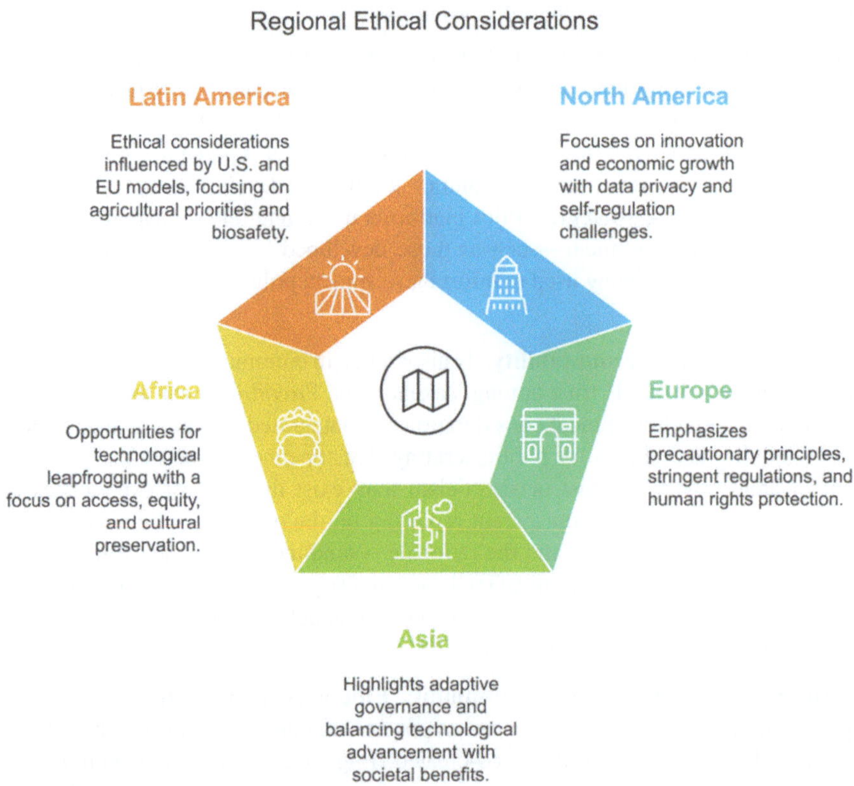

Fig. 6.7 Regional ethical perspectives in AI and synthetic biology: this pentagon diagram illustrates how diverse regions prioritize ethical issues in the governance of emerging technologies. North America emphasizes innovation alongside challenges in data privacy and self-regulation. Europe applies the precautionary principle, focusing on stringent regulations and human rights. Asia balances rapid technological advancement with collective welfare and societal benefits. Africa highlights opportunities for technological leapfrogging while ensuring equity, cultural preservation, and environmental considerations. Latin America draws from both U.S. and EU influences, emphasizing agricultural priorities and biosafety

aligning universal principles (like human rights) with deeply local concerns (such as environmental stewardship or communal decision-making). Existing measures—for instance, UNESCO's encouragement of local ethics committees and nation-specific guidelines—help safeguard cultural pluralism. Ultimately, building ethically robust AI–SB governance requires more than high-level declarations—it demands locally grounded frameworks, inclusive participation, and mechanisms that can evolve with both the technology and the societies it serves (Fig. 6.8).

6.5 Conclusion

Adaptive, culturally aware governance frameworks are essential to effectively manage the multifaceted ethical complexities arising from AI–SB convergence, acknowledging that globally shared guidelines require context-specific adaptations to remain legitimate and inclusive. Looking ahead, interdisciplinary collaboration emerges as a cornerstone for developing comprehensive ethical guidelines—yet in the context of ethical and cultural diversity, such collaboration must go beyond merely assembling experts from different fields. It should proactively incorporate a range of cultural perspectives and local knowledge systems, fostering genuine dialogue among scientists, ethicists, policymakers, technologists, social scientists, and community representatives. By recognizing and respecting how beliefs, values, and social norms vary across regions, these teams can anticipate diverse viewpoints early in the innovation cycle, thereby designing solutions that are both innovative and ethically sound in various cultural settings.

In parallel, active public participation—informed by transparent communication and genuine dialogue rather than mere information campaigns—offers a crucial mechanism to capture the on-the-ground concerns of those most affected by AI–SB applications. This engagement ensures that governance structures remain responsive to both broad ethical principles (e.g., human rights, ecological integrity) and local nuances (e.g., religious constraints, communal decision-making).

By committing to adaptive, culturally sensitive governance, interdisciplinary collaboration, and robust public engagement, we can guide the development of AI and Synthetic Biology toward a future reflecting widely endorsed ethical aspirations—such as human dignity, environmental sustainability, and social equity—while still allowing each society to embed its own values and priorities. These shared ideals are not intended as a single universal prescription; rather, they represent common ground among many global actors that can be further honed and adapted at regional and local levels. Striking this balance between global coherence and contextual flexibility will be key in leveraging AI–SB's transformative potential responsibly and equitably across diverse cultural landscapes.

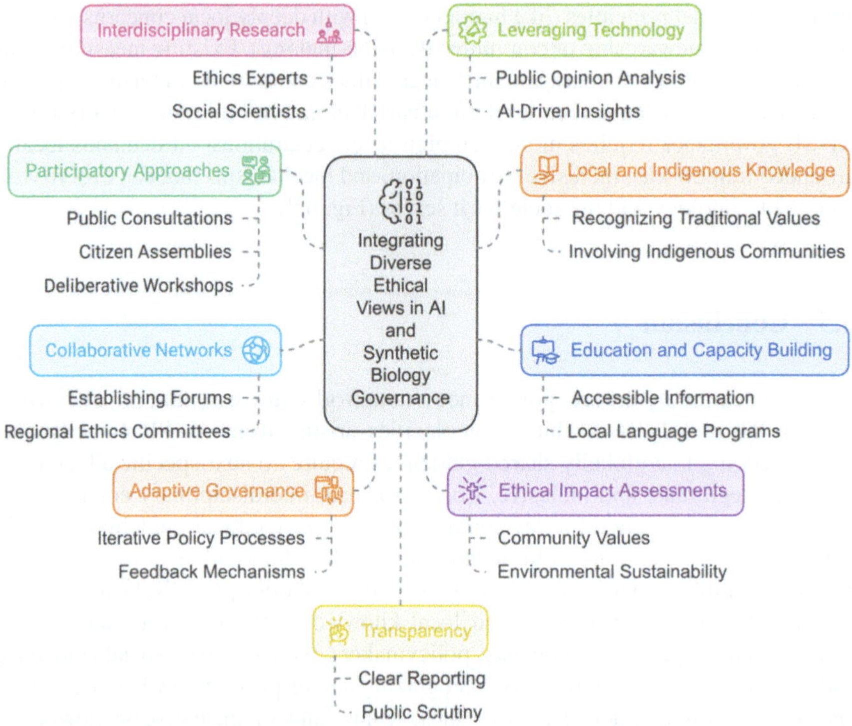

Fig. 6.8 Navigating ethical complexity in AI and synthetic biology governance: this comprehensive framework highlights the importance of integrating diverse ethical perspectives for responsible technological advancement. The central focus is on a holistic approach that encompasses interdisciplinary research and collaborative networks, ensuring input from ethics experts, social scientists, and regional committees to create inclusive policies. Participatory approaches, such as citizen assemblies and public consultations, are key in capturing a broad range of societal views, while leveraging technology like AI-driven insights ensures that public opinion is analyzed efficiently. The framework also values local and indigenous knowledge, recognizing the critical role of traditional values and actively involving these communities in decision-making processes. Education and capacity building provide accessible information and local language programs to empower all stakeholders, promoting widespread understanding and engagement. Ethical impact assessments are used to consider community values and environmental sustainability, maintaining alignment with societal well-being. Adaptive governance, with iterative policy processes and feedback mechanisms, ensures that governance evolves in response to new challenges, while transparency is upheld through clear reporting and public scrutiny. Altogether, this framework promotes a governance model that is adaptable, inclusive, and ethically informed to meet the demands of emerging AI and synthetic biology technologies

References

1. Topol E (2019) Deep medicine: how artificial intelligence can make healthcare human again. Basic Books
2. Floridi L (2014) The fourth revolution: how the infosphere is reshaping human reality. Oxford University Press

3. Char DS, Shah NH, Magnus D (2018) Implementing machine learning in health care—addressing ethical challenges. N Engl J Med 378(11):981–983
4. Carolan M (2021) The sociology of food and agriculture. Routledge
5. Kloppenburg J (2014) Re-purposing the master's tools: the open source seed initiative and the struggle for seed sovereignty. J Peasant Stud 41(6):1225–1246
6. Redford KH, Adams W, Mace GM (2013) Synthetic biology and conservation of nature: wicked problems and wicked solutions. PLoS Biol 11(4):e1001530
7. Sandler R (2017) Environmental ethics. Oxford University Press
8. Stilgoe J, Owen R, Macnaghten P (2013) Developing a framework for responsible innovation. Res Policy 42(9):1568–1580
9. Guston DH (2014) Understanding 'anticipatory governance.' Soc Stud Sci 44(2):218–242
10. Whittlestone J, Nyrup R, Alexandrova A, Dihal K, Cave S (2019) Ethical and societal implications of algorithms, data, and artificial intelligence: a roadmap for research. Nuffield Foundation
11. Kaebnick GE, Gusmano MK, Murray TH (2014) The ethics of synthetic biology: next steps and prior questions. Hastings Cent Rep 44(S5)
12. Taddeo M, Floridi L (2018) How AI can be a force for good. Science 361(6404):751–752
13. Boenink M, Swierstra T, Stemerding D (2010) Anticipating the interaction between technology and morality: a scenario study of experimenting with humans in bionanotechnology. Stud Ethics Law Technol 4(2)
14. Capurro R (2018) Intercultural information ethics. In: Localizing the internet. Brill | Fink, pp 19–38
15. Jasanoff S (2011) Constitutional moments in governing science and technology. Sci Eng Ethics 17(4):621–638
16. Jasanoff S (2003) Technologies of humility: citizen participation in governing science. Minerva 41(3):223–244
17. Balmer A, Calvert J, Marris C, Molyneux-Hodgson S, Frow E, Kearnes M, Bulpin K, Schyfter P, Mackenzie A, Martin P (2015) Taking roles in interdisciplinary collaborations: reflections on working in post-ELSI spaces in the UK synthetic biology community. Sci Technol Stud
18. Costanza-Chock S (2018) Design justice: towards an intersectional feminist framework for design theory and practice. In: Proceedings of the DRS conference
19. Zou J, Huss M, Abid A, Mohammadi P, Torkamani A, Telenti A (2019) A primer on deep learning in genomics. Nat Genet 51(1):12–18
20. Pauwels E (2011) The value of science and technology studies (STS) to sustainability research: a critical approach toward synthetic biology promises. In: European research on sustainable development. Springer Berlin Heidelberg, pp 111–135
21. UNESCO (2005) Universal declaration on bioethics and human rights
22. CIOMS (2016) 2016 CIOMS international ethical guidelines for health-related research involving humans
23. Have H (2006) The activities of UNESCO in the area of ethics. Kennedy Inst Ethics J 16(4):333–351
24. European Parliament (2020) The impact of the General Data Protection Regulation (GDPR) on artificial intelligence
25. High-Level Expert Group on Artificial Intelligence (2019) Ethics guidelines for trustworthy AI
26. NIH (2024) NIH guidelines for research involving recombinant or synthetic nucleic acid molecules (NIH guidelines)
27. Bohua L, Yuexin W, Yakun O, Kunlan Z, Huan L, Ruipeng L (2023) Ethical framework on risk governance of synthetic biology. J Biosaf Biosecur 5(2):45–56
28. Tang L, Kuzma J, Zhang X, Song X, Li Y, Liu H, Hu G (2023) Synthetic biology and governance research in China: a 40-year evolution. Scientometrics 128(9):5293–5310
29. Trump BD (2017) Synthetic biology regulation and governance: lessons from TAPIC for the United States, European Union, and Singapore. Health Policy 121(11):1139–1146

30. Millett P, Alexanian T (2021) Implementing adaptive risk management for synthetic biology: lessons from iGEM's safety and security programme. Eng Biol 5(3):64–71
31. Undheim TA (2024) The whack-a-mole governance challenge for AI-enabled synthetic biology: literature review and emerging frameworks. Front Bioeng Biotechnol 12
32. Heavey P (2013) Synthetic biology ethics: a deontological assessment. Bioethics 27(8):442–452
33. Kronberger N, Holtz P, Kerbe W, Strasser E, Wagner W (2009) Communicating synthetic biology: from the lab via the media to the broader public. Syst Synth Biol 3(1–4):19–26
34. Nesbeth DN, Zaikin A, Saka Y, Romano MC, Giuraniuc CV, Kanakov O, Laptyeva T (2016) Synthetic biology routes to bio-artificial intelligence. Essays Biochem 60(4):381–391
35. Eriksson M, Brown WT, Gordon LB, Glynn MW, Singer J, Scott L, Erdos MR, Robbins CM, Moses TY, Berglund P, Dutra A, Pak E, Durkin S, Csoka AB, Boehnke M, Glover TW, Collins FS (2003) Recurrent de novo point mutations in lamin A cause Hutchinson-Gilford progeria syndrome. Nature 423(6937):293–298
36. Dias R, Torkamani A (2019) Artificial intelligence in clinical and genomic diagnostics. Genome Med 11(1):70
37. Heidari Feidt R, Ienca M, Elger BS, Folcher M (2019) Synthetic biology and the translational imperative. Sci Eng Ethics 25(1):33–52
38. Kraft SA, Cho MK, Gillespie K, Halley M, Varsava N, Ormond KE, Luft HS, Wilfond BS, Lee SS-J (2018) Beyond consent: building trusting relationships with diverse populations in precision medicine research. Am J Bioeth 18(4):3–20
39. Baylis F, Robert JS (2004) The inevitability of genetic enhancement technologies. Bioethics 18(1):1–26
40. Wee R, Henaghan M, Winship I (2013) Ethics: dynamic consent in the digital age of biology: online initiatives and regulatory considerations. J Prim Health Care 5(4):341
41. Rothstein M (2014) Privacy and technology in the twenty-first century
42. Burall S (2018) Rethink public engagement for gene editing. Nature 555(7697):438–439
43. Potrykus I (2001) Golden rice and beyond. Plant Physiol 125(3):1157–1161
44. Macknight RC, Laing WA, Bulley SM, Broad RC, Johnson AA, Hellens RP (2017) Increasing ascorbate levels in crops to enhance human nutrition and plant abiotic stress tolerance. Curr Opin Biotechnol 44:153–160
45. Amin L, Azlan NAA, Ahmad R (2013) Ethical perception of human gene in transgenic banana. Afr J Biotechnol
46. Gaskell G, Stares S, Allansdottir A, Allum N, Corchero C, Fischler C, Jackson J, Kronberger N, Hampel J, Mejlgaard N, Quintanilha A, Rammer A, Revuelta G, Stoneman P, Torgersen H, Twardowski T, Wagner W (2010) Europeans and biotechnology in 2010: winds of change? Public Underst Sci
47. Mampuys R (2018) Socio-economic considerations in regulatory decision-making on genetically modified crops
48. Macovei O (2022) The ethics of synthetic biology—at the confluence of ecoethics and technoethics. Postmod Open 13(3):234–250
49. Jin S, Clark B, Kuznesof S, Lin X, Frewer LJ (2019) Synthetic biology applied in the agrifood sector: public perceptions, attitudes and implications for future studies. Trends Food Sci Technol 91:454–466
50. Nielsen AAK, Voigt CA (2018) Deep learning to predict the lab-of-origin of engineered DNA. Nat Commun 9(1):3135
51. ÓhÉigeartaigh SS, Whittlestone J, Liu Y, Zeng Y, Liu Z (2020) Overcoming barriers to cross-cultural cooperation in AI ethics and governance. Philos Technol 33(4):571–593
52. Bensaude Vincent B (2013) Ethical perspectives on synthetic biology. Biol Theory 8(4):368–375
53. Deplazes-Zemp A, Gregorowius D, Biller-Andorno N (2015) Different understandings of life as an opportunity to enrich the debate about synthetic biology. NanoEthics 9(2):179–188
54. Schenker JG (2000) Women's reproductive health: monotheistic religious perspectives. Int J Gynecol Obstet 70(1):77–86

55. Noble SU (2018) Algorithms of oppression. NYU Press
56. Metcalf J, Moss E, Boyd D (2019) Owning ethics: corporate logics, silicon valley, and the institutionalization of ethics. Soc Res 86(2):449–476
57. Douglas CM, Stemerding D (2014) Challenges for the European governance of synthetic biology for human health. Life Sci Soc Policy 10(1):6
58. Wareham C, Nardini C (2015) Policy on synthetic biology: deliberation, probability, and the precautionary paradox. Bioethics 29(2):118–125
59. Heyd D (2012) Is there anything unique in the ethics of synthetic biology? Perspect Biol Med 55(4):581–589
60. Anyshchenko A, Yarnold J (2021) From 'mad cow' crisis to synthetic biology: challenges to EU regulation of GMOs beyond the European context. Int Environ Agreem 21(3):391–404
61. Sundaram LS, Ajioka JW, Molloy JC (2023) Synthetic biology regulation in Europe: containment, release and beyond. Synth Biol 8(1)
62. Pei L, Gaisser S, Schmidt M (2012) Synthetic biology in the view of European public funding organisations. Public Underst Sci 21(2):149–162
63. Carter SR, Rodemeyer M, Garfinkel MS, Friedman RM (2014) Synthetic biology and the U.S. biotechnology regulatory system: challenges and options
64. Levidow L, Bijman J (2002) Farm inputs under pressure from the European food industry. Food Policy 27(1):31–45
65. Bioethics Advisory Committee (2021) Ethics guidelines for human biomedical research
66. Whyte K (2017) Indigenous climate change studies: indigenizing futures, decolonizing the Anthropocene. Engl Lang Notes 55(1–2):153–162
67. United Nations (2007) United Nations declaration on the rights of indigenous peoples
68. Eriksson D, Kershen D, Nepomuceno A, Pogson BJ, Prieto H, Purnhagen K, Smyth S, Wesseler J, Whelan A (2019) A comparison of the EU regulatory approach to directed mutagenesis with that of other jurisdictions, consequences for international trade and potential steps forward. New Phytol 222(4):1673–1684
69. Laird SA, Wynberg RP (2016) Locating responsible research and innovation within access and benefit sharing spaces of the convention on biological diversity: the challenge of emerging technologies. NanoEthics 10(2):189–200
70. Macnaghten P, Guivant JS (2011) Converging citizens? Nanotechnology and the political imaginary of public engagement in Brazil and the United Kingdom. Public Underst Sci 20(2):207–220
71. Dietz T, Ostrom E, Stern PC (2003) The struggle to govern the commons. Science 302(5652):1907–1912
72. Vanclay F (2003) International principles for social impact assessment. Impact Assess Proj Apprais 21(1):5–12
73. Frodeman R, Klein J, Dos Santos Pacheco R (2019) The Oxford handbook of interdisciplinarity
74. Ignatidou S (2019) AI-driven personalization in digital media: political and societal implications
75. Fisher E, Rip A (2013) Responsible innovation: multi-level dynamics and soft intervention practices. In: Responsible innovation. Wiley, pp 165–183
76. Chilvers J, Kearnes M (2016) Remaking participation: science, environment and emergent publics

Chapter 7
Ethical Implications and Governance Solutions for the Convergence of AI and Synthetic Biology in Healthcare and Agriculture

Juan C. Cruz, Ruth Mampuys, German Reyes, and Sarah Dryhurst

Abstract The convergence of Artificial Intelligence (AI) and Synthetic Biology (SB) offers transformative potential in healthcare and agriculture but introduces significant ethical, social, and ecological challenges. This chapter comprehensively analyzes these dualities, addressing critical issues such as data privacy violations, algorithmic bias exacerbating inequities, diluted accountability in complex systems, adverse socioeconomic impacts on vulnerable populations, and environmental integrity concerns. In healthcare, AI-driven diagnostics and personalized therapies raise concerns over patient data security, equitable access, and the opacity of decision-making. In agriculture, AI-SB innovations challenge farmer data ownership, socioeconomic fairness, and ecological safety. To navigate these complexities, the chapter advocates for robust, adaptive governance solutions grounded in interdisciplinary collaboration, stakeholder participation, and the foundational TAPIC principles (Transparency, Accountability, Participation, Integrity, Capacity). It presents a structured framework comprising eight core governance pillars—spanning enhanced data protection, algorithmic transparency, ethical protocols for AI-generated biological systems, ecological impact evaluation, and collaborative pilot programs. These pillars are operationalized through five key data governance domains (privacy,

J. C. Cruz (✉) · G. Reyes
Department of Biomedical Engineering, Universidad de los Andes, Bogotá, Colombia
e-mail: jc.cruz@uniandes.edu.co

G. Reyes
e-mail: g.reyes@uniandes.edu.co

R. Mampuys
The Netherlands Commission on Genetic Modification (COGEM), Bilthoven, Utrecht, The Netherlands
e-mail: mampuys@wrr.nl

S. Dryhurst
Institute for Risk and Disaster Reduction, University College London, London, UK
e-mail: s.dryhurst@ucl.ac.uk

C. L. Cummings et al. (eds.), *Biotechnology and AI: Technological Convergence and Information Hazards*, NATO Science for Peace and Security Series A: Chemistry and Biology, https://doi.org/10.1007/978-3-032-05246-9_7

sharing, standardization, benefit-sharing, monitoring). This comprehensive, proactive approach aims to ensure that AI-SB innovations are developed and deployed responsibly, aligning technological advancements with societal values and mitigating risks to foster an equitable, sustainable, and just future.

Keywords AI and synthetic biology convergence · Ethical implications · Governance frameworks · Healthcare applications · Agricultural applications

7.1 Introduction

The integration of Artificial Intelligence (AI) with Synthetic Biology (SB) is fundamentally transforming several fields, notably healthcare and agriculture, creating both remarkable possibilities and distinct ethical challenges. In healthcare, AI-driven diagnostics and personalized therapies show promise for improving patient outcomes, yet they also raise concerns about privacy, bias, and equitable access [1]. Additionally, reliance on algorithmic decision-making calls into question the role of human judgment in clinical practice [2]. In agriculture, AI and SB are revolutionizing farming, enhancing productivity and food security. However, the use of genetically modified organisms (GMOs) and data-centric farming brings ethical implications—from risks to small-scale farmers to ecological impacts [3]. Concerns also arise about data ownership, potential corporate monopolies, and the clash with traditional practices. Later sections explore how policy formulation and inclusive governance can address these conflicts [4]. Beyond technological innovation, this convergence also reframes longstanding bioethical principles—shifting justice from resource allocation toward addressing algorithm-driven inequities in care.

Justice in bioethics—once primarily about resource allocation—now must account for algorithmic bias and unequal access introduced by AI-SB convergence [5]. While AI-driven therapies can improve accuracy and efficiency, they risk exacerbating health inequalities if not properly regulated. Section 7.2 delves into data bias and equitable access as central concerns. Similarly, the principle of non-maleficence faces challenges as AI and SB introduce risks of unforeseen harm—topics explored in Sects. 7.2.4 and 7.3.3. Autonomy is also at stake, as AI and SB complicate informed consent and challenge participants' comprehension of novel technologies. See Sects. 7.2.1 and 7.5.1 for a more detailed discussion.

Given these emerging ethical tensions, adaptive governance is essential to ensure oversight keeps pace with innovation. Section 7.4.4 discusses how anticipatory, flexible policies can better align AI–SB innovation with ethical oversight [6].

This chapter explores not only the transformative potential of these innovations but also the profound ethical implications they bring. By combining the analytical power of AI with the design capabilities of SB, entirely new possibilities emerge for personalized medicine, enhanced diagnostics, sustainable farming, and beyond. However, these breakthroughs also introduce significant challenges, such as data privacy risks, algorithmic biases, inequitable access, and ecological concerns, all of which demand rigorous ethical scrutiny and adaptive governance.

Our discussion is framed around a dual focus: first, identifying the ethical challenges inherent in this technological convergence; and second, proposing governance strategies that can effectively mitigate these risks. Although our primary focus is on healthcare and agriculture, many of the insights and solutions presented here extend to broader applications within the realm of biotechnology. For example, over-reliance on AI recommendations risks diminishing clinician autonomy—and shifts liability for diagnostic errors toward developers and institutions.

7.2 Ethical Implications in Precision Health

In recent years, healthcare has undergone a profound shift due to the convergence of AI and SB, enabling innovations such as AI-driven genomics and engineered biological interventions. These approaches enhance precision diagnostics and open pathways to personalized therapies that were once purely speculative.

Machine learning–based diagnostic tools can detect disease markers earlier and more accurately, while engineered biological systems (e.g., customized gene therapies) target previously untreatable conditions. Despite these advances, the AI–SB convergence in healthcare raises intertwined ethical concerns, from privacy and bias to accountability and human oversight.

AI-based clinical tools depend on large genetic and medical datasets, creating risks of re-identification and unauthorized data use. Potential harms include discrimination, psychological distress, or infringements on patient autonomy. While measures such as dynamic consent and advanced encryption offer some protections, their implementation varies widely across jurisdictions. Additionally, algorithmic bias remains a pressing challenge; models that lack diverse training data can yield skewed diagnoses and further entrench health disparities. For deeper analysis of these issues, see Sects. 7.2.1 and 7.2.2. Mitigating bias calls for more inclusive datasets, robust oversight, and transparent audits. Nevertheless, broader social determinants can still amplify inequities, underscoring the need for holistic solutions that go beyond technical fixes (see also Sect. 7.2.2 on equity and fair access).

Incorporating AI into clinical workflows also raises concerns about accountability and patient autonomy. Overreliance on automated outputs may undermine clinical judgment if not carefully managed, emphasizing the importance of explainable AI and participatory governance (Fig. 7.1).

While the convergence of AI and SB in healthcare presents substantial ethical challenges—such as data privacy risks, algorithmic bias, and questions of accountability—several real-world initiatives offer promising examples of how these issues are being addressed in practice. One notable approach is the implementation of dynamic consent models in genomic research. In these projects, digital platforms enable patients to monitor and modify their consent preferences over time as new research applications arise. This not only reinforces individual autonomy but

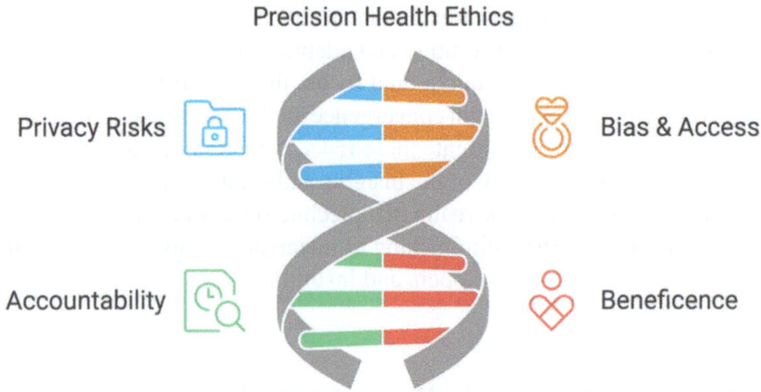

Fig. 7.1 Precision health ethics framework. This figure illustrates the four fundamental ethical dimensions of precision health through a DNA helix metaphor, emphasizing their intrinsic connection to genetic and personalized medicine. The central double helix structure, with its color-coded segments, represents how these ethical considerations are woven into the very fabric of precision health. Privacy risks (blue, represented by a locked folder icon) addresses data protection and consent management; bias and access (orange, shown with a target heart icon) focuses on healthcare equity and resource distribution; accountability (green, depicted with a document magnifying glass icon) emphasizes transparency and responsibility; and beneficence (red, illustrated with a caring heart icon) represents patient benefits and risk management

also builds trust between researchers and participants by providing continuous transparency regarding data use [7].

In parallel, initiatives such as the Open Algorithms (OPAL) project are pioneering frameworks to improve algorithmic transparency. By enabling third-party audits and employing explainable AI methods—such as Local Interpretable Model-Agnostic Explanations (LIME) and SHapley Additive exPlanations (SHAP)—these projects ensure that clinical decision-making is both interpretable and accountable [8, 9]. Such efforts are critical for supporting clinicians in understanding AI-generated recommendations and for maintaining robust oversight of complex algorithms. Furthermore, several hospitals and research centers have begun integrating explainable AI systems into their clinical workflows. Early pilot programs demonstrate that when healthcare providers have access to interpretability tools, they are more likely to trust and effectively use AI as an adjunct to their clinical expertise. These programs illustrate a successful melding of cutting-edge technology with practical clinical needs. See Sects. 7.5.1 and 7.5.2 for more on these governance and technical solutions.

Regulations such as GDPR guide data protection for genetic and clinical information [10], complemented by measures like differential privacy [11] and homomorphic encryption [12]. These legal and technical frameworks pave the way for responsible data handling, while Sects. 7.2.1 and 7.5.1 elaborate on specific policy solutions.

7.2.1 Privacy Risks

This section focuses exclusively on the ethical implications of privacy risks in AI-SB convergence; governance and technical solutions are discussed in Sect. 7.5. Patient privacy in AI-enhanced healthcare extends beyond individual data to potentially affect entire families [13]. Healthcare systems often handle genetic and clinical records that, if exposed, could lead to discrimination or long-term psychological harm. While technical safeguards (e.g., encryption, secure data management) and legal measures (e.g., GDPR) offer baseline protection, rapid AI developments demand more nuanced solutions—such as dynamic consent—to maintain robust privacy over time. For instance, breaches of genomic databases have led to inadvertent disclosure of familial disease predispositions, raising ethical concerns about intergenerational harm. In sum, the risk of unauthorized re-identification and familial disclosure underscores how AI-driven precision health can breach core privacy rights—extending harm beyond individual patients to entire families. Section 7.5.1 details governance strategies such as dynamic consent models, enhanced encryption standards, and transparent data-sharing agreements designed to uphold data sovereignty and minimize re-identification risks.

7.2.2 Bias and Access

This section focuses exclusively on the ethical implications of bias and access in AI-SB convergence; governance and technical solutions are discussed in Sect. 7.5. Morley et al. emphasize that AI algorithms trained on biased datasets can perpetuate and amplify disparities in healthcare delivery [14]. As Wahl et al. demonstrate, AI systems and clinical trials predominantly sourced from high-income countries and specific demographic groups may perform poorly or even cause harm when applied to underrepresented communities or different genetic, environmental, and sociocultural contexts [15]. It is crucial to prioritize fair and equal access to AI-powered healthcare solutions across different populations, ensuring both the availability and effectiveness of these technologies for all demographic groups. Developing inclusive databases that accurately reflect the full spectrum of human diversity is essential. Additionally, robust oversight mechanisms are needed to detect and mitigate biases in AI healthcare systems to preserve fairness. At a global level, promoting international collaboration is vital to ensure equitable distribution of these technologies' benefits among nations and socioeconomic classes, preventing the exacerbation of global health disparities [16].

Baylis and Robert caution against the potential for improved genetic therapies to generate a novel manifestation of genetic disparity, wherein only the affluent possess the means to benefit from transformative medical interventions [17]. Despite comprehensive efforts, significant gaps remain in addressing issues such as intellectual property, particularly in disadvantaged settings. Patents for AI algorithms and synthetic biological components may hinder access to essential innovations, limiting participation for researchers and institutions in developing nations and consolidating control in technologically advanced countries [18]. This raises concerns about technological colonialism, where advanced nations exert control over global markets and resource distribution.

Access and equity concerns remain central to the ethical implementation of AI and SB in healthcare. As Rajkomar et al. emphasize, careful adaptation of these technologies to diverse contexts, combined with equitable implementation strategies, is essential to ensure that AI and SB innovations reduce rather than exacerbate global health disparities [19]. Challenges such as resource limitations, data biases, and lack of transparency must be addressed to prevent a bifurcated health system that favors affluent populations. Ultimately, algorithmic bias in AI-enabled diagnostics threatens to perpetuate and deepen existing health disparities, denying underrepresented groups equitable access to life-saving innovations. Section 7.5.2 outlines concrete solutions—ranging from inclusive dataset mandates and bias auditing protocols to tiered implementation subsidies—that aim to ensure fair deployment of precision health technologies.

7.2.3 Accountability

This section focuses exclusively on the ethical implications of accountability in AI-SB convergence; governance and technical solutions are discussed in Sect. 7.5. Achieving accountability and transparency in AI-driven synthetic biology is inherently complex, given the multitude of stakeholders—developers, researchers, regulators, and end-users—whose roles and responsibilities often overlap [20]. This can create "diluted accountability," where no single entity is clearly held responsible for errors or unethical outcomes. The fundamental complexity of AI systems, especially those utilizing deep learning methods, poses a substantial obstacle to achieving transparency. Karim et al. emphasize that while AI systems have attained exceptional performance in several fields, their decision-making processes often remain opaque even to developers [21]. This "black box" nature makes it particularly challenging for non-technical stakeholders to scrutinize and understand how these systems arrive at their decisions, especially in critical applications like healthcare or financial services. Addressing these issues requires algorithmic impact assessments, third-party audits, and the adoption of explainable AI techniques.

In the field of SB, transparency issues have a distinct dimension. Pauwels cautions that the complex genomic alterations associated with SB can present challenges in tracking and monitoring [22]. Given the intricate nature of these biological systems, even minor alterations can have extensive and often unforeseen repercussions. Achieving transparency in this context necessitates not only the unrestricted availability of data but also the capacity to understand and analyze intricate biological information. A lack of openness in SB procedures can give rise to substantial ethical and safety issues, especially when considering potential environmental consequences or biosecurity hazards.

Karim et al. highlight that although AI systems have attained exceptional performance in several fields, this often comes at the expense of interpretability, making it difficult for users to fully comprehend the underlying decision-making processes [23]. Cutting-edge machine learning models, particularly deep neural networks, are intrinsically complex, characterized by millions of parameters and intricate architectures that resist straightforward explanation. This complexity presents a substantial obstacle for end-users and regulators who must comprehend the reasoning behind AI decisions, especially in critical sectors such as healthcare or financial services. In SB, the challenge of achieving transparency is further complicated by the vast and intricate nature of biological data. Alterovitz et al. emphasize the challenges associated with controlling and interpreting the large volumes of data produced by SB studies [24]. Ensuring the accessibility and comprehensibility of this data for all stakeholders necessitates sophisticated data management infrastructure and powerful visualization technologies. The multidisciplinary character of SB adds to this problem, as it requires expertise from several disciplines to fully evaluate and analyze the data.

Moreover, the international scope of AI and SB research and applications conflicts with the context-specific nature of regulatory frameworks, posing considerable difficulties in maintaining uniform accountability and transparency norms. Pang et al. highlight disparities in legislation among various countries and regions, which can result in a fragmented approach to governance [25]. Although a patchwork of legislative frameworks is an inevitable feature in nearly every field, this fragmentation can obstruct international collaboration and create potential vulnerabilities that might be exploited to bypass ethical norms or transparency obligations. Furthermore, the rapid rate of innovation in these domains frequently surpasses the pace at which legal structures are updated. It is therefore wise to strive for as much uniformity as possible in accountability and transparency norms, even if a globally uniform legislative framework remains elusive. Kiseleva et al. observe that this delay—often referred to as 'the pacing problem'—contributes to a regulatory void where emerging ethical concerns may not be sufficiently addressed [20]. The inherent dynamism of AI and SB necessitates that legal frameworks be both broad and flexible. However, the demand for adaptability must be balanced with the need for stability and predictability in regulatory contexts.

Pang et al. suggest collaborative initiatives to standardize global legislation for AI and SB as a means to address the challenges posed by inconsistent regulations [25]. This strategy involves working with international organizations to establish standardized frameworks that guarantee uniform implementation of transparency and accountability measures across diverse legal systems. However, given the significant ethical and cultural differences among regions, truly uniform global legislation may be an aspirational goal. Instead, the objective should be to identify a core set of common principles—such as accountability, transparency, robust data protection, and informed consent—that can serve as a consistent legal foundation while still allowing for necessary local adaptations. Such harmonization could effectively mitigate regulatory arbitrage and foster a fairer environment for ethical innovation.

Kiseleva et al. advocate for the establishment of adaptive legal frameworks capable of promptly responding to new developments in AI and SB [20]. These frameworks should possess sufficient adaptability to integrate emerging ethical issues and technological advancements while ensuring a consistent legal foundation. By 'consistent legal foundation,' we mean a core set of stable principles—such as data protection, accountability, and transparency—that remain unchanged over time, even as modular guidelines are updated to address new challenges like AI interpretability or synthetic biology safety. This approach ensures that while the regulatory framework can evolve in response to technological advances, its foundational ethical commitments remain constant. This may entail implementing systems for periodic evaluation and revision of regulations, along with establishing expert advisory bodies to provide continuous guidance on technological progress.

Pauwels underscores the need for comprehensive stakeholder engagement in AI and SB governance [22]. This involves not only public consultations and participatory governance models but also clear dissemination of information regarding potential hazards, advantages, and uncertainties. As Bubela et al. advocate, providing unambiguous and easily understandable information to non-expert audiences promotes universal comprehension and acceptance of these technologies, while fostering meaningful dialogue about their ethical implications and societal impacts [26].

Moreover, the implementation of public databases that provide controlled access to information about biotechnological systems has the potential to greatly improve openness and accountability. Wittner suggests a method in which different degrees of access are designated to various stakeholders, enabling thorough examination of the impacts of biotechnological advancements while safeguarding confidential data [27]. This approach empowers academics, policymakers, and the public to engage in more informed decision-making on biotechnology applications.

Participatory accountability systems involve the active involvement of stakeholders, including the public, in the implementation and oversight of biotechnology innovations. Keller advocates for the necessity of public involvement in transparency procedures, proposing that watchdog groups and individual citizens should be granted enhanced access rights to information regarding algorithmic and biotechnological systems [28]. This approach goes beyond mere transparency by actively engaging the public in decision-making processes and oversight. For example, one method is to assemble citizen panels to evaluate proposed biotechnological applications, thereby ensuring that diverse viewpoints are considered. However, while stakeholder participation can enrich decision-making, it is not a panacea. In practice, these participatory processes can sometimes lead to deadlock when consensus is hard to achieve, potentially obstructing timely decisions. While stakeholder participation is valuable for ensuring diverse viewpoints in decision-making, it can also risk obstructing the process if those involved are unable to reach consensus. In practice, such consultative approaches have sometimes led to deadlock, reminding us that these solutions are not a panacea. To counterbalance this potential drawback, incorporating watchdog organizations into regulatory procedures can provide an extra level of rigor and expertise, thereby augmenting overall accountability.

The application of the Accountability for Reasonableness (A4R) concept by Maluka et al. in healthcare priority-setting provides valuable insights for improving accountability in biotechnological developments [29]. This approach emphasizes four essential criteria: relevance, publicity, appeals/revision, and enforcement. In biotechnology, applying the A4R framework could ensure that decision-making processes are transparent, rely on pertinent criteria, allow for modification, and are enforceable. For instance, when evaluating the authorization of a new gene therapy, the A4R framework would necessitate explicit communication of the criteria used for decision-making, public disclosure of the reasoning behind decisions, procedures for stakeholders to contest decisions, and strict enforcement of established standards. By employing A4R principles, biotechnological advancements can achieve enhanced legitimacy, equity, and stakeholder involvement.

Achieving accountability and transparency in AI and SB is a complex but essential endeavor. The multiplicity of stakeholders, inherent opacity of technologies, and technological and legislative obstacles present significant challenges. However, by developing explainable AI models, standardizing global legislation, fostering public participation, and utilizing advanced methods such as reverse engineering, QII metrics [30], collaborative projects like OPAL [31], and frameworks like A4R, these challenges can be addressed. Implementing these strategies requires continuous vigilance and flexibility, given the rapidly evolving nature of AI and SB. By enhancing transparency and accountability, we can ensure that these powerful technologies are developed and deployed ethically, responsibly, and in a manner that earns public trust. Ultimately, fostering a transparent and accountable environment will contribute to the responsible advancement of AI and SB, benefiting society as a whole. The diffuse responsibilities across developers, clinicians, and institutions create a "diluted accountability" landscape in AI-SB applications, complicating recourse for error and eroding trust in clinical decision-making. Section 7.5.2 presents governance mechanisms—including mandatory algorithmic impact assessments, third-party auditing frameworks, and clear liability pathways—that restore accountability and transparency in AI-driven care.

7.2.4 Beneficence

This section focuses exclusively on the ethical implications of beneficence in AI-SB convergence; governance and technical solutions are discussed in Sect. 7.5. Mitigating unintended consequences in SB and AI requires a comprehensive approach to risk management. Chu and Agapito-Tenfen highlight how off-target effects in genome editing can introduce deleterious mutations with serious health implications [32]. To address these risks, Trump et al. advocate for international cooperation in developing standardized regulatory standards, given the transboundary nature of potential risks associated with synthetic organisms and AI systems [33]. By employing these strategies, society can harness the potential benefits of these technologies while minimizing risks to the public health. A proactive and collaborative effort is essential to navigate the ethical and ecological challenges posed by advanced genetic engineering tools, ensuring responsible innovation that aligns with societal values.

Effective governance requires both robust oversight and public engagement. Gottwald proposes independent ethical review boards to evaluate SB projects comprehensively [34], while Bensaude Vincent emphasizes the importance of educational initiatives that extend beyond technical explanations to include discussions of ethical implications, potential risks, and governance challenges [35]. While AI-SB convergence holds promise for unprecedented patient benefits, it also introduces risks of unintended harms—from off-target gene edits to unpredictable treatment outcomes—that challenge the principle of "do no harm." Section 7.5.4 proposes targeted safeguards such as Safety-by-Design protocols, dual-use risk assessments, and iterative pilot testing to proactively balance innovation with patient safety.

7.3 Ethical Implications in Sustainable Agriculture

The integration of AI and synthetic biology in sustainable agriculture brings about nuanced ethical challenges that reach well beyond those in conventional farming. As the industry increasingly incorporates advanced technologies, four core ethical areas come to the forefront: privacy and data ownership, socioeconomic fairness, environmental safety, and the preservation of farmer autonomy. Together, these considerations lay the ethical groundwork essential for agricultural innovation. A thoughtful approach to these dimensions is vital to ensuring that new technologies in agriculture not only support sustainability but also safeguard farmer rights and uphold ecological integrity (Fig. 7.2).

The integration of AI into agriculture heralds promising advancements in productivity, efficiency, and sustainability, while simultaneously raising significant concerns about equity and fair access, particularly for small-scale farmers in developing regions. Wolfert et al. highlight how large agribusinesses, equipped with substantial resources, can readily invest in advanced technologies like precision farming equipment, drones, and sophisticated data analytics platforms [36]. This technological advantage, as Rotz et al. observe, creates a widening digital divide that threatens to deepen existing inequalities within the agricultural sector [37]. Carolan emphasizes how the concentration of agricultural power among a select few corporations can lead to the marginalization of smallholders, as AI-enhanced large farms gain competitive advantages through increased efficiency and profitability [3]. This dynamic particularly affects developing regions, where agriculture serves as the primary source of income and employment for rural communities. Wang further notes how the deployment of AI technologies can disrupt traditional agricultural practices deeply embedded in cultural heritage and local knowledge systems, potentially undermining both cultural identity and crucial agrobiodiversity [38].

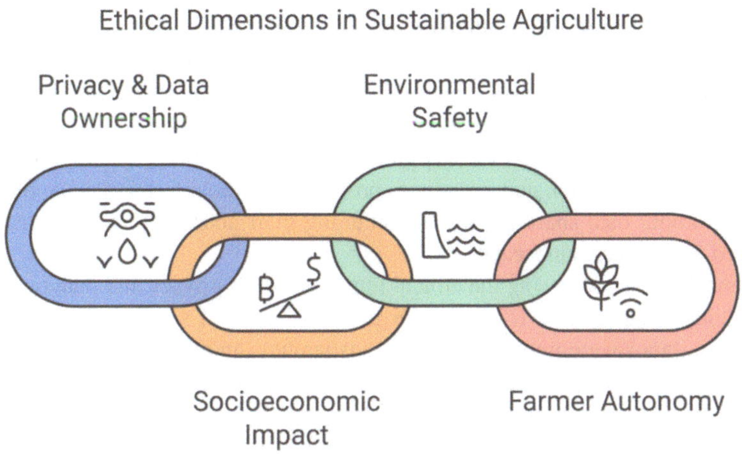

Fig. 7.2 Ethical dimensions in sustainable agriculture. This figure illustrates the interconnected ethical dimensions in AI-driven sustainable agriculture through a chain metaphor, emphasizing their interdependence and collective importance. Each link represents a critical ethical consideration: privacy and data ownership (blue link, with a drone and data icon) addresses agricultural data protection and farmer control over information; socioeconomic impact (orange link, with a balance scale icon) represents the economic fairness and resource distribution aspects; environmental safety (green link, with an environmental monitoring icon) focuses on ecological protection and biodiversity preservation; and farmer autonomy (red link, with a smart farming icon) emphasizes independent decision-making and rights preservation

7.3.1 Privacy and Data Ownership

This subsection examines the ethical implications of privacy and data ownership in AI-driven agriculture; governance and mitigation strategies appear in Sect. 7.5. Digital technologies, particularly AI-driven analytics, have transformed agriculture into a data-centric enterprise. Bronson and Knezevic highlight how small-scale farmers, often lacking technical expertise and legal resources, are particularly vulnerable in this landscape [39]. The risks extend beyond basic privacy concerns—sensitive information about farming practices, crop yields, and land usage can be exploited by competitors or companies to manipulate input prices based on detailed knowledge of farmers' operations [40]. For instance, competitors might use this data to gain a market advantage, or companies could manipulate input prices based on detailed knowledge of a farmer's operations. Such scenarios can lead to economic disadvantages for small-scale farmers, threatening their livelihoods. Moreover, the absence of robust legal frameworks governing agricultural data exacerbates these vulnerabilities. Wiseman and Sanderson note how legislation has not kept pace with technological advancements, creating a regulatory void where farmers' data ownership, consent, and usage rights remain inadequately protected [41]. This gap allows corporations to establish terms that favor their interests, often at the expense of the farmers who generate the data.

To address these ethical challenges, a multi-faceted approach is essential. Carolan advocates for data sovereignty through collective ownership models and cooperative platforms, empowering farmers with greater agency over their data [42]. Kaur et al. emphasize the importance of transparent data use practices, including clear agreements that outline data usage, access rights, and farmer benefits [43]. Supporting these measures with education and awareness programs can help farmers make informed decisions about data sharing while maintaining their autonomy in an increasingly digitized agricultural landscape.

In conclusion, addressing the ethical implications of privacy and data ownership in sustainable agriculture is critical for the well-being of small-scale farmers. By promoting data sovereignty, enhancing transparency, and implementing supportive policies, it is possible to harness the benefits of technological advancements while safeguarding the rights and interests of farmers. The asymmetry of data control in AI-powered farming exposes smallholders to risks of exploitation and loss of autonomy, making data sovereignty an ethical imperative. Section 7.5.1 details governance measures—such as farmer-led data cooperatives, transparent licensing agreements, and robust consent protocols—to protect agricultural data rights.

7.3.2 Socioeconomic Impact

This subsection examines the ethical implications of socioeconomic impact in AI-driven agriculture; governance and mitigation strategies appear in Sect. 7.5. Modern agriculture's socioeconomic landscape is characterized by concentrated corporate power that fundamentally alters traditional agricultural frameworks. Howard documents how multinational corporations have established unprecedented control through ownership of genetically modified seeds, proprietary technologies, and agricultural data [44]. This consolidation, as Kloppenburg notes, creates substantial barriers for farmers, particularly affecting traditional practices like seed saving and reuse [45]. Shiva documents how the shift towards industrialized agriculture has precipitated a concerning erosion of indigenous knowledge systems and biodiversity [46]. Traditional farming methods, which historically balanced sustainability with ecosystem health, are increasingly supplanted by yield-maximizing practices that often overlook environmental considerations. This transformation, as Patel observes, particularly impacts small-scale farmers who face mounting financial pressures from compulsory purchases of patented seeds and inputs [47].

The resulting socioeconomic disparities, as analyzed by McMichael, manifest in the increasing concentration of wealth and resources among corporate entities and large-scale producers, while smallholders face diminishing opportunities [48]. This imbalance often catalyzes rural depopulation as farmers abandon their agricultural heritage in search of alternative livelihoods, fundamentally altering rural community structures. Altieri emphasizes the ethical imperative of ensuring equitable resource access and protecting cultural diversity alongside environmental sustainability [49]. González de Molina et al. propose alternative approaches through agroecology and food sovereignty movements, offering potential pathways to redistribute agricultural power and prioritize local needs [50]. Effective solutions encompass comprehensive policy interventions to curtail monopolistic practices, deliberate support for traditional farming methods, and community empowerment initiatives such as cooperatives and participatory decision-making processes. This complex challenge requires carefully balancing technological advancement with the preservation of traditional agricultural practices, emphasizing solutions that protect vulnerable farming communities while promoting sustainable agricultural development.

This complex challenge requires a delicate balance between technological advancement and the preservation of traditional agricultural practices, emphasizing the need for solutions that protect vulnerable farming communities while promoting sustainable agricultural development. The path forward necessitates careful consideration of how to maintain agricultural innovation while ensuring equitable access and preserving valuable traditional farming knowledge and practices. Corporate consolidation of AI-SB tools threatens to widen socioeconomic divides by limiting smallholder access to innovation and eroding traditional practices. Section 7.5.7 proposes policy levers—like tiered technology subsidies, cooperative business models, and antitrust safeguards—to foster equitable participation in the agricultural bioeconomy.

7.3.3 Environmental Safety

This subsection examines the ethical implications of environmental safety in AI-driven agriculture; governance and mitigation strategies appear in Sect. 7.5. The deployment of genetically modified organisms (GMOs) into natural ecosystems presents complex ecological challenges that require nuanced analysis. Tsatsakis et al. demonstrate how Bt crops can affect non-target insects, including beneficial pollinators and natural predators, with impacts propagating through food chains in ways that are challenging to predict or reverse [51]. Ellstrand further highlights how gene flow from GMOs to wild relatives can lead to the emergence of superweeds or invasive species resistant to conventional control methods, thereby threatening native plant populations [52]. Additionally, AI-optimized crops, enhanced for traits such as increased yield or drought resistance, may outcompete traditional varieties, potentially reducing genetic diversity that is crucial for resilience against pests and diseases [36]. Concerns have also been raised about the development of resistance in target pest populations; in some regions, the extensive cultivation of Bt crops has led to the evolution of resistant insect populations. This resistance not only undermines the effectiveness of GMOs but can also lead to increased reliance on chemical pesticides, potentially escalating environmental and health risks [51].

Significant scientific debate exists on these issues. While some studies report adverse non-target effects [51], others have found minimal or inconsistent impacts [53, 54]. This divergence in findings underscores that ecological consequences of GMO deployment are influenced by factors such as environmental context, agricultural practices, and the specific traits of the modified organisms.

To mitigate risks associated with the release of GMOs, scientists have developed advanced biocontainment techniques. For example, in industrial applications of synthetic biology, one promising approach is the use of synthetic auxotrophies—engineering GMOs to depend on artificial nutrients or compounds that do not occur naturally. Mandell et al. demonstrated the effectiveness of this method by creating E. coli strains that rely on synthetic amino acids for survival, achieving escape frequencies below measurable thresholds [55]. However, in the context of agricultural biotechnology, such biocontainment strategies are not typically employed as primary safety measures. Instead, most countries require rigorous environmental risk analyses prior to market approval, ensuring that any GM crop with scientifically demonstrated risks is not commercialized. Effective management of the risks associated with SB and GMOs also necessitates the establishment of comprehensive regulatory frameworks. This includes thorough risk assessment procedures, consideration of potential ecological impacts, and a commitment to minimizing unintended health effects. Given the global scope of biotechnology research and trade, international cooperation in establishing and standardizing regulatory parameters is crucial.

Effective management of the risks associated with SB and GMOs also necessitates the establishment of comprehensive regulatory frameworks. This includes thorough risk assessment procedures, consideration of potential ecological impacts, and a commitment to minimizing unintended health effects. Given the global scope of biotechnology research and trade, international cooperation in establishing and standardizing regulatory parameters is crucial [22]. For an overview of existing regulatory frameworks and their key features, see Table 7.1. While these frameworks provide important safeguards, gaps remain that call for further harmonization and adaptive governance.

Post-market monitoring of GM crops is both vital for detecting unforeseen repercussions and a mandatory component of regulatory oversight, particularly in regions like the European Union. Under EU regulations, extensive long-term ecological monitoring programs track transgene survival, spread, and effects on non-target species. Moe-Behrens et al. advocate for adaptive monitoring systems involving repeated cycles of assessment, implementation, and evaluation [56]. This approach is complemented by the 'Safety-by-Design' (SbD) concept, which Asín-García et al. emphasize for incorporating genetic safeguards into synthetic organisms' design protocols [57]. In contrast to industrial applications—where physical containment measures, such as bioreactors equipped with redundant safety mechanisms, provide an additional layer of protection—in agricultural biotechnology, risk management is primarily achieved through rigorous pre-market environmental risk assessments, mandatory post-market monitoring, and regulatory reviews rather than through physical containment strategies.

Furthermore, the engagement of stakeholders, including scientists, policymakers, and the public, is crucial in addressing environmental and ecological safety concerns. Transparent communication about the benefits and risks associated with GMOs and AI-optimized crops can enhance public trust and facilitate informed decision-making

Table 7.1 Overview of existing regulatory frameworks for biotechnology risk management

Region/entity	Key regulations/guidelines	Comment
European Union	– GMO Directive (2001/18/EC) –Regulation (EC) No 1829/2003 (genetically modified food/feed) –Regulation (EC) No 1830/2003 (traceability and labeling)	Provides comprehensive regulation for GMO release, traceability, and labeling
International agreements	– Cartagena protocol on biosafety (2000)	Focuses on safe handling, transport, and transboundary movements of living modified organisms
OECD guidelines	OECD recommendations on risk assessment and harmonization	Offers guidance to member countries to align risk assessment procedures and regulatory practices
United States	Oversight by USDA, FDA, and EPA	Multiple agencies cover environmental, food, and health safety aspects; regulatory responsibilities are shared
Other regions	Various national/regional frameworks	Regulations vary widely in stringency and scope, often reflecting local priorities and capacities

[58]. The release of AI-optimized organisms into complex ecosystems poses unpredictable ecological harms that challenge non-maleficence and biodiversity preservation. Section 7.5.5 introduces adaptive risk assessments, post-market ecological monitoring, and Safety-by-Design protocols to anticipate and mitigate environmental risks.

7.3.4 Farmer Autonomy

This subsection examines the ethical implications of farmer autonomy in AI-driven agriculture; governance and mitigation strategies appear in Sect. 7.5. The integration of AI-driven agricultural technologies represents a profound transformation of farming landscapes. As Rotz et al. document, this shift fundamentally changes how farming knowledge is acquired, valued, and transmitted across generations [37]. While these advances promise heightened efficiency and productivity, they simultaneously pose significant risks to farmers' self-determination and traditional agricultural wisdom, affecting operations of all sizes. Chen highlights how large agritech corporations' proprietary technologies, protected by stringent intellectual property rights, often bind farmers to restrictive licensing agreements, limiting their ability to repair, modify, or fully comprehend the very tools that have become essential to their daily operations [59]. Bronson and Knezevic introduce the compelling concept of "digital feudalism" to characterize this emerging dynamic, drawing parallels between modern farmers and medieval tenants who worked land without owning their essential tools [39]. This power imbalance is further intensified by corporate data monopolies, which, as Wiseman et al. observe, accumulate and control vast quantities of agricultural data generated by farmers without providing adequate transparency or compensation, effectively diminishing farmers' marketplace bargaining power [41].

Eastwood et al. note how this technological shift threatens not only cultural heritage but potentially compromises agricultural systems' resilience and sustainability by promoting standardized practices that may overlook regional ecological diversity [60]. Van der Burg et al. emphasize the critical issue of data privacy, highlighting farmers' limited control over their data's usage, sharing, and commercialization [61]. This lack of data sovereignty raises significant ethical concerns about consent and exploitation, particularly regarding the potential manipulation of market prices or targeted product marketing that could disadvantage small-scale operators.

To address these challenges while preserving farmer autonomy, several complementary approaches have emerged. Kloppenburg advocates for open-source technologies enabling farmers to modify and adapt tools to their specific needs [45]. This aligns with Chen support for right-to-repair legislation and Wiseman et al.'s emphasis on transparent data practices [41, 59]. Eastwood et al. highlight the value of collaborative and participatory approaches in technology design [60], while Rotz et al. underscore the importance of enhanced technological literacy among farmers to bridge the knowledge gap between technology providers and users [37].

This complex intersection of AI technology and agriculture necessitates careful consideration of how to harness technological benefits while preserving farmer autonomy, traditional knowledge, and community well-being. The path forward requires balancing innovation with preservation of agricultural heritage, ensuring that technological advancement serves rather than subjugates the farming community's interests. Dependency on proprietary AI tools risks undermining farmer decision-making and traditional knowledge, raising concerns about digital feudalism. Section 7.5.7 outlines governance strategies—including open-source technology mandates, right-to-repair policies, and participatory oversight—that safeguard farmer autonomy.

7.4 Discussion

7.4.1 Balancing Innovation and Ethical Responsibility

The rapid convergence of AI and synthetic biology (SB) is transforming healthcare and agriculture, offering unprecedented opportunities for precision diagnostics, personalized therapies, and sustainable food production. Yet this acceleration also outpaces the ethical and regulatory frameworks designed to govern these innovations, creating a governance paradox whereby technological capabilities evolve faster than our capacity to manage associated risks [6, 62]. As a result, without proactive oversight, AI-SB convergence risks exacerbating health disparities, undermining patient autonomy, and causing unintended ecological harms [1, 42].

At the heart of this paradox lies a tension between core bioethical principles and emerging technological realities. Autonomy is challenged when clinicians and farmers become over-reliant on opaque AI recommendations, shifting liability away from human decision-makers [2]. Justice is strained as algorithmic bias and concentrated data ownership threaten equitable access to AI-enabled interventions across diverse populations [14, 19]. Non-maleficence and beneficence are complicated by unpredictable off-target effects in engineered organisms and algorithmic errors that may inflict harm at scale [32, 51].

Emerging governance approaches address these complex challenges through three interlocking strategies. Interdisciplinary collaboration brings together technologists, ethicists, policymakers, and affected communities to ensure diverse perspectives shape ethical guidelines [62, 63]. Adaptive regulatory frameworks—or anticipatory governance—enable modular policy updates that respond quickly to novel AI-SB developments while embedding core TAPIC principles (Transparency, Accountability, Participation, Integrity, Capacity) [6, 20]. Finally, technical safeguards such as explainable AI, dynamic consent, and Safety-by-Design integrate ethical considerations directly into system architectures [9, 55, 64]. Together, these strategies form a coherent governance ecosystem capable of aligning innovation with societal values and mitigating the risks posed by AI-SB convergence.

7.4.2 Adaptive Regulatory Frameworks

This section turns to the governance mechanisms that can address the ethical dilemmas discussed above before they escalate. Rather than detailing the problems again, the following paragraphs outline anticipatory governance approaches—spanning flexible regulations, stakeholder participation, and technological safeguards—that proactively respond to the rapid evolution of AI–SB convergence [6, 62]. By focusing on forward-looking policies, we can align innovation with societal values and ensure that core principles such as autonomy, justice, and accountability remain upheld. One of the most pressing needs, given the rapid pace of AI–SB development, is a regulatory structure capable of regular updates. Modular policy design offers a practical way to integrate incremental changes without overhauling entire legal frameworks each time a novel tool or genetic breakthrough surfaces. For instance, adding new guidelines to address emerging forms of data misuse can be done swiftly if regulatory components are designed as "plug-and-play" modules [65]. This flexibility helps bridge the gap between inherited regulations—such as those crafted for traditional GMOs—and the dynamic needs of AI-augmented synthetic biology.

Crucially, dynamic consent and collective data governance can fill the privacy and ownership gaps described in Sect. 7.2.1 and 7.3.1. In healthcare contexts, dynamic consent portals enable patients to update preferences as novel diagnostic techniques arise—preserving autonomy without stalling innovation [64]. Similarly, agricultural data cooperatives empower small-scale farmers, giving them collective bargaining power to negotiate fair terms for data sharing [42]. Both strategies reinforce data sovereignty, ensuring that stakeholders remain active participants rather than passive data sources. Besides policy instruments, advanced technologies also offer potent solutions to privacy and equity concerns. Blockchain-based ledgers can audit data flows in real time, minimizing information asymmetry and preventing the unapproved reuse of sensitive records [66]. Likewise, explainable AI frameworks [8, 9] can illuminate algorithmic decision-making, mitigating hidden biases that disproportionately affect underserved groups. By embedding these technical safeguards into anticipatory governance, policymakers ensure that AI–SB tools remain both innovative and responsible in practice.

A layered approach to risk management can further strengthen anticipatory governance (see Fig. 7.3). This figure illustrates four concentric circles, each representing a stage in the governance process—Risk Identification, Risk Assessment, Risk Analysis, and Mitigation Strategies—with outer to inner layers showing how initial detection funnels into structured assessments and culminates in actionable remedies. By visually mapping potential harms at each stage, stakeholders can direct resources where they matter most, identifying vulnerabilities early and applying context-specific policies or technical solutions. This structured process helps avert the "pacing problem," where the rapid rise of AI–SB capabilities outstrips the slower evolution of ethical and legal norms [62]. Even the most robust frameworks and cutting-edge tools risk failing if stakeholders are not actively engaged. Diverse participation is crucial: ethicists and regulators bring legal safeguards, industry players contribute real-world feasibility, while affected communities—patients, farmers, local citizens—validate whether these frameworks reflect lived realities. Methods like public consultations, citizen juries, and multi-stakeholder review boards offer ways to continuously integrate a wide range of experiences and values [20]. This inclusive oversight not only improves policy outcomes but also boosts public trust in AI–SB innovations.

Ultimately, anticipatory governance in AI–SB must remain agile to address emerging frontiers such as AI-driven gene editing or self-evolving synthetic organisms. Legislators and regulatory agencies can solidify this agility by periodically reviewing policies, updating guidelines, and employing scenario planning to forecast potential ethical concerns. When backed by transparent data practices, broad-based stakeholder input, and layered risk management methods (Fig. 7.3), these efforts form a coherent strategy for ensuring that AI–SB advances deliver societal benefits without compromising fundamental rights or ecological integrity.

In addition to early risk identification and scenario planning, ethical forecasting plays a crucial role in anticipatory governance by enabling proactive anticipation of future ethical challenges. To further enhance anticipatory governance, serves as a proactive strategy to predict forthcoming ethical dilemmas emerging from advancements in AI and SB. This process encompasses several essential stages:

- Trend Analysis: By analyzing current technical paths, research trends, funding patterns, and emerging technologies, stakeholders can forecast hypothetical future

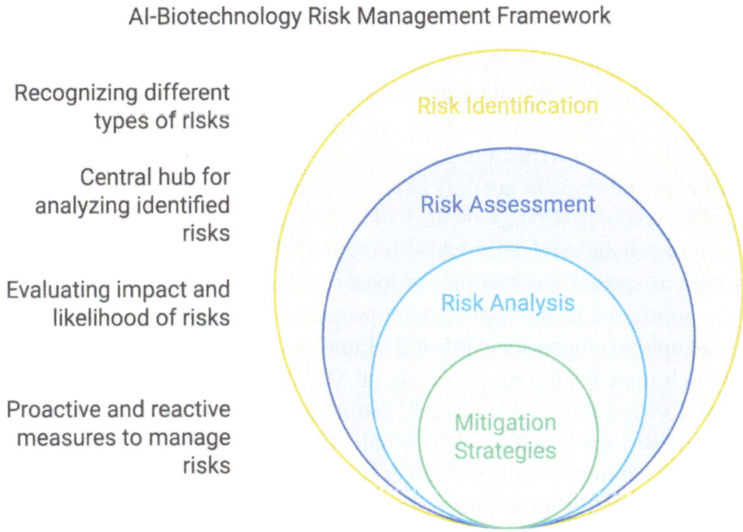

Fig. 7.3 AI-biotechnology risk management framework. This diagram illustrates the layered approach to risk management in AI-biotechnology, progressing from outer to inner layers. The outermost yellow circle represents risk identification, where different types of risks are initially recognized. Moving inward, the blue layer shows risk assessment, serving as a central hub for analyzing identified risks. The light blue circle indicates risk analysis, where the impact and likelihood of risks are evaluated. At the core, the green circle represents mitigation strategies, where proactive and reactive measures are developed to manage risks

advancements and pinpoint potential ethical issues [67, 68]. This involves examining areas of immediate expansion and identifying areas where ethical concerns may arise.

- Impact Assessment: This stage evaluates the possible ethical consequences of projected technologies on individuals, society, and the environment [69]. It considers both direct effects and second- and third-order impacts that may not be immediately apparent.
- Multidisciplinary Collaboration: Engaging specialists from various disciplines ensures that ethical forecasting comprehensively identifies gaps and unforeseen ethical dilemmas that may result from the convergence of multiple technologies or fields of study [62].
- Continuous Monitoring: Implementing continuous monitoring technologies, such as AI-driven systems that analyze scientific literature, news sources, and social media, enables real-time tracking of evolving ethical concerns. This facilitates the timely identification and response to emerging issues.

Anticipatory governance also benefits from collaborative arrangements that bridge government, industry, and research institutions. Well-structured public–private partnerships (PPPs) can align commercial interests with socially responsible objectives by tying regulatory incentives—such as expedited approvals or tax breaks—to compliance with robust ethical standards [70]. This strategy encourages private actors to integrate governance "by design," implementing transparency tools and data safeguards from the outset. When industry stakeholders actively participate in drafting and testing governance models, policy measures become more feasible and agile, reducing the risk of superficial or inconsistent adoption.

Governance frameworks are only as strong as their capacity to deploy them effectively. In low-resource settings, the lack of technical expertise and limited regulatory infrastructure can make advanced AI–SB regulations impractical. Capacity-building efforts, such as specialized training for local regulators, the creation of open educational resources, and technology transfer programs, are critical for ensuring that all regions can uphold ethical standards [63]. Equally important is addressing the digital divide, which impedes the equitable use of AI-based tools—particularly in rural or underserved communities. Dedicated funding and international collaboration can help close these gaps, ensuring that compliance and benefit-sharing do not remain privileges of wealthier nations and larger institutions.

Integrating TAPIC principles—Transparency, Accountability, Participation, Integrity, and Capacity—into adaptive policies enhances the effectiveness and ethical grounding of governance frameworks [71] (Fig. 7.4):

- Transparency: Implement open disclosure of AI algorithms used in biotechnology analysis and create public registries for AI-biotechnology projects and results.
- Accountability: Establish explicit lines of accountability for AI-powered decision-making, accompanied by tailored auditing procedures to evaluate adherence and effectiveness.
- Participation: Develop multi-stakeholder forums to facilitate ongoing discussions on governance matters and employ participatory technology evaluation mechanisms.
- Integrity: Establish specialized ethical review committees and develop AI training modules focused on integrity for academics and practitioners.
- Capacity: Invest in education and training programs and create centers of excellence for AI-biotechnology governance research and implementation.

The swift rate of progress in AI and biotechnology demands governance systems that remain flexible and adaptive. In the immediate future, this means instituting prompt processes to address emerging ethical concerns and conducting periodic evaluation cycles to gauge the effectiveness of policies. Over the longer term, it requires scenario planning, horizon scanning, and the modular design of regulatory frameworks—all of which can accommodate steady technological advances without necessitating a full overhaul. In domains as dynamic and influential as AI and SB, proactive governance also entails establishing multidisciplinary research institutes that continuously assess potential long-term effects. Notably, AI itself can help identify deficiencies in governance and suggest remedies, offering a novel opportunity to refine regulatory structures even as new capabilities emerge [62, 69].

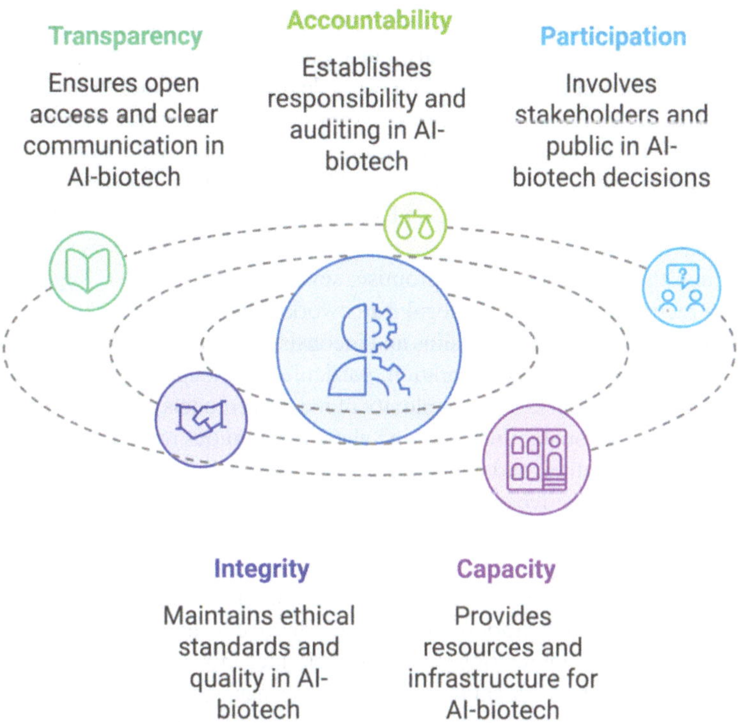

Fig. 7.4 Framework for AI-biotech governance: the TAPIC approach. This figure illustrates the five interconnected principles of the TAPIC framework essential for effective AI-biotechnology governance. At its center, it places AI-biotech development (represented by the human–machine interface icon) as the core focus around which all principles revolve. The framework consists of: transparency (green), ensuring open access and clear communication in AI-biotech developments; accountability (lime), establishing clear lines of responsibility and auditing mechanisms; participation (blue), involving diverse stakeholders and the public in decision-making processes; integrity (purple), maintaining ethical standards and quality assurance; and capacity (pink), providing necessary resources and infrastructure

The successful implementation of anticipatory governance for AI and SB hinges critically on inclusive stakeholder engagement [62]. Given the complex implications of these emerging technologies, diverse perspectives—ranging from technologists and policymakers to marginalized communities—ensure ethical frameworks address both comprehensiveness and equity [72]. Mechanisms such as public consultations, expert panels, participatory design workshops, and citizen juries bridge the gap between specialized scientific expertise and broader societal values [73, 74]. By actively involving affected groups in ethical guideline development, governance models gain both legitimacy and resilience.

Emerging technological tools are increasingly reshaping these governance frameworks. AI can rapidly process vast datasets to detect anomalies or potential policy breaches, enabling timely interventions [62, 75]. Blockchain's decentralized, tamper-proof ledger [76, 77] offers traceability and a reliable audit trail, strengthening transparency and accountability by recording every data access or transaction [66]. Meanwhile, natural language processing (NLP) automates the interpretation of complex legal documents, keeping governance frameworks updated with evolving regulations and reducing administrative overhead [78, 79]. Together, these innovations can bolster compliance, enhance trust, and minimize human error across high-stakes applications in AI and SB.

Although these strategies show promise, several barriers persist. Regulatory lag remains a recurring challenge, as legal frameworks often struggle to keep pace with rapid developments. Institutional silos and inconsistent stakeholder participation can further stymie meaningful collaboration, particularly among marginalized communities. Moreover, the black-box complexity of many AI models complicates efforts at transparency and accountability. Looking ahead, continuous review cycles, renewed international cooperation, and expanded capacity-building initiatives will be essential to maintaining effective oversight [6, 11, 62]. Balancing innovation with ethical responsibility must therefore be viewed as an ongoing process, receptive to new challenges as they emerge.

Developing adaptable ethical strategies for AI and SB is not just an academic exercise; it will shape the trajectory of these transformative technologies. By employing dynamic frameworks, integrating diverse stakeholder input, and adopting forward-looking methods such as ethical forecasting and scenario planning, we can establish a responsive ethical ecosystem that evolves alongside rapid technological advances. Ethics in such cutting-edge fields is never a fixed endpoint but an ongoing process—one that demands continual vigilance, flexibility, and a readiness to reassess our positions in light of fresh data and insights. This collective effort—spanning disciplines, cultures, and interests—aims to ensure that AI and SB deliver on their extraordinary promise while upholding our most cherished values for a more equitable, responsible, and beneficial future.

7.4.3 Interdisciplinary Collaboration

The convergence of AI and SB presents complex ethical challenges requiring insights from data science, bioethics, policy, agriculture, and beyond. Traditional frameworks often lack the agility to address rapid technological developments, leaving gaps in areas such as consent, privacy, and ecological impact [62, 80]. Engaging environmental scientists, geneticists, agronomists, and ethicists in a collaborative process ensures that new organisms, AI-driven diagnostics, and other emerging solutions are evaluated from multiple angles. This comprehensive perspective not only enhances the robustness of governance structures but also demystifies complex technologies, making them more accessible to broader society [42].

Delgado and Åm highlight how joint research initiatives among diverse stakeholders offer a more holistic understanding of AI–SB's broad impacts [63]. Policymakers, scientists, ethicists, and public representatives can develop solutions more likely to gain widespread acceptance. This approach is echoed by the European Commission's High-Level Expert Group on AI [70], which underscores the importance of open dialogue to secure equitable, well-informed policies. Blasimme et al. further recommend interdisciplinary educational programs to break down disciplinary silos [81], while Floridi advocates for stakeholder engagement platforms that help build trust and legitimacy [62].

Embedding TAPIC principles within interdisciplinary collaboration strengthens governance frameworks [82]. Figure 7.5 illustrates how each stakeholder group contributes to Transparency, Accountability, Participation, Integrity, and Capacity:

- Transparency: Foster transparency by ensuring that all stakeholders have access to information regarding AI-biotechnology projects and their governance.
- Accountability: Promote accountability by defining clear roles and responsibilities within collaborative efforts, ensuring that each stakeholder is answerable for their contributions.
- Participation: Encourage active participation from a diverse range of stakeholders, including marginalized communities, to ensure that multiple perspectives are considered in governance decisions.
- Integrity: Maintain integrity by adhering to ethical standards and best practices in all collaborative endeavors, ensuring that governance decisions are made with honesty and moral uprightness.
- Capacity: Enhance capacity by providing stakeholders with the necessary training and resources to effectively contribute to governance discussions and implementations.

To translate the conceptual model shown in Fig. 7.5 into concrete action, Table 7.2 maps each stakeholder to its governance role and primary TAPIC contribution.

Early risk identification is a core benefit of integrated perspectives: involving ethicists, ecologists, and data scientists can uncover unforeseen ecological or social harms—such as the spread of engineered genes or hidden algorithmic biases—before wide deployment [6, 58]. Domain specialists also help refine governance strategies so they align with both innovation and public interest [83], ensuring privacy safeguards in AI diagnostics [19] and culturally appropriate solutions in agriculture [42].

Effective governance also hinges on inclusive engagement. Interdisciplinary working groups, symposia, and online platforms unite AI experts, biotechnologists, ethicists, regulators, and affected communities—fostering knowledge exchange, transparent decision-making, and co-creation of policy [63, 81]. Public consultations, citizen science, and educational initiatives further ground governance in lived experience, build trust, and ensure AI-SB developments reflect shared societal values [84, 85].

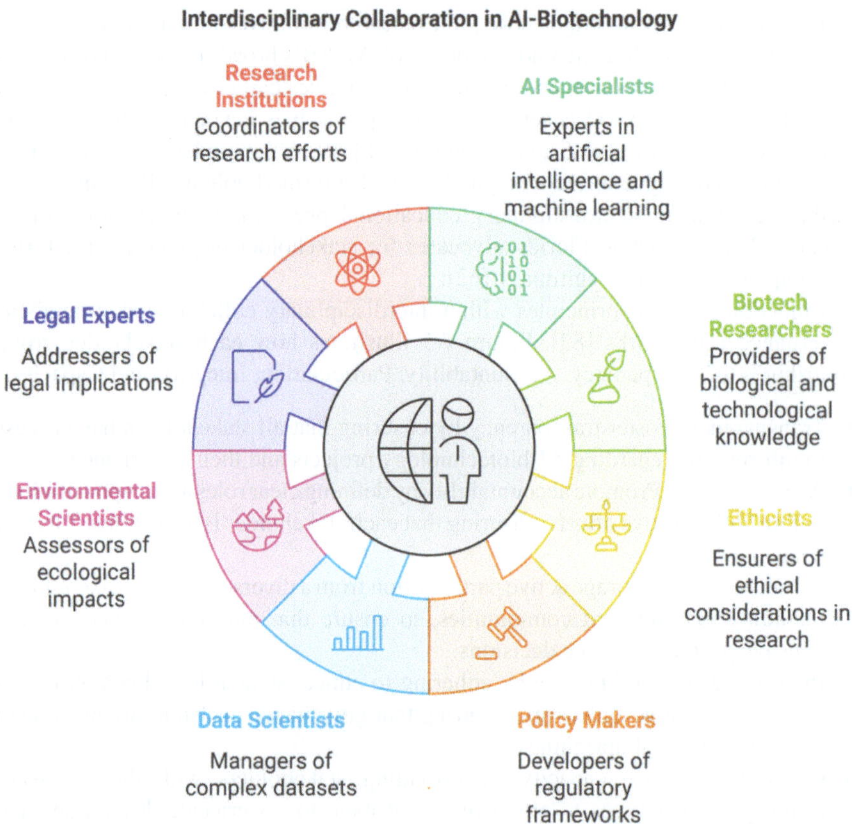

Fig. 7.5 Interdisciplinary collaboration in AI-biotechnology. This diagram illustrates the integrated network of eight essential stakeholder groups required for comprehensive AI-biotechnology governance. At the center, a globe-and-human icon represents the global human impact of AI-biotech developments. Radiating outward in distinct colored segments are the key stakeholders: AI specialists (mint green), who provide expertise in artificial intelligence and machine learning; biotech researchers (lime), who contribute biological and technological knowledge; ethicists (yellow), who ensure ethical considerations in research; policy makers (orange), who develop regulatory frameworks; data scientists (blue), who manage complex datasets; environmental scientists (pink), who assess ecological impacts; legal experts (purple), who address legal implications; and research institutions (red), who coordinate research efforts

7.5 Governance Strategy: Eight Pillars for AI in Biotechnology

This section presents eight core governance pillars designed to ensure responsible development, deployment, and oversight of AI-driven biotechnology. Each pillar is framed by (1) a concise description of the ethical or regulatory challenge; (2) a proposed solution; and (3) key limitations or considerations that must be addressed during implementation.

Table 7.2 Stakeholder roles and TAPIC contributions

Stakeholder	Role in governance	Key TAPIC contributions
AI specialists	Develop transparent, explainable models and implement bias audits	Transparency
Biotech researchers	Design safety-by-design protocols and conduct risk assessments	Integrity
Ethicists	Identify and articulate ethical risks; guide normative frameworks	Accountability
Policymakers	Draft adaptive regulations; enable modular policy updates	Capacity
Farmers and patients	Provide lived experience to shape context-appropriate governance	Participation
Industry	Embed best practices; commit to open reporting and third-party audits	Transparency
NGOs/Public	Monitor compliance; advocate for underserved voices	Accountability
Regulators	Enforce standards; conduct ongoing oversight and evaluation	Integrity

7.5.1 Broadening Data Protection Rules

Expand existing data protection regulations to specifically include biological data created and processed by AI. This ensures that sensitive genetic and biological information is adequately safeguarded against misuse and unauthorized access.

The Challenge. The rapid growth of AI-driven analytics in biotechnology has outpaced existing data protection laws, such as the General Data Protection Regulation (GDPR) in the European Union, which do not explicitly address AI-derived genetic and biological data. While these regulations offer baseline safeguards (e.g., requiring consent for data collection), they often do not account for the heightened re-identification risks posed by machine learning techniques that can cross-reference anonymized datasets with other sources. Additionally, questions about data ownership—particularly in cases where commercial entities aggregate and monetize health or agricultural data—remain unresolved. Individuals, small-scale farmers, or research participants may unknowingly relinquish control over their genetic information, leading to potential misuse, discrimination, or long-term privacy violations that can affect not just the individual but also biological relatives [86].

Proposed Solutions.

- Explicit AI-Focused Legislation. Expand existing legal frameworks (e.g., GDPR, HIPAA) to specifically include AI-processed biological data, clarifying ownership rights and lawful usage.

- Dynamic Consent Models. Implement continuous or "dynamic" consent so individuals can update their data-sharing preferences over time—especially critical when new AI applications or research aims emerge [64].
- Differential Privacy and Homomorphic Encryption. Apply privacy-preserving technologies that allow computations on encrypted genetic data without exposing raw identifiers [11, 12]. This reduces re-identification risks in collaborative research environments.
- Secure Data Platforms and Audits. Mandate rigorous auditing of AI data flows, including secure data repositories with robust encryption at rest and in transit, plus regular third-party security assessments to ensure compliance.
- Transparent Licensing and Data-Sharing Agreements. Require that data holders provide clear contracts outlining how genetic or agricultural data can be used, shared, or sold, ensuring that contributors remain informed about commercial activities.

Critical Reflection. Though these technical and legal proposals are promising, barriers to implementation persist. Advanced encryption techniques (e.g., fully homomorphic encryption) can be resource-intensive, raising cost and computational overhead concerns for smaller labs or developing countries. Dynamic consent frameworks, while empowering, may lead to "consent fatigue," where individuals are overwhelmed by frequent requests to update preferences. Further, cross-border data transfers pose significant compliance challenges; laws vary widely by jurisdiction, potentially creating regulatory gaps. Finally, some stakeholders (especially large commercial entities) may have minimal incentive to adopt stricter data controls unless enforcement is both clear and consistent.

Existing Frameworks and Real-World Examples.

- GDPR/HIPAA. Provide a foundational legal framework for personal data protection, but do not explicitly address AI-driven re-identification threats or the complexities of aggregated genetic data ownership.
- Genetic Information Nondiscrimination Act (GINA). In the United States, GINA prohibits genetic discrimination in health insurance and employment; however, it does not cover other areas (e.g., life insurance) nor address newly emerging AI methodologies.
- Dynamic Consent in Genomic Research. Projects like the UK's Genomics England pilot dynamic consent portals, giving participants ongoing control over how their genetic data is used.
- Blockchain-Based Data Sharing. Some research consortia experiment with blockchain for tamper-proof data logs, enhancing traceability. While not yet widespread, these pilots highlight possible paths for secure, auditable data-sharing.

7.5.2 Algorithmic Transparency

Integrate measures that enforce transparency in AI systems used for biological data analysis. This includes open disclosure of AI algorithms, methodologies, and decision-making processes to allow for external scrutiny and accountability.

The Challenge. A key hurdle in AI-driven biotechnology is the "black box" nature of many machine learning models, particularly deep neural networks. The complexity and opacity of these systems make it difficult for users, regulators, and even developers to understand how inputs are transformed into outputs [21]. This lack of transparency can undermine user trust—for instance, in healthcare diagnostics, physicians may be reluctant to rely on AI recommendations they cannot interpret. Additionally, it can hamper accountability, since pinpointing responsibility for errors or biases becomes challenging if the underlying decision logic is inaccessible [20]. In biotechnology, these concerns extend to AI-assisted gene editing tools or synthetic organisms, where the stakes of inadvertent mistakes can be exceptionally high.

Proposed Solutions.

- Explainable AI (XAI) Methods. Approaches such as Local Interpretable Model-Agnostic Explanations (LIME) [8], SHapley Additive exPlanations (SHAP) [9], and Quantitative Input Influence (QII) [30] can be embedded in AI pipelines to provide end-users and stakeholders with insights into how decisions are made.
- Mandatory Auditing and "Algorithmic Impact Assessments." Requiring third-party audits or internal "impact assessments" ensures that AI developers document design choices, training data sources, and known limitations [31]. This clarifies the scope of an algorithm's usage and reveals potential biases.
- Open Algorithms Initiatives. Projects like the Open Algorithms (OPAL) project promote collaborative frameworks for algorithmic development, offering privacy-preserving data access without divulging raw datasets. This setup preserves accountability by allowing external experts to evaluate code performance and fairness [31].
- Interpretability by Design. From the inception of AI system development, incorporate model architectures that prioritize transparency—such as simpler models, rule-based systems, or hybrid solutions that blend deep learning with interpretable layers [23].

Critical Reflection. While explainable AI tools can improve interpretability, they do not fully resolve the trade-off between transparency and performance: highly accurate models (e.g., deep neural networks) often remain inherently complex. Moreover, auditing large-scale AI systems demands specialized skills and can be resource-intensive, raising equity concerns for smaller labs and low-resource institutions. The intellectual property dimension also complicates open disclosure—companies may be reluctant to reveal proprietary models. Additionally, interpretability methods like LIME or SHAP may offer only partial, post-hoc explanations rather than truly transparent decision rules. Without robust regulatory pressure, many organizations have minimal incentive to invest in thorough auditing or open-source practices.

Existing Frameworks and Real-World Examples.

- GDPR's "Right to Explanation." Although not perfectly defined in practice, the GDPR hints that individuals have the right to understand significant factors behind automated decisions. This concept has spurred policy debates on enforcing algorithmic transparency in Europe [10].
- OPAL Project. By allowing external audits on aggregated data, OPAL provides a model for how private companies and public agencies could share insights from AI systems without compromising sensitive data [31].
- Institutional Review Boards (IRBs). In some cases, IRBs have begun reviewing algorithmic protocols in clinical settings, requiring documentation that clarifies how AI systems handle bias or missing data.
- Emerging Legal Proposals (e.g., EU AI Act). The proposed EU AI Act classifies certain AI applications (e.g., in healthcare or biotech) as high-risk, requiring more stringent transparency and oversight standards.

7.5.3 *Federated Learning Criteria*

Develop precise criteria for the application of federated learning in sensitive biological data scenarios. Federated learning enables decentralized data processing, enhancing privacy by keeping data localized while still allowing AI models to learn from diverse datasets.

The Challenge. Federated learning (FL) allows AI models to train across decentralized datasets without requiring raw data to be gathered in a single location. While this approach offers robust privacy advantages—particularly valuable for sensitive genetic or clinical data—it also introduces technical and governance challenges. One major issue is data heterogeneity: local datasets may vary greatly in size, quality, or labeling, potentially skewing model performance [87]. Additionally, implementing FL at scale demands significant infrastructure, including reliable network bandwidth and on-device computational power. From a governance perspective, lack of clear criteria for how and when to deploy FL, especially in biotech research, can lead to inconsistencies. For instance, small labs may struggle to maintain secure local data storage or comply with multiple institutional review board (IRB) requirements across regions.

Proposed Solutions.

- Standardized FL Protocols. Develop explicit guidelines outlining minimum data quality, device security, and encryption standards for participating institutions. These could draw from existing frameworks in privacy-sensitive sectors such as healthcare, ensuring uniform levels of protection and interoperability.

- Robust Encryption and Secure Aggregation. Combine federated learning with differential privacy or secure multi-party computation techniques (e.g., homomorphic encryption) to safeguard intermediate model updates [88].
- Hierarchical FL Models. Implement hierarchical architectures that cluster similar datasets for localized training, reducing the impact of outliers and data imbalance [89].
- Consent and IRB Coordination. Clearly define how each local dataset is acquired and ensure dynamic consent processes remain valid even if local data is kept on-site. IRBs should coordinate to streamline multi-site approvals.
- Federated Learning Audits. Enforce periodic third-party audits of federated model performance, focusing on bias detection and data misuse prevention.

Critical Reflection. Though federated learning promises enhanced privacy, practical bottlenecks remain. Smaller institutions may lack the infrastructure or funding to implement the secure hardware required for FL at scale. Heterogeneous data can degrade overall model accuracy, especially if certain sites have extremely limited samples. Cross-border issues add complexity, as local regulations or differing consent guidelines might conflict. Additionally, even if raw data stays local, metadata and model updates can still leak sensitive information under certain conditions [90, 91]. Without consistent standards and enforcement mechanisms, FL could inadvertently become a fragmented system where only well-resourced entities reap the benefits.

Existing Frameworks and Real-World Examples.

- Healthcare Federated Learning Initiatives. Projects such as the Federated Tumor Segmentation (FeTS) Challenge demonstrate FL in multi-institutional medical imaging research, showing feasibility for privacy-sensitive applications [92].
- Big Tech Implementations. Companies like Google have piloted FL on user devices for predictive text, offering partial models for biotech to emulate [87].
- Ongoing Regulatory Discussions. While no formal global standard exists for FL in biotech, proposals under the EU's emerging AI regulations hint at stricter oversight of cross-border collaborations and data flows.
- Secure Enclaves. Some academic consortia experiment with secure enclaves—trusted environments that handle model updates—to mitigate data leakage risks further.

7.5.4 Ethical Protocols for AI-Generated Biological Systems

Establish well-defined protocols that guide the ethical creation and implementation of AI-generated biological entities. These protocols should address potential ethical dilemmas, safety concerns, and long-term ecological impacts.

The Challenge. Advances in AI are making it possible to design novel genetic pathways or organisms with unprecedented speed and complexity [80]. While such innovations carry promise for healthcare (e.g., customized gene therapies) and agriculture (e.g., pest-resistant crops), the ethical and safety concerns associated with engineering living systems are profound. Dual-use risks—where the same techniques applied for therapeutic benefits can be misused for harmful purposes—complicate governance [34]. Additionally, unforeseen ecological impacts may arise if AI-designed organisms are released into the environment without robust containment strategies [93]. Traditional regulatory structures often lag behind these breakthroughs, leaving gaps where new types of engineered organisms do not neatly fit existing oversight categories [35].

Proposed Solutions.

- Standardized Ethical Review Protocols. Expand the role of institutional or independent ethical review boards (ERBs) and biosafety committees to evaluate AI-driven synthetic biology proposals [34]. These bodies should have guidelines specific to AI-augmented design processes.
- Safety-by-Design (SbD) Frameworks. Require incorporation of genetic safeguards—such as kill switches or synthetic auxotrophies—to minimize ecological and public-health risks [55, 57].
- Dual-Use Compliance Mechanisms. Introduce mandatory dual-use risk assessments, accompanied by secure data-sharing platforms that track where and how AI tools are used. Align these measures with international biosecurity norms to prevent weaponization of engineered organisms [33].
- Stakeholder and Community Engagement. Engage farmers, patient advocacy groups, and local communities in protocol development, ensuring that those directly affected by AI-designed organisms have a voice in release and monitoring [35].
- Iterative Pilot Programs. Before large-scale deployment, conduct controlled pilot tests with comprehensive risk assessments and monitoring to gather empirical data on ecological effects [56].

Critical Reflection. Even with defined ethical protocols, genuine predictive certainty about AI-engineered organisms remains elusive—particularly when considering long-term ecological impacts. Genetics can exhibit complex interactions with other species, possibly triggering unexpected evolutionary changes. Moreover, commercial pressures may incentivize rapid product development, potentially undermining thorough risk reviews if not rigorously enforced. Ethical review boards, while crucial, can face resource limitations and may struggle to keep pace with the volume and sophistication of AI-driven projects. Lastly, dual-use governance must balance scientific openness with security measures, a tension that can stifle international collaboration if not carefully managed.

Existing Frameworks and Real-World Examples.

- NIH Guidelines for Recombinant DNA Research. Although primarily designed for traditional genetic engineering, these guidelines can partially extend to AI-generated modifications, offering a foundation for biosafety classification.

- iGEM (International Genetically Engineered Machine) Competition. Though educational, iGEM teams often pioneer synthetic biology projects. Many are now adopting expanded safety checklists that factor in AI-assisted design processes.
- Cartagena Protocol on Biosafety. Provides an international framework for the transboundary movement of living modified organisms, but its scope does not explicitly address advanced AI design tools.
- Safety-by-Design Initiatives. Synthetic auxotrophy studies illustrate how built-in containment measures can help mitigate ecological risks, though implementation varies widely by jurisdiction [55].

7.5.5 Evaluating Ecological Consequences

Implement frameworks to assess and mitigate the long-term ecological consequences of biotechnological advancements powered by AI. This involves conducting thorough environmental impact assessments and developing strategies to minimize negative ecological outcomes.

The Challenge. AI-driven biotechnology—ranging from genetically modified crops to engineered microorganisms—poses significant ecological uncertainties. While AI may enhance precision in gene editing (e.g., CRISPR-based methods) and crop optimization, the potential for unintended ecological impacts remains high [51]. These impacts can include gene flow to wild relatives [52], the emergence of pest or weed resistance, and reduced genetic diversity in crops if monocultures are encouraged. Moreover, small-scale pilot projects rarely capture the long-term, wide-ranging environmental interactions that occur once new organisms are deployed at scale. Without rigorous and continuous ecological monitoring, unforeseen harms to pollinators, soil microbiomes, or broader ecosystems may only surface when they are difficult—if not impossible—to reverse.

Proposed Solutions.

- Adaptive Risk Assessments. Move beyond static, one-time evaluations; require ongoing, adaptive risk assessments that evolve with ecological feedback and updated AI models [56].
- Post-Market Environmental Monitoring. Establish continuous monitoring systems for AI-optimized crops or engineered organisms after commercial release, with mandatory reporting of anomalies or environmental disturbances [54, 94].
- Biocontainment Measures. Incorporate physical or genetic safeguards (e.g., synthetic auxotrophies, kill switches) to mitigate the risks of unintended spread [55].
- Multi-stakeholder Oversight. Involve ecologists, local communities, and environmental NGOs in both the risk assessment and post-deployment monitoring processes, ensuring ecological expertise shapes regulatory decisions.

- Scenario Planning and Modeling. Use AI-driven scenario simulations to predict potential evolutionary pathways or ecological interactions under varied environmental conditions, improving proactive risk identification.

Critical Reflection. Even the most stringent monitoring frameworks can struggle with long latency periods: ecological disruptions may take multiple growing seasons or generational cycles to manifest. Commercial pressures to roll out new biotech solutions rapidly can undermine thorough evaluations, especially if monitoring protocols aren't enforced or well-funded. Additionally, model bias—due to limited data on biodiversity or marginal ecosystems—can skew risk assessments. In regions with less regulatory infrastructure, there is a risk that AI-designed organisms will be tested or deployed without robust oversight, disproportionately impacting vulnerable ecosystems and local populations. Global coordination is also crucial: organisms do not respect borders, and transboundary movement may bypass localized regulations.

Existing Frameworks and Real-World Examples.

- Cartagena Protocol on Biosafety. Offers international guidelines for living modified organisms, but doesn't fully integrate AI-driven enhancements or ongoing dynamic risk assessments [22].
- Long-Term Ecological Research (LTER) Programs. Some countries maintain LTER sites to study ecosystem changes over decades, providing a potential model for continuous monitoring of AI-driven GMO impacts.
- EU Directives on GMOs. European regulations require post-market monitoring and renewal of approvals every 10 years, a system that could be expanded to accommodate AI-augmented organisms and more frequent reviews (Regulation (EC) No 1829/2003).
- Community-Based Monitoring Initiatives. Pilot programs in agriculture, such as citizen-science efforts to report pest and pollinator changes, show promise for capturing local ecosystem data that official channels might overlook.

7.5.6 Pilot Implementation Programs

Launch pilot programs to test and evaluate the effectiveness of refined governance frameworks. These programs provide practical insights and data to inform further improvements and adjustments.

The Challenge. Implementing large-scale reforms in AI-driven biotechnology often requires real-world testing before establishing permanent policies or regulations [63]. Relying solely on theoretical models or small-scale laboratory experiments can overlook crucial variables that arise in practical settings, such as the complexity of clinical workflows or diverse farming conditions. Additionally, stakeholders—ranging from healthcare providers to small-scale farmers—may be hesitant to adopt emerging AI solutions without tangible demonstrations of safety, efficacy, and cost-effectiveness.

Consequently, the gap between research innovation and policy implementation can persist, delaying or undermining the responsible roll-out of promising technologies.

Proposed Solutions.

- Targeted Pilot Projects. Launch well-structured pilot programs that test new governance frameworks or AI applications in controlled yet realistic conditions [14]. These can occur in select hospitals, farms, or research centers to gather feedback on technical feasibility and ethical compliance.
- Multi-stakeholder Consortia. Form public–private partnerships (PPPs) or consortia that pool resources and expertise, ensuring transparency and balanced oversight [70]. This approach encourages collaboration among technologists, ethicists, patient or farmer advocacy groups, and regulatory bodies.
- Phased Scale-Up. Begin with small, region-specific trials to address local needs, then expand gradually to broader populations. This phased approach highlights emergent challenges—like infrastructure gaps or cultural barriers—before wider adoption.
- Evaluation Metrics and Audit Trails. Define clear success metrics (e.g., reductions in diagnostic errors, yield improvements without ecological harm), along with robust auditing mechanisms to track compliance with governance protocols.
- Knowledge-Sharing Platforms. Create open-access repositories documenting pilot results (both successes and failures), enabling continuous learning and refinements to governance frameworks.

Critical Reflection. Pilot programs can be time-consuming and resource-intensive, posing barriers for smaller organizations or low-income regions with limited budgets. Even well-designed pilots may inadvertently overlook longer-term effects, especially if testing periods are too short. There is also a risk of pilot inflation—where organizations repeatedly conduct pilots without ever moving to full implementation, leading to "pilot purgatory." Additionally, private actors might selectively share only positive outcomes, skewing the evidence base for policy decisions [4]. Sustained political and financial support is vital to prevent pilot projects from stalling or failing to translate into effective long-term solutions.

Existing Frameworks and Real-World Examples.

- Dynamic Consent Pilots. Some hospitals and genomic research initiatives have run pilot programs testing dynamic consent portals, revealing practical guidelines for user interfaces, consent updates, and data-sharing preferences [64].
- Agricultural 'Living Labs'. In regions like the EU, living labs on smart farms explore AI's impact on crop management, generating real-time feedback on ecological factors such as soil health and biodiversity [36].
- OPAL Project and FeTS Challenge. While these primarily address data privacy or federated learning in healthcare, they demonstrate how time-bounded, multi-site pilots can guide future regulatory decisions [92].

- Regional Policy Testbeds. Several national agencies (e.g., in Singapore or the UK) have established AI "regulatory sandboxes" where new governance rules can be trialed at small scale and refined before nationwide rollout [62].

7.5.7 *Collaborative Endeavors*

Foster cooperative efforts between regulatory agencies and research institutions. Collaborative initiatives ensure that governance frameworks are informed by the latest scientific developments and practical experiences.

The Challenge. Regulating AI-driven biotechnology often requires input from a broad range of stakeholders—including scientists, ethicists, policymakers, industry partners, civil society groups, and local communities [62]. However, each group has different priorities and operating styles, which can lead to siloed efforts or conflicting agendas [63]. Coordinating across disparate sectors is further complicated by varied regulatory landscapes in different countries, making global alignment on standards difficult. Inadequate collaboration can delay necessary reforms or produce piecemeal regulations that fail to keep pace with rapidly evolving AI and synthetic biology innovations.

Proposed Solutions. Building genuine collaboration in AI–SB requires practical, well-supported strategies that unite diverse fields. Below are actionable measures—illustrated by real-world examples—to foster effective interdisciplinary partnerships (Fig. 7.6):

- Interdisciplinary Research Hubs. Establish dedicated centers or working groups that unite computer scientists, biologists, ethicists, and policymakers under one organizational umbrella [70]. Such hubs can help break down silos and ensure holistic perspectives inform AI–SB projects from the start.
- Public–Private Partnerships (PPPs). Encourage structured partnerships between academia, government agencies, non-governmental organizations, and industry to share resources and expertise. Joint funding and governance arrangements can foster transparency and balanced decision-making [4].
- Organizing Cross-Disciplinary Workshops and Conferences: Events that gather professionals from multiple disciplines foster knowledge exchange and build collaborative networks.
- Developing Interdisciplinary Educational Programs and Training: Interdisciplinary training equips future researchers and practitioners with the skills necessary for effective collaboration.
- Providing Funding for Interdisciplinary Research: Designated grants for interdisciplinary projects encourage collaboration across traditional research boundaries.
- Showcasing Successful Collaborative Frameworks: Highlighting effective interdisciplinary collaborations can serve as models for new initiatives.

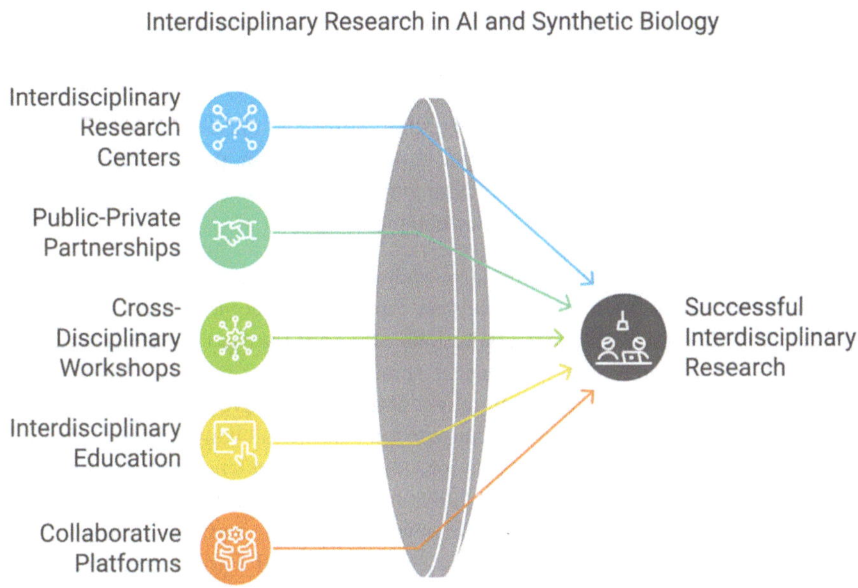

Fig. 7.6 Pathways to successful interdisciplinary research in AI and synthetic biology. This figure illustrates the five key pathways that contribute to successful interdisciplinary research in AI and synthetic biology. The left side presents the essential components: Interdisciplinary Research Centers (blue), represented by a network node icon, exemplified by initiatives like the UNESCO International Bioethics Committee; public–private partnerships (green), shown with a handshake icon, as demonstrated by the global partnership on AI; cross-disciplinary workshops (lime), depicted by a collaborative network icon, such as those hosted by the Synthetic Biology Leadership Council; interdisciplinary education (yellow), represented by an interactive learning icon, like MIT Schwarzman College of Computing's integrated programs; and collaborative platforms (orange), shown with a people-and-gears icon, exemplified by the open synthetic biology project. The gray portal-like structure in the center represents the integration and transformation of these components, leading to the ultimate goal of successful interdisciplinary research (dark gray icon)

- Multi-stakeholder Review Boards. Expand or reform existing boards—like Institutional Review Boards (IRBs) or Ethics Committees—to include representatives from civil society, patient or farmer advocacy groups, and policy experts. This ensures diverse viewpoints when evaluating new technologies or research proposals.
- International Alignment Efforts. Create global task forces or align existing treaties (e.g., involving WHO, FAO, OECD) to harmonize AI–SB standards and reduce regulatory fragmentation [22].
- Leveraging Advanced Technologies for Collaboration: Using tools like AI-driven project management, virtual reality, and blockchain can improve the effectiveness of interdisciplinary partnerships.

- Developing Policies that Encourage Interdisciplinary Research: Policies that support interdisciplinary work motivate institutions and researchers to engage in cross-disciplinary collaborations.
- Online Collaborative Platforms. Use digital tools (e.g., open-source repositories, shared data portals) that facilitate real-time collaboration, particularly for cross-border research or AI model development. This approach also helps maintain transparency around data use and ownership.

Critical Reflection. While collaborative endeavors promise more inclusive governance, power imbalances among partners can skew outcomes. Larger corporate entities or well-funded research institutions often hold greater influence in PPPs, potentially overshadowing the voices of smaller labs, low-resource countries, or community representatives. Additionally, confidentiality agreements—common in industry partnerships—may restrict knowledge sharing, slowing broader scientific progress. Aligning international regulations also proves challenging due to geopolitical frictions and cultural differences. Sustained funding is crucial to keep interdisciplinary centers or multi-stakeholder review boards active and effective over time. Without robust oversight, collaboration can devolve into a token gesture, where diverse stakeholders are named but not genuinely empowered to shape decisions.

Existing Frameworks and Real-World Examples.

- European Commission High-Level Expert Group on AI. Brings together policy, academia, industry, and civil society, providing guidelines like "Ethics Guidelines for Trustworthy AI" and serving as a model for multi-stakeholder engagement [72].
- The Global Partnership on AI (GPAI). Involves multiple governments, private sector partners, and research organizations collaborating on policy recommendations and AI ethics.
- iGEM (International Genetically Engineered Machine) Foundation. Encourages interdisciplinary teams—from biology, engineering, and social sciences—to address safety, ethics, and public engagement in synthetic biology competitions.
- Collaborative Online Platforms (e.g., Open Synthetic Biology Project). Provide repositories for sharing protocols, AI models, and ethical guidelines, fostering a culture of transparency in biotech innovation.

7.5.8 Periodic Evaluation and Improvement

Establish procedures for the regular evaluation and enhancement of governance frameworks based on actual results and emerging challenges. This continuous improvement process ensures that governance remains relevant and effective in addressing new ethical and technical issues.

The Challenge. Technologies in AI and synthetic biology evolve at unprecedented speed, outstripping the pace of traditional lawmaking and oversight [6]. Even well-structured governance frameworks risk obsolescence if they rely on static rules that fail to anticipate new breakthroughs or shifting public concerns [62, 95]. Implementing one-time regulations without scheduled reviews can lead to regulatory lag, where existing policies become increasingly unfit for purpose. Additionally, resource constraints and political inertia often impede the creation of mechanisms for continuous or periodic policy updates.

Proposed Solutions.

- Mandatory Review Cycles. Require periodic re-examination (e.g., every 2–3 years) of regulatory frameworks and AI–SB projects, aligned with technological milestones or funding cycles [65, 96].
- Adaptive Regulatory Sandboxes. Establish "sandbox" environments where new policies can be tested in real-world conditions, with built-in protocols for feedback loops and iteration [63].
- Expert Advisory Panels. Form permanent, interdisciplinary advisory bodies that meet regularly to evaluate emerging AI–SB developments, recommend updates, and foresee risks [62].
- Horizon-Scanning Programs. Allocate resources to prospective technology analysis—scanning scientific literature, patents, and public discourse to identify early signals of disruptive innovations.
- Public Consultation and Accountability Mechanisms. Involve civil society, advocacy groups, and industry stakeholders in scheduled reviews. Publicly disclose findings and action items so the process remains transparent and inclusive.

Critical Reflection. Even with structured review cycles, political and budgetary constraints may curtail the practical impact of iterative governance. Some institutions may resist frequent policy revisions if it demands additional compliance burdens or threatens entrenched interests. Moreover, long feedback loops and bureaucratic processes can slow responsiveness when urgent changes are needed. International coordination remains a challenge: while adaptive review might work in one jurisdiction, it may not align easily with another's legislative timelines. Balancing stability (so stakeholders can plan confidently) with flexibility (to adapt to rapid innovation) requires a delicate policy design that may not always be straightforward to implement [20].

Existing Frameworks and Real-World Examples.

- EU GMO Approvals Renewal Process. GMO authorizations in the EU undergo renewal every 10 years, necessitating updated reviews of risk assessments. This model could be extended to AI-based biotech products with shorter intervals.
- IEEE Standards Association. Some emerging standards programs employ iterative processes for updating guidelines on AI ethics, holding annual or biennial public consultations [97].

- UK Regulatory Horizons Council. This advisory body scans for innovations that require new or revised regulation, offering a potential blueprint for scheduled horizon-scanning and adaptive policy.
- Pilot Policy Feedback Loops (see Sect. 7.5.6). Pilot programs often incorporate iterative improvements based on real-world data, providing a microcosm of how periodic evaluation can succeed before scaling up.

7.6 Governance Frameworks and Data Management

The convergence of AI and SB in both healthcare and agriculture has given rise to a range of ethical challenges—from data privacy risks and algorithmic bias in clinical settings to data ownership and socioeconomic inequalities in farming. Yet these seemingly distinct domains also exhibit overlapping concerns. Personalized medicine and enhanced diagnostics bring issues of patient autonomy to the fore, while digital farming raises questions about environmental impact and small-scale producer rights.

Despite the varied contexts, the governance solutions share common ground. Technical safeguards like differential privacy [11] and homomorphic encryption [12] protect sensitive health or agricultural data. Dynamic consent [64] and transparent data-sharing agreements bolster trust, and adaptive regulatory frameworks [6, 62] consistently aim to align innovation with ethical principles. This chapter explores how these strategies can be systematically applied across both health and agricultural sectors to foster responsible AI–SB advancements.

Section 7.6 synthesizes data governance strategies introduced in Sect. 7.5, demonstrating how robust data models, privacy safeguards, and benefit-sharing mechanisms operationalize each governance pillar. We organize this section by five core data domains—Privacy Protection, Data Sharing, Standardization, Benefit-Sharing, and Monitoring—each mapped to the relevant governance pillar(s).

Table 7.3 summarizes the five core data governance domains, links each to the corresponding governance pillar(s) introduced in Sect. 7.5, and indicates the subsection where each domain is discussed in detail.

7.6.1 Privacy Protections in Genetic Data Management

Privacy protection encompasses both legal and technical safeguards designed to ensure that sensitive biological data remains under the control of its rightful owners. Key strategies include data minimization (collecting only what is strictly necessary), robust de-identification protocols, and the application of differential privacy—which adds calibrated noise to datasets to prevent re-identification while preserving analytical utility [98, 99]. Homomorphic encryption further strengthens protection by allowing computation on encrypted data without exposing raw values [12], and

Table 7.3 Mapping of data governance domains to governance pillars

Data domain	Description	Linked governance pillar(s)	Section for details
Privacy protection	Techniques to secure genetic and agronomic data (e.g., dynamic consent, encryption)	7.5.1 Broadening data protection rules	7.6.1
Data sharing models	Multi-tier access models, cooperatives, blockchain audit trails	7.5.7 Collaborative endeavors; 7.5.1	7.6.2
Standardization and ontologies	FAIR principles, metadata standards, ontologies (GO, ChEBI)	7.5.2 Algorithmic transparency	7.6.3
Equitable benefit-sharing	Tiered licensing, capacity building, Nagoya Protocol alignment	7.5.6 Pilot implementation; 7.5.7	7.6.4
Testing and monitoring	High-throughput safety testing, post-market environmental monitoring	7.5.5 Evaluating ecological consequences; 7.5.8 Periodic evaluation	7.6.5

dynamic consent portals empower individuals to update sharing preferences as research evolves [64]. Figure 7.7 illustrates the four-stage Data Privacy Protection Process—moving from data minimization through anonymization, encryption, and strict access controls—to demonstrate how layered safeguards collectively reduce privacy risks. Continuous impact assessments and breach response protocols complete this framework, ensuring that privacy measures adapt alongside technological advancements. See Sect. 7.5.1 for detailed governance strategies on implementing and enforcing these protections.

7.6.2 Data Sharing Models in AI-Biotechnology

Effective data sharing models balance the imperative for open scientific collaboration with the need to protect contributor rights and maintain trust. Multi-tier access frameworks range from fully public, anonymized datasets to controlled environments where only vetted researchers can access identifiable information under strict governance agreements [100]. Cooperative data platforms and data trusts give contributors collective bargaining power to negotiate terms of use, benefit-sharing, and revenue distribution, while blockchain-based audit trails provide immutable records of every transaction—ensuring transparency and accountability. Dynamic consent mechanisms can be integrated into sharing platforms to allow real-time updates to participant preferences, reinforcing autonomy and trust. By clearly defining usage

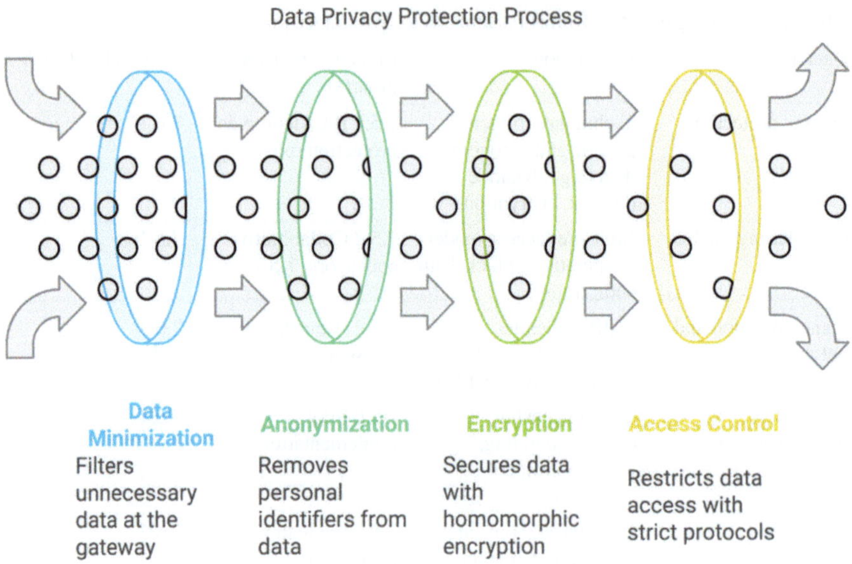

Fig. 7.7 Data privacy protection process in AI-biotechnology. This figure illustrates the four-stage data privacy protection process essential for securing sensitive biological and genetic information in AI-biotechnology applications. The process flows through sequential privacy-enhancing stages: data minimization (blue), which filters unnecessary data at the entry point; anonymization (green), which removes personal identifiers from the retained data; encryption (lime), which secures data using homomorphic encryption techniques; and access control (yellow), which implements strict protocols for data access

rights, access protocols, and benefit-sharing arrangements, these models help prevent exploitation, support equitable participation across stakeholders, and align with governance pillars of collaborative oversight (Sect. 7.5.7) and privacy protection (Sect. 7.5.1).

7.6.3 Data Standardization and Ontologies in AI-Biotechnology

Standardization and ontology development ensure that biological data is findable, interoperable, and reusable—key prerequisites for trustworthy AI in biotechnology. By adopting FAIR principles [99] and establishing ontologies such as the Gene Ontology (GO) and Chemical Entities of Biological Interest (ChEBI), researchers can harmonize data formats, metadata schemas, and terminologies across disciplines, reducing misinterpretation and facilitating cross-study validation [101, 102]. Consistent data structures not only improve reproducibility and comparability but also enable automated quality checks and bias detection within AI pipelines. See

Sect. 7.5.2 for governance strategies linking data standardization to algorithmic transparency and accountability.

7.6.4 *Equitable Benefit-Sharing Mechanisms*

Equitable benefit-sharing mechanisms allocate both monetary and non-monetary gains from AI-biotechnology innovations to all stakeholders, particularly historically marginalized communities. Tiered licensing models adjust fees based on economic context, while capacity-building programs transfer skills and technology to under-resourced regions, promoting local autonomy and reducing dependency [4, 103]. Legally binding benefit-sharing agreements—aligned with international frameworks such as the Nagoya Protocol—ensure transparent reporting of profits, co-ownership of IP, and reinvestment into community priorities. See Sects. 7.5.6 and 7.5.7 for detailed governance approaches that operationalize fair distribution of AI-biotech benefits.

7.6.5 *High-Throughput Testing and Monitoring Frameworks*

High-throughput testing and continuous monitoring are critical to ensure that AI-designed biological systems meet rigorous safety, efficacy, and environmental standards. Rigorous pre-market evaluations—combined with ongoing post-market monitoring—allow researchers to detect unintended consequences (such as off-target effects in gene editing or unforeseen ecological impacts) and provide data to iteratively refine both the technology and its governing policies [56, 94]. Adaptive testing protocols, including those implemented in controlled laboratory settings and long-term ecological research (LTER) programs, help identify potential hazards before wide-scale deployment. Moreover, regulatory frameworks—such as those stipulated in EU Directives on GMOs—mandate periodic safety assessments and environmental monitoring, ensuring that risk management measures remain responsive to evolving scientific evidence. See Sects. 7.5.5 and 7.5.8 for detailed governance strategies and periodic evaluation processes that support a proactive testing and monitoring regime.

7.7 Conclusions and Pathways Forward

The convergence of artificial intelligence (AI) and synthetic biology (SB) heralds a new era of innovation capable of transforming healthcare, agriculture, and environmental management—offering solutions for personalized medicine, sustainable food systems, and ecological restoration. Yet these advances also amplify ethical, social, and ecological risks if governance does not keep pace. This chapter has presented eight governance pillars (Sect. 7.5) as a cohesive framework to translate ethical principles into actionable policy, technical, and institutional strategies. Section 7.6 demonstrated how five core data governance domains (privacy protection, data sharing, standardization, benefit-sharing, testing and monitoring) operationalize these pillars across contexts.

7.7.1 Key Takeaways

- Privacy and Equity: AI-SB tools magnify data sensitivity and bias risks in precision health and small-holder agriculture; robust privacy safeguards and inclusive data governance are non-negotiable.
- Accountability and Transparency: Diluted responsibility in multi-actor systems demands mandatory impact assessments, third-party audits, and explainable AI to ensure trustworthy decision-making.
- Safety and Stewardship: Adaptive risk assessments, Safety-by-Design genetic safeguards, and post-market ecological monitoring form a layered defense against unintended harms.
- Participation and Integrity: Genuine stakeholder engagement—from patient panels to farmer cooperatives—fosters legitimacy, addresses power imbalances, and grounds innovation in lived experience.

7.7.2 Unique Convergence Opportunities

AI's predictive analytics combined with SB's design capabilities can accelerate novel therapies and climate-resilient crops. By integrating ethics-by-design early in R&D, stakeholders can preempt harm while maximizing societal benefit. Cross-sectoral learning—for example, applying dynamic consent models developed in genomics to agricultural data cooperatives—illustrates the multiplier effect of convergence.

7.7.3 Priority Research and Policy Agenda

- Adaptive governance evaluation: Launch multi-jurisdictional regulatory sandbox pilots to measure responsiveness, enforcement efficacy, and public trust impacts.
- Participatory governance experiments: Conduct longitudinal studies comparing dynamic consent, data cooperatives, and citizen juries to identify scalable best practices for stakeholder empowerment.

- Long-term impact modeling: Develop AI-enabled scenario planning platforms that simulate ecological, economic, and social trajectories of AI-SB interventions under diverse conditions.
- Capacity building and equity: Invest in technical and regulatory infrastructure in low-resource settings to prevent widening global divides in biotech capabilities.
- Global policy harmonization: Align national standards via international bodies (e.g., WHO, FAO, OECD) to reduce regulatory fragmentation and enable safe cross-border collaboration.

By embedding these strategies within the TAPIC framework—Transparency, Accountability, Participation, Integrity, Capacity—policymakers, researchers, and practitioners can ensure that AI-SB innovation remains ethically grounded, socially equitable, and ecologically sound. Ultimately, responsible governance will determine whether AI and synthetic biology deliver on their extraordinary promise: creating a healthier, more sustainable, and just future for all.

References

1. Topol E (2019) Deep medicine: how artificial intelligence can make healthcare human again. Basic Books
2. Char DS, Shah NH, Magnus D (2018) Implementing machine learning in health care—addressing ethical challenges. N Engl J Med 378(11):981–983
3. Carolan M (2021) The sociology of food and agriculture. Routledge
4. Kloppenburg J (2014) Re-purposing the master's tools: the open source seed initiative and the struggle for seed sovereignty. J Peasant Stud 41(6):1225–1246
5. Daniels N (2001) Justice, health, and healthcare. Am J Bioeth 1(2):2–16
6. Guston DH (2014) Understanding 'anticipatory governance.' Soc Stud Sci 44(2):218–242
7. Kaye J, Stranger M (2009) The end of individual control over health information: promoting fair information practices and the governance of biobank research. In: Governing biobanks
8. Ribeiro MT, Singh S, Guestrin C (2016) Why should I trust you?: explaining the predictions of any classifier
9. Lundberg S, Lee S-I (2017) A unified approach to interpreting model predictions
10. European Union (2016) General data protection regulation (GDPR)
11. Dwork C, McSherry F, Nissim K, Smith A (2006) Calibrating noise to sensitivity in private data analysis, pp 265–284
12. Gentry C (2009) Fully homomorphic encryption using ideal lattices. Stanford University
13. Görsoy G, Doerr M, Wilbanks J, Wagner J, Tang H, Brenner S (2020) Navigating ethical quandaries with the privacy dilemma of biomedical datasets. In: Pacific symposium on biocomputing
14. Morley J, Machado CCV, Burr C, Cowls J, Joshi I, Taddeo M, Floridi L (2020) The ethics of AI in health care: a mapping review. Soc Sci Med 260:113172
15. Wahl B, Cossy-Gantner A, Germann S, Schwalbe NR (2018) Artificial intelligence (AI) and global health: how can AI contribute to health in resource-poor settings? BMJ Glob Health 3(4):e000798
16. Beauchamp T, Childress J (2013) Principles of biomedical ethics. Oxford University Press
17. Baylis F, Robert JS (2004) The inevitability of genetic enhancement technologies. Bioethics 18(1):1–26
18. (2009) Recent patent applications in synthetic biology. Nat Biotechnol 27(12):1127

19. Rajkomar A, Hardt M, Howell MD, Corrado G, Chin MH (2018) Ensuring fairness in machine learning to advance health equity. Ann Intern Med
20. Kiseleva A, Kotzinos D, De Hert P (2022) Transparency of AI in healthcare as a multilayered system of accountabilities: between legal requirements and technical limitations. Front Artif Intell 5
21. Karim MdR, Shajalal M, Graß A, Döhmen T, Chala SA, Boden A, Beecks C, Decker S (2023) Interpreting black-box machine learning models for high dimensional datasets. In: 2023 IEEE 10th international conference on data science and advanced analytics (DSAA), pp 1–10
22. Pauwels E (2013) Public understanding of synthetic biology. Bioscience 63(2):79–89
23. Karim MR, Islam T, Shajalal M, Beyan O, Lange C, Cochez M, Rebholz-Schuhmann D, Decker S (2023) Explainable AI for bioinformatics: methods, tools and applications. Brief Bioinform 24(5)
24. Alterovitz G, Muso T, Ramoni MF (2010) The challenges of informatics in synthetic biology: from biomolecular networks to artificial organisms. Brief Bioinform 11(1):80–95
25. Pang S, Lee SY, Seul JY (2017) Policy challenges and ethical issues with the breakthrough technology: the case of synthetic biology. Sci Technol Soc 22(3):455–472
26. Bubela T, Hagen G, Einsiedel E (2012) Synthetic biology confronts publics and policy makers: challenges for communication, regulation and commercialization. Trends Biotechnol 30(3):132–137
27. Wittner F (2019) A public database as a way towards more effective algorithm regulation and transparency, pp 175–192
28. Keller P (2019) Participatory accountability at the dawn of artificial intelligence. SSRN Electron J
29. Maluka S, Kamuzora P, SanSebastián M, Byskov J, Ndawi B, Olsen ØE, Hurtig A-K (2011) Implementing accountability for reasonableness framework at district level in Tanzania: a realist evaluation. Implement Sci 6(1):11
30. Datta A, Sen S, Zick Y (2016) Algorithmic transparency via quantitative input influence: theory and experiments with learning systems. In: 2016 IEEE symposium on security and privacy (SP), pp 598–617
31. Lepri B, Oliver N, Letouzé E, Pentland A, Vinck P (2018) Fair, transparent, and accountable algorithmic decision-making processes. Philos Technol 31(4):611–627
32. Chu P, Agapito-Tenfen SZ (2022) Unintended genomic outcomes in current and next generation GM techniques: a systematic review. Plants 11(21):2997
33. Trump BD, Galaitsi S, Appleton E, Bleijs DA, Florin M, Gollihar JD, Hamilton RA, Kuiken T, Lentzos F, Mampuys R, Merad M, Novossiolova T, Oye K, Perkins E, Garcia-Reyero N, Rhodes C, Linkov I (2020) Building biosecurity for synthetic biology. Mol Syst Biol 16(7)
34. Gottwald F-T (2017) Integrity at risk: potentials and dangers of synthetic biology and how to govern with integrity. In: The role of integrity in the governance of the commons, pp 23–35
35. Bensaude Vincent B (2013) Ethical perspectives on synthetic biology. Biol Theory 8(4):368–375
36. Wolfert S, Ge L, Verdouw C, Bogaardt M-J (2017) Big data in smart farming—a review. Agric Syst 153:69–80
37. Rotz S, Gravely E, Mosby I, Duncan E, Finnis E, Horgan M, LeBlanc J, Martin R, Neufeld HT, Nixon A, Pant L, Shalla V, Fraser E (2019) Automated pastures and the digital divide: how agricultural technologies are shaping labour and rural communities. J Rural Stud 68:112–122
38. Wang X (2022) Artificial intelligence in the protection and inheritance of cultural landscape heritage in traditional village. Sci Program 2022:1–11
39. Bronson K, Knezevic I (2016) Big data in food and agriculture. Big Data Soci 3(1)
40. Fleming A, Jakku E, Lim-Camacho L, Taylor B, Thorburn P (2018) Is big data for big farming or for everyone? Perceptions in the Australian grains industry. Agron Sustain Dev 38(3):24
41. Wiseman L, Sanderson J (2019) Farms create lots of data, but farmers don't control where it ends up and who can use it. Phys.org
42. Carolan M (2018) 'Smart' farming techniques as political ontology: access, sovereignty and the performance of neoliberal and not-so-neoliberal worlds. Sociol Rural 58(4):745–764

43. Kaur J, Hazrati Fard SM, Amiri-Zarandi M, Dara R (2022) Protecting farmers' data privacy and confidentiality: recommendations and considerations. Front Sustain Food Syst 6
44. Howard PH (2015) Intellectual property and consolidation in the seed industry. Crop Sci 55(6):2489–2495
45. Kloppenburg J (2010) Impeding dispossession, enabling repossession: biological open source and the recovery of seed sovereignty. J Agrar Chang 10(3):367–388
46. Shiva V (2020) Reclaiming the commons: biodiversity, traditional knowledge, and the rights of Mother Earth
47. Patel R (2010) The value of nothing: how to reshape market society and redefine democracy
48. Mcmichael P (2013) Value-chain agriculture and debt relations: contradictory outcomes. Third World Q 34(4):671–690
49. Altieri MA, Farrell JG, Hecht SB, Liebman M, Magdoff F, Murphy B, Norgaard RB, Sikor TO (2018) Agroecology. CRC Press
50. González de Molina M, López-García D, Guzmán G (2017) Politizando el consumo alimentario: estrategias para avanzar en la transición agroecológica. Redes
51. Tsatsakis AM, Nawaz MA, Kouretas D, Balias G, Savolainen K, Tutelyan VA, Golokhvast KS, Lee JD, Yang SH, Chung G (2017) Environmental impacts of genetically modified plants: a review. Environ Res 156:818–833
52. Ellstrand NC (2018) "Born to run"? Not necessarily: species and trait bias in persistent free-living transgenic plants. Front Bioeng Biotechnol 6
53. Marvier M, McCreedy C, Regetz J, Kareiva P (2007) A meta-analysis of effects of Bt cotton and maize on nontarget invertebrates. Science 316(5830):1475–1477
54. Romeis J, Meissle M, Bigler F (2006) Transgenic crops expressing Bacillus thuringiensis toxins and biological control. Nat Biotechnol 24(1):63–71
55. Mandell DJ, Lajoie MJ, Mee MT, Takeuchi R, Kuznetsov G, Norville JE, Gregg CJ, Stoddard BL, Church GM (2015) Biocontainment of genetically modified organisms by synthetic protein design. Nature 518(7537):55–60
56. Moe-Behrens GHG, Davis R, Haynes KA (2013) Preparing synthetic biology for the world. Front Microbiol 4
57. Asin-Garcia E, Kallergi A, Landeweerd L, Martins dos Santos VAP (2020) Genetic safeguards for safety-by-design: so close yet so far. Trends Biotechnol 38(12):1308–1312
58. Hartley S, Gillund F, van Hove L, Wickson F (2016) Essential features of responsible governance of agricultural biotechnology. PLoS Biol 14(5):e1002453
59. Chen T, Lv L, Wang D, Zhang J, Yang Y, Zhao Z, Wang C, Guo X, Chen H, Wang Q, Xu Y, Zhang Q, Du B, Zhang L, Tao D (2025) Empowering agrifood system with artificial intelligence: a survey of the progress, challenges and opportunities. ACM Comput Surv 57(2):1–37
60. Eastwood C, Turner JA, Romera A, Selbie D, Henwood R, Espig M, Wever M (2023) A review of multi-scale barriers to transitioning from digital agriculture to a digital bioeconomy. CABI Rev
61. van der Burg S, Wiseman L, Krkeljas J (2021) Trust in farm data sharing: reflections on the EU code of conduct for agricultural data sharing. Ethics Inf Technol 23(3):185–198
62. Floridi L (2018) Soft ethics, the governance of the digital and the general data protection regulation. Philos Trans R Soc A: Math Phys Eng Sci 376(2133):20180081
63. Delgado A, Åm H (2018) Experiments in interdisciplinarity: responsible research and innovation and the public good. PLoS Biol 16(3):e2003921
64. Kaye J, Whitley EA, Lund D, Morrison M, Teare H, Melham K (2015) Dynamic consent: a patient interface for twenty-first century research networks. Eur J Hum Genet 23(2):141–146
65. van der Heijden AHC (1996) Two stages in visual information processing and visual perception? Vis Cogn 3(4):325–362
66. Trapscott D, Trapscott A (2016) Blockchain revolution: how the technology behind bitcoin is changing money, business, and the world. Penguin
67. Börjeson L, Höjer M, Dreborg K-H, Ekvall T, Finnveden G (2006) Scenario types and techniques: towards a user's guide. Futures 38(7):723–739

68. Schwartz P (1996) The art of the long view: planning for the future in an uncertain world
69. Porter M, Kramer M (2006) Strategy and society: the link between competitive advantage and corporate social responsibility
70. European Commission (2020) Public-private partnerships in transport research
71. Trump B, Cummings C, Klasa K, Galaitsi S, Linkov I (2023) Governing biotechnology to provide safety and security and address ethical, legal, and social implications. Front Genet 13
72. High-Level Expert Group on Artificial Intelligence (2019) Ethics guidelines for trustworthy AI
73. Rowe G, Frewer LJ (2000) Public participation methods: a framework for evaluation. Sci Technol Human Values 25(1):3–29
74. Smith G, Wales C (2000) Citizens' juries and deliberative democracy. Polit Stud 48(1):51–65
75. Goodman B, Flaxman S (2017) European Union regulations on algorithmic decision making and a "right to explanation." AI Mag 38(3):50–57
76. Nakamoto S (2008) Bitcoin: a peer-to-peer electronic cash system
77. Zheng Z, Xie S, Dai HN, Chen X, Wang H (2018) Blockchain challenges and opportunities: a survey. Int J Web Grid Serv 14(4):352
78. Katz DM, Bommarito MJ, Blackman J (2017) A general approach for predicting the behavior of the Supreme Court of the United States. PLoS ONE 12(4):e0174698
79. Susskind R, Susskind D (2017) The future of the professions: how technology will transform the work of human experts. Oxford University Press
80. Benner SA, Sismour AM (2005) Synthetic biology. Nat Rev Genet 6(7):533–543
81. Blasimme A, Vayena E, Hafen E (2018) Democratizing health research through data cooperatives. Philos Technol 31(3):473–479
82. Greer SL, Wismar M, Figueras J (2016) Strengthening health system governance: better policies, stronger performance. Open University Press
83. Marchant GE, Wallach W (2015) Coordinating technology governance. Iss Sci Technol
84. Burgess A, van Diggele C, Mellis C (2018) Mentorship in the health professions: a review. Clin Teach 15(3):197–202
85. Stahl BC, Wright D (2018) Ethics and privacy in AI and big data: implementing responsible research and innovation. IEEE Secur Priv 16(3):26–33
86. Gymrek M, McGuire AL, Golan D, Halperin E, Erlich Y (2013) Identifying personal genomes by surname inference. Science 339(6117):321–324
87. Kairouz P, McMahan HB, Avent B, Bellet A, Bennis M, Bhagoji AN, Bonawitz K, Charles Z, Cormode G, Cummings R et al (2021) Advances and open problems in federated learning. Found Trends Mach Learn 14(1–2):1–210
88. Yang Q, Liu Y, Chen T, Tong Y (2019) Federated machine learning: concept and applications. ACM Trans Intell Syst Technol 10(2)
89. Li T, Sahu AK, Talwalkar A, Smith V (2020) Federated learning: challenges, methods, and future directions. IEEE Signal Process Mag 37(3):50–60
90. Salem A, Bhattacharya A, Backes M, Fritz M, Zhang Y (2019) Updates-leak: data set inference and reconstruction attacks in online learning. Mach Learn
91. Zhu L, Liu Z, Han S (2019) Deep leakage from gradients. Cryptogr Secur
92. Sheller MJ, Edwards B, Reina GA, Martin J, Pati S, Kotrotsou A, Milchenko M, Xu W, Marcus D, Colen RR, Bakas S (2020) Federated learning in medicine: facilitating multi-institutional collaborations without sharing patient data. Sci Rep 10(1):12598
93. Wallach W, Saner M, Marchant G (2018) Beyond cost-benefit analysis in the governance of synthetic biology. Hastings Center Rep 48(S1)
94. Naranjo SE (2009) Impacts of Bt crops on non-target invertebrates and insecticide use patterns. CABI Rev 1–11
95. Floridi L (2014) The fourth revolution: how the infosphere is reshaping human reality. Oxford University Press
96. van der Heijden K (2004) Scenarios: the art of strategic conversation, 2nd edn
97. IEEE (2019) AI standards: best practices for ethical systems

98. Erlich Y, Narayanan A (2014) Routes for breaching and protecting genetic privacy. Nat Rev Genet 15(6):409–421
99. Wilkinson MD, Dumontier M, Aalbersberg IJ, Appleton G, Axton M, Baak A, Blomberg N, Boiten J-W, da Silva Santos LB, Bourne PE et al (2016) The FAIR guiding principles for scientific data management and stewardship. Sci Data 3(1):160018
100. Ozercan HI, Ileri AM, Ayday E, Alkan C (2018) Realizing the potential of blockchain technologies in genomics. Genome Res 28(9):1255–1263
101. Ashburner M, Ball CA, Blake JA, Botstein D, Butler H, Cherry JM, Davis AP, Dolinski K, Dwight SS, Eppig JT et al (2000) Gene ontology: tool for the unification of biology. Nat Genet 25(1):25–29
102. de Matos P, Alcántara R, Dekker A, Ennis M, Hastings J, Haug K, Spiteri I, Turner S, Steinbeck C (2010) Chemical entities of biological interest: an update. Nucleic Acids Res 38(suppl_1):D249–D254
103. Carbonell I (2016) The ethics of big data in big agriculture. Internet Policy Rev 5(1)

Chapter 8
Assessing Governance and Regulatory Frameworks for Converging Technologies: The Case of Artificial Intelligence in Biological Engineering and Design Technologies

Gregory P. Nichols

Abstract This chapter explores gaps in governance for addressing converging technologies. One emerging example of convergence is the overlap of artificial intelligence (AI) with biological technologies and engineering. While biotechnologies are somewhat better regulated and monitored compared to other technologies, artificial intelligence is not, and the combination of these two is certainly not explicitly or fully covered by most existing regulations or risk governance framework. An overview of existing or relevant regulations for AI and biological technologies is discussed, including how regulations and laws are being used to address contemporary needs and where actual and perceived shortfalls exist for regulating a convergence of the two technologies. Additionally, there is a discussion providing an analysis of regulatory approaches from US, EU, and other international perspectives regarding how this current challenge is being viewed and especially the characteristics of each approach based on risk-tolerance, social/cultural norms, historical sensitivities, and regional differences. This widening gap between technology implementation and appropriate oversight poses a unique challenge regarding how to keep the global population safe and well during this technology boom while the growing specter of converging technologies and emerging technological systems remains a relatively understudied and misunderstood aspect of global uncertainty.

Keywords Artificial intelligence · Biotechnology · Technological convergence · Governance · Regulation

G. P. Nichols (✉)
Oak Ridge Associated Universities (ORAU), Oak Ridge, TN, USA
e-mail: Greg.Nichols@orau.org

C. L. Cummings et al. (eds.), *Biotechnology and AI: Technological Convergence and Information Hazards*, NATO Science for Peace and Security Series A: Chemistry and Biology, https://doi.org/10.1007/978-3-032-05246-9_8

8.1 Introduction

The advent of biotechnology arguably begins over 7,000 years ago with the discovery of fermentation by the Chinese to create beer. Artificial intelligence (AI) has roots back to Ancient Greece, where myths and stories relayed adventures of heroes who fought automatons that exhibited human thought and reason. Although these technologies in some ways have been with humanity longer than most, they still bring awe and wonder, surprise and shock, and even fear and trembling at times. With all the excitement and uncertainty surrounding both biotechnology and AI, their full potentials are most likely still not achieved, and this often brings a certain amount of nervousness to humans, as it begs the age-old question: "what if?" It is a fair question to ask, both from the positive (what if AI and/or biotech could lead to achievements that greatly benefit human beings) and the negative (what if one or both technologies cause irreparable harm to our environment or to humanity).

Given the insatiable curiosity of humans, this question is tested ad nauseam, which can often lead to even more complicated and complex situations. An emerging example would be the concept of technology convergence, where two or more individual technologies are intentionally or unintentionally intertwined, leading to some new hybrid technology that may or may not keep any resemblance of the original underlying technologies. Herein lies the major issue at hand: if one technology itself opens the door for uncertainty and ambiguity, and is not yet clearly defined, then what happens whenever another technology, which also brings its own ambiguity and uncertainty, is added to the mix? Does the new creation resemble a neatly aligned wall, where each brick fits snugly with one another, or does it take on the shape of something more sinister, akin to Frankenstein's monster? Or does it, more realistically, fall somewhere in between, reaching a gray area that is much harder to define and control?

This concept of convergence has overwhelmingly hit a fever-pitch with the contemporary notion of AI somehow creating or enhancing breakthroughs in biological technology. However, these two veins of technology have always been on a collision course, in that efforts to control potential risks while pushing the boundaries of new developments appear to mirror one another. Modern developments within each technology have roots in the post-World War II era. Artificial intelligence was arguably born from efforts to decode the German Enigma code during the War, with the spark ignited in 1956 at the seminal Dartmouth Summer Research Project on Artificial Intelligence. One of the earliest successes in using AI for real-world applications was the ground-breaking MYCIN, a program developed at Stanford University in the early 1970s, which identified pathogenic bacteria and recommended the proper antibiotic and therapeutic dose for treating related infections [1]. Contemporary notions of biotechnology most likely were made possible through discovering the structure of deoxyribonucleic acid (DNA) in 1952 [2]. This discovery arguably opened the door for all related biotechnology research and discoveries to follow, leading to breakthroughs for CRISPR, cloning, mapping the human genome, mRNA vaccine technology, and the advent of synthetic biology.

8.2 Why This Topic and Why Now?

It may seem rather odd to pair two technologies that seem so opposed to one another—the rather soft-feeling and life-supporting biological technologies and engineering rooted in DNA and other organic molecules compared with AI, a technology generally more thought of as creating malevolent robots doomed to take over humanity. However, recent trends in science, technology, and societal actions demonstrate that not only are there distinct physical developments of the convergence of these technologies, but also conceptually, the ideas are merging. The use of software to predict biological outputs in general has been a common process for nearly 20 years, beginning with Rosetta@home, a platform managed at the University of Washington that predicts protein interactions and protein design [3]. Recent developments have created more powerful iterations of these capabilities. In 2018, Google DeepMind launched AlphaFold, an AI tool that predicts the structures of proteins so accurate that they cannot be distinguished from actual experimentally observed structures [4]. In 2022, OpenAI launched its open-sourced generative AI model known as ChatGPT, which revolutionized the way that the average person can access complex AI platforms and use them for everyday use, prompting warnings from experts, including Eric Schmidt, founder and former CEO of Google, that these new platforms threaten biological weapons treaties and potentially open the door to the development of super-weapons, like killer plagues [5]. Also, in 2022, the AI lab at Meta released the ESM Metagenomic Atlas, a database containing the structures of over 600 million proteins [6]. Following on the heels of key breakthroughs, Microsoft launched BioGPT in 2023, a generative program relying on deep learning to answer complex biomedical inquires [7]. Much of this work led to another venture in 2024, EvolutionaryScale, a startup funded by Amazon and NVDIA, which uses large language models to identify and predict protein structures with a tool known as ESM3 [8]. A recent publication reveals that ESM3 has been able to model 500 million years of evolution [9]. DeepMind also released a new project in 2024—AlphaProteo—an AI platform that designs novel proteins for use in biological and health research [10]. A broad list of AI tools and methods currently used in biotechnology is found in Table 8.1.

The COVID-19 Pandemic in many ways helped to cement the usefulness of AI in biotech development, especially with protein and pharmaceutical design. AI algorithms and supercomputing resources helped slog through the massive amount of data collected from SARS-CoV-2, including genomic information, patient data, and other biological information, to create possibilities for new types of effective treatments. While groundbreaking discoveries were made with AI during the Covid-19 era, this new potential threat prompted research institutions, like the Johns Hopkins University Center for Health Security, to begin work in earnest aimed at understanding the new health and biosecurity threats of the integration of AI and biotechnology. At the same time, the dual nature of this new convergence was made clear, as projected threats gave way to the acknowledgement of scientific achievements. The 2024 Nobel Prize in Chemistry was awarded to David Baker "for computational protein design" and Demis Hassabis and John Jumper "for protein structure prediction," and the

Table 8.1 AI-powered tools and techniques that enhance biotechnologies [11]

AI tool	Function	Application in biotechnology
XGBoost	Efficient prediction	Drug discovery and development
Random Forest	Pattern recognition	Drug discovery and development
Artificial Neural Network	Precise diagnoses	Personalized medicine
SignatraX	Drug target identification	Transcriptomic
CRISPR-P	Design high-fidelity guide RNAs	CRISPR technology
Generative Deep Neural Networks (GDNN)	Molecule design	Drug design
Alpha Fold	Protein structure prediction	Drug discovery and development
Deep Chem	Molecular property prediction	Drug design
Soft Sensors	Estimate bioprocess parameters	Bioprocessing
Deep Variant	Genomic variation analysis	Genomics
Convolutional neural networks (CNN)	Protein structure prediction	Proteomics
DeepCpfR	Predicting off-target effects of guide RNA sequences	CRISPR technology
Docking Simulations	Computer-aided drug discovery	Bio-molecular engineering
Lifecycle Analysis	Assess the environmental impacts	Bioplastics

2024 Nobel Prize in Physics to John Hopfield and Geoffrey Hinton "for foundational discoveries and inventions that enable machine learning with artificial neural networks" [12].

8.3 Technological Convergence and Governance

Technological convergence is the general trend for two or more seemingly unrelated technologies to become more conjoined and begin to operate as a more cohesive, integrated technological system or platform. Helbing and colleagues have identified three distinctive characteristics of converging technologies [13]:

1. "The increasing ubiquity and pervasive distribution of sensing and computing capabilities across both physical objects and biological organisms;
2. The erosion of a clear separation between the physical, digital, and biological domains due to emerging technologies such as artificial intelligence (AI), gene editing, nanotechnology, biomedical engineering, neurotechnology and robotics;

3. The increasingly frequent co-occurrence of the technologies listed above and their large-scale spreading in ways that may be hard to detect, protect from, and manage."

The latest item that can be added to the list is AIxBio (Artificial Intelligence and Biotechnology). Although the use of AI is not limited to biotechnology per se (it could include other related technologies like gene-editing, synthetic biology, biological weapons development, and biological design) most of what this chapter covers can fit into that broad category.

Governance of converging technologies can be quite difficult given a few things for consideration. First, some sort of oversight for one or more parent technologies may already exist, and that underlying regulation or law may or may not be adequate, but if it does exist, it provides foundational or partial oversight to some degree. As a corollary, that underlaying foundation may also provide loopholes for the new converging technology, which in a way could make it even more difficult to manage. Next, even if some sort of regulation is in place, it does not mean that it will sufficiently cover the daughter technologies or platforms.

One of the key points to make in regulating converging or converged technologies is derived from a philosophical principle known as the sorites paradox. The paradox, also known as the pile of sand problem, begins with a pile of sand of no discernable size. Imagine that a single grain of sand is removed from the pile, repeatedly, until at some point no pile of sand remains. The question at the core of the paradox is "When does the pile of sand cease to exist?" That is, at what point can the pile of sand no longer be called a pile of sand? Thus lies the crux of governance in converging technologies: At what point does the original regulation(s) no longer cover the new thing, because that thing is now so different, it cannot be sufficiently covered by any of the original regulations, written as intended.

Figure 8.1 provides an overview of some of the ways in which digital and AI-related technologies are merging with biotechnologies in the life sciences. While no one can predict exactly what will happen and how, this outlay provides a reasonable roadmap in understanding what could be expected as convergence of AI and biotechnology continues to develop.

8.4 Asilomar—Then and Now

For whatever reason, it seems that all roads regarding pivotal decisions on emerging concepts appear to consolidate at the Asilomar State Beach and Conference Grounds in Pacific Grove, California, near Monterey on the Central California Coast. Since 1972, seven major conferences have been held here, including a pivotal meeting in 1975 regarding an emerging technology at the time: Recombinant DNA. Co-organized by future Nobel Laureate, Paul Berg, the Asilomar Conference on Recombinant DNA was a multi-day gathering of leading experts in DNA technology, with the goal of evaluating potential safety hazards to determine how best to move forward

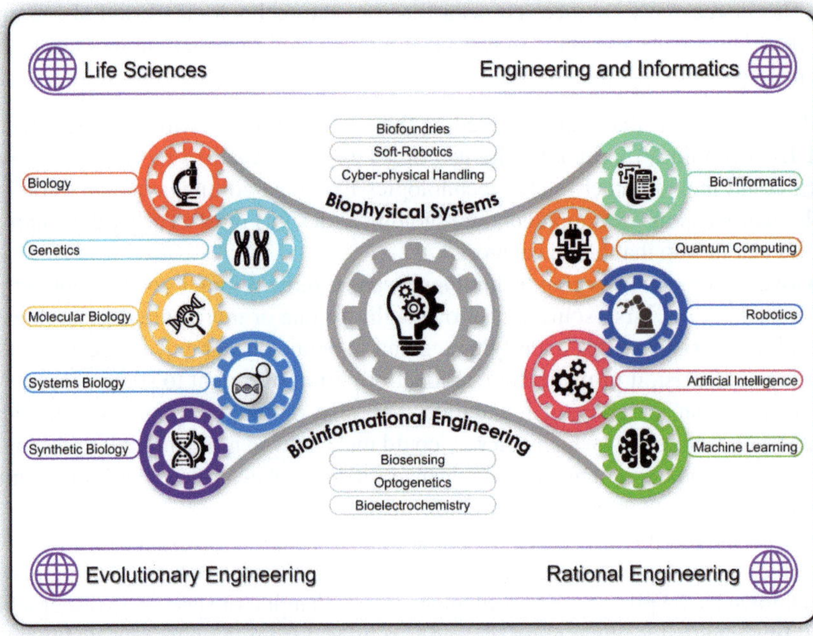

Fig. 8.1 Associated sciences and technologies enabling engineering biology [14] *Source* Used under the creative common license: http://creativecommons.org/licenses/by/4.0/

in a responsible way [15]. It is generally accepted that the Asilomar Conference on Recombinant DNA most likely paved the way for an era of expansion in DNA technology research, leading to the mapping of the complete human genome in 2003 and eventually new gene-editing technologies like CRISPR (clustered regularly interspaced short palindromic repeats).

The 1972 conference in many ways can be seen as a model for how to organize experts around a central concern of risk for an emerging technology. In fact, this very concept was revived again in January 2017, when a group of more than 100 experts, organized by the Future of Life Institute, met at Asilomar to discuss the development and future of AI (Fig. 8.2). The outcome of the conference was a list of 23 principles (known as the Asilomar AI Principles) meant to govern the responsible use and development of AI [16]:

Research Issues

1. Research Goal: The goal of AI research should be to create not undirected intelligence, but beneficial intelligence
2. Research Funding: Investments in AI should be accompanied by funding for research on ensuring its beneficial use, including thorny questions in computer science, economics, law, ethics, and social studies, such as:

Fig. 8.2 Participants at the Asilomar conference on beneficial AI [17] *Source* Used under creative common license: Deed - Attribution-ShareAlike 4.0 International – Creative Commons

a. How can we make future AI systems highly robust, so that they do what we want without malfunctioning or getting hacked?
b. How can we grow our prosperity through automation while maintaining people's resources and purpose?
c. How can we update our legal systems to be more fair and efficient, to keep pace with AI, and to manage the risks associated with AI?
d. What set of values should AI be aligned with, and what legal and ethical status should it have?

3. Science-Policy Link: There should be constructive and healthy exchange between AI researchers and policy-makers.
4. Research Culture: A culture of cooperation, trust, and transparency should be fostered among researchers and developers of AI.
5. Race Avoidance: Teams developing AI systems should actively cooperate to avoid corner-cutting on safety standards.

Ethics and Values

6. Safety: AI systems should be safe and secure throughout their operational lifetime, and verifiably so where applicable and feasible.
7. Failure Transparency: If an AI system causes harm, it should be possible to ascertain why.
8. Judicial Transparency: Any involvement by an autonomous system in judicial decision-making should provide a satisfactory explanation auditable by a competent human authority.
9. Responsibility: Designers and builders of advanced AI systems are stakeholders in the moral implications of their use, misuse, and actions, with a responsibility and opportunity to shape those implications.
10. Value Alignment: Highly autonomous AI systems should be designed so that their goals and behaviors can be assured to align with human values throughout their operation.
11. Human Values: AI systems should be designed and operated so as to be compatible with ideals of human dignity, rights, freedoms, and cultural diversity.

12. Personal Privacy: People should have the right to access, manage and control the data they generate, given AI systems' power to analyze and utilize that data.
13. Liberty and Privacy: The application of AI to personal data must not unreasonably curtail people's real or perceived liberty.
14. Shared Benefit: AI technologies should benefit and empower as many people as possible.
15. Shared Prosperity: The economic prosperity created by AI should be shared broadly, to benefit all of humanity.
16. Human Control: Humans should choose how and whether to delegate decisions to AI systems, to accomplish human-chosen objectives.
17. Non-subversion: The power conferred by control of highly advanced AI systems should respect and improve, rather than subvert, the social and civic processes on which the health of society depends.
18. 1AI Arms Race: An arms race in lethal autonomous weapons should be avoided.

Longer-Term Issues

19. Capability Caution: There being no consensus, we should avoid strong assumptions regarding upper limits on future AI capabilities.
20. Importance: Advanced AI could represent a profound change in the history of life on Earth, and should be planned for and managed with commensurate care and resources.
21. Risks: Risks posed by AI systems, especially catastrophic or existential risks, must be subject to planning and mitigation efforts commensurate with their expected impact.
22. Recursive Self-improvement: AI systems designed to recursively self-improve or self-replicate in a manner that could lead to rapidly increasing quality or quantity must be subject to strict safety and control measures.
23. Common Good: Superintelligence should only be developed in the service of widely shared ethical ideals, and for the benefit of all humanity rather than one state or organization.

The 1975 and 2017 Asilomar conferences demonstrate the concern that experts in emerging fields do have at times for the safe and responsible use of innovations in their fields. These two particular conferences demonstrate one way that experts and scientists can police or self-regulate their own actions. While these conferences were independent of one another, in both subject and time (separated by greater than 40 years), it is somewhat ironic that now, the two technologies they represent (biotechnology and AI), are merging.

8.5 Tools for Technology Oversight

8.5.1 Background

The approach to developing regulatory guidance and policy for any subject or emerging field is typically governed by not just existing science but also the socio-cultural values of a particular nation and its collective body-politic, which is usually governed by the level of perceived necessity, appetite for risk, and alignment with societal norms. These drivers often differ from region to region, both within countries and between countries, which can make achieving national and global agreement challenging, as a one-size-fits-all approach may not be feasible in all cases. However, patterns typically emerge in the regulatory front, and these historical trends can be used to anticipate the governance approaches that may be taken for new technologies.

The idea of providing guidance for technology development and use is not a new concept. Many different types of rules related to technological oversight currently exist. The challenge comes with designing or developing rules for new technologies and certainly for those combinations of new technologies that may or may not even have structured oversight. The real issue is when a technology becomes so different that it no longer fits into the standing regulatory scheme of the parent technologies (Fig. 8.3)—that identification is more art that science at times.

There are four interrelated tools of oversight in technology, and these apply to converging technologies as well: Laws, regulations, policies, and governance. Each offers a different goal with different purposes. Laws are generally a broad set of rules created by a legislative body of some kind that oversee behavior and impose penalties for the violation of that set of rules. Regulations are usually developed within

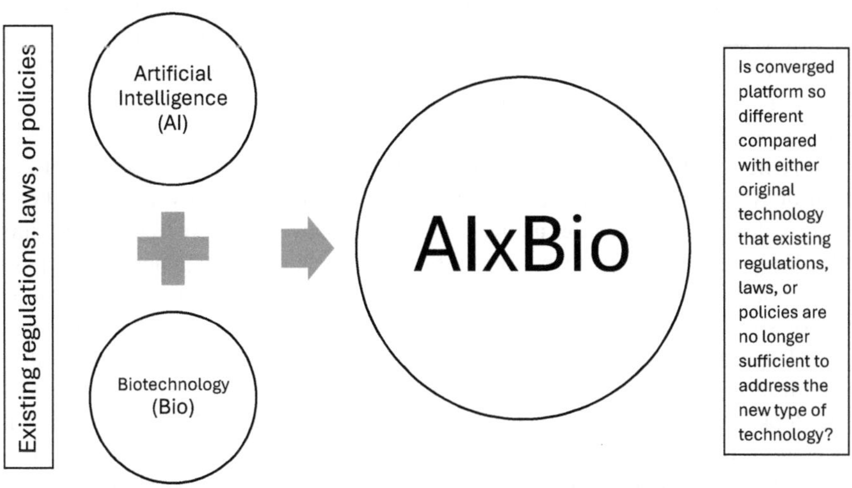

Fig. 8.3 Scheme for transitioning from governing individual technologies to governing converging technologies, specifically for AI, Biotech, and AIxBio

an executive agency and focus on enforcing specific aspects of laws or directives within that agency and can help strengthen the standardization of specific aspects of a particular target. Policies are sets of principles or plans that outline the specific actions a governing body will take in its approach toward overseeing or managing an issue or area of focus. Finally, governance refers to the overarching approach to controlling and managing resources; it goes hand-in-hand with creating and utilizing laws, regulations, and policies to achieve some desired outcome.

While all of these concepts cover different aspects, they are related and often interconnected. However, the general approach contains some shared elements. First, there is some designated authority that assumes responsibility of control for overseeing and enforcing the process. Second, there is a specific identified target of that particular authority (e.g., a group of people, an organization, an issue, a technology, a process, etc.). Next, there is a command or set of commands that the target must follow. Finally, if the target of one of these tools fails to comply, then there is some sort of consequence or set of consequence related to any such violation. These tools work in tandem towards achieving a specific goal of some kind, whether that is reducing risk, increasing productivity, promoting innovation, supporting economic development, or protecting a specific group or region.

Governance of converging technologies does not typically evolve directly to specific rules. As noted in Fig. 8.3, there may come a natural tipping point when it is sufficiently clear that existing rules no longer apply to the developing converged technology. That moment does not always come in a clearly defined manner, but if and when that moment occurs, then some group of governing bodies or interested parties will begin the process of developing those new rules. However, until that moment does occur, a vacuum often exists where a gray or fuzzy area emerges regarding what to do, if anything in the meantime. This can be a dangerous time where undefined or unclear risks and other unintended consequences can begin stirring. It can also be a time to apply lessons learned from other successful implementation strategies. While this gray area exists, many overarching and underlaying laws and codes related to basic health protections, safety, security, and privacy, still typically apply. Other general principles may also be applicable, such as responsible innovation and risk management.

Specific to AI and related technologies that include AI, lies the concept of any form of AI being trustworthy, safe, and secure. While not necessarily specific to AIxBio, this concept provides an overarching foundation of where to begin building rules for governing AI and anything in-between. Related is the general concept of the *precautionary principle*, typically more European-centric, but the idea that all new or significantly different technologies should be treated as if they are inherently dangerous or risky and need added protections and specific governance unless or until proven otherwise. This differs from a more US-based approach that typically looks at economics and necessity in terms of regulation. Neither approach is necessarily right or wrong, but each do offer their own sets of benefits and challenges (For an example of how an approach to risk for another emerging technology, nanotechnology, was evaluated in the US versus Europe, see *Bridging international approaches on nanoEHS* [18]). In general, almost all technologies impact the same

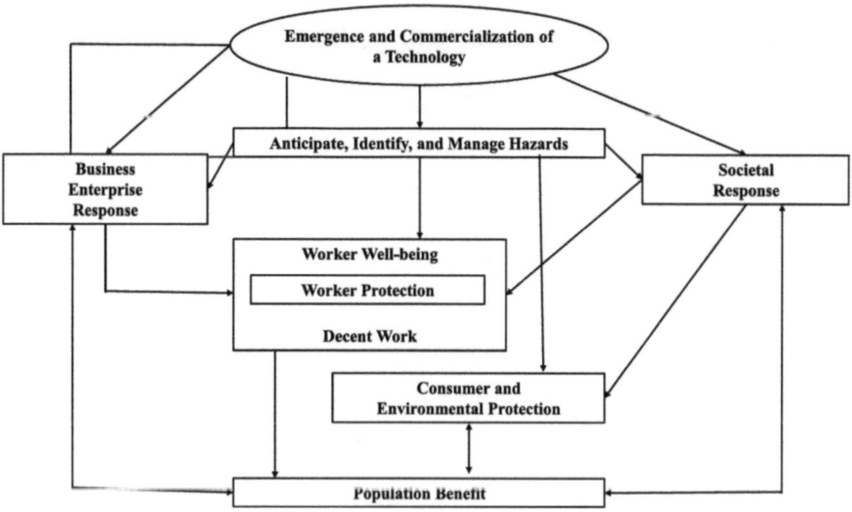

Fig. 8.4 Pathways for responsible development of emerging technologies [19]

groups eventually, even if it is not always in the same way; therefore, historical trends can be used to evaluate a potential impact from AI, biotechnology or in this case AIxBio. Herein lies the concept of responsible development that is often used as an overarching method of anticipating risks and controlling the safe and purposeful development of a technology (see Fig. 8.4).

8.6 Integration

It may be helpful to think of AIXBio as a series of events in a chain that creates a system, and regulations or laws can be added as checks-and-balances at multiple points along this system. Consider an approach from quality management known as a supplier-input-process-output-customer (SIPOC) analysis. SIPOC is used to understand a process and the requirements related to each step. In the case of AIxBio, SIPOC can be used to map out a potential process of using AI with biotechnology to create a more realistic picture of what could be the most likely scenarios for use (Fig. 8.5).

The process begins with suppliers, which in this case range from publicly funded researchers to corporations to even private citizens. These are the individuals or groups that supply data and software that are used to ultimately create some sort of output. The next component, where the actual data that set the basis for using AI-related tools comes into play, provides the first point where some sort of oversight could be implemented. Restrictions can be placed on what types of data are available for use and by whom. The next point is related and focuses on the users of that

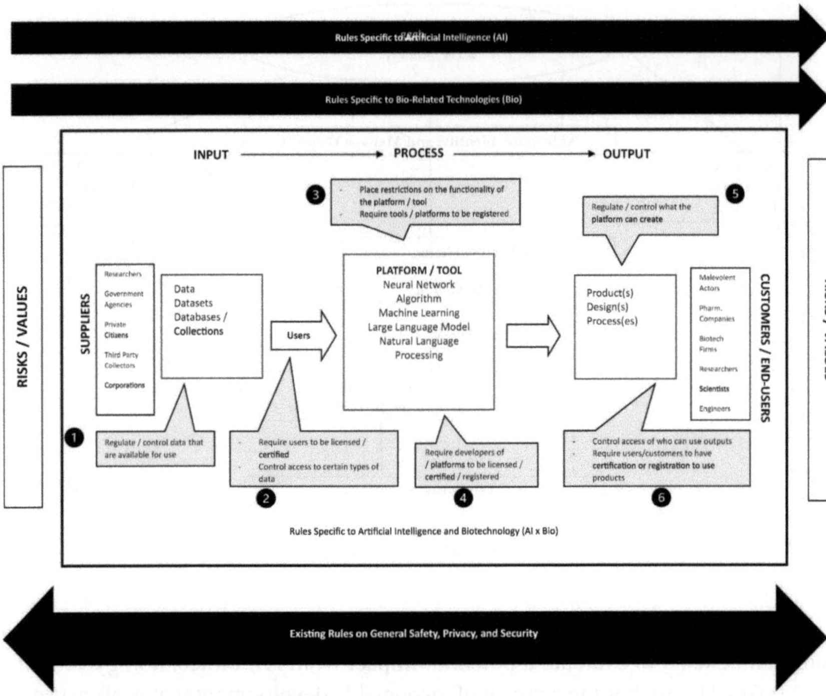

Fig. 8.5 Supplier-Input-Process-Output-Customer (SIPOC) analysis for the convergence of artificial intelligence and biotechnology (AIxBio) with checkpoints for possible integration of regulatory or legal action to minimize illegitimate actions and adverse impact

data. Some current efforts aim to require users of this data to be licensed or certified somehow so that their activities can be tracked and controlled easier and/or they will be trained to use the data properly, hence minimizing the likelihood of unethical or negligent misuse. The next area with multiple checkpoints is the actual Process area itself, where data are used in a tool to create some sort of output. It is here, where the bulk of the misuse could occur and suggestions range from placing restrictions on the functionality of platforms (i.e., requiring laws that limit what tasks certain models can actually perform) to requiring tools and platforms to be registered and listed in some sort of clearinghouse, almost like the lists used to ensure the safe use of nuclear materials. Also, at this point, the concept of licensure or certification has been brought up—this time to register developers of these platforms. Once the process has created an output, some approaches aim to limit what the AI platforms can actually be used to create, therefore potentially limiting products. For example, using AI to create biological weapons is a concern, and requiring manufacturers or developers to program AI tools to refuse to create bioweapons based on key word use or certain triggers could be a requirement of legitimate and public tools. Finally, at the last stage where customers or end-users have the opportunity to use the approaches, several approaches can be taken to possibly minimize illegitimate or nefarious use. Again, common themes such as controlling access to who can use the outputs and requiring end-users to register or have a certification that authenticates their legitimate claim to using the output.

One important thing to keep in mind is that regulations and laws to define specific uses of AIxBio may be necessary in the long run, but there are guardrails in place from general laws that still apply to most scenarios. For example, in looking at Fig. 8.5 again, existing rules on safety, privacy, and security would still apply to any use of data going into any of the models and how the outputs of those models can be used. Additionally, existing regulations for biotechnology and AI would still apply to whatever parts of the new technology integration would apply and be relevant. It is not the case that AIxBio is completely unbound; it is just possible that loopholes may show in existing laws and regulations and could need to be bolstered by more specific actions to address those gaps.

A final piece to developing any kind of regulation, but more specifically where it pertains to AIxBio, is a set of overarching values and/or risk tolerance used to define what types of behavior would be inappropriate or risky. These conditions would vary from place to place and agency to agency. However, almost universally there is a prohibition on the use of AIxBio to create biological weapons or weapons of mass destruction, especially in a civilian context or by nefarious non-state actors, so this could be a good place to start and ensure that existing guidelines are robust enough to prevent this outcome, and if not, that the necessary regulations or laws are created, aimed at this particular aspect.

8.7 Overview of Current Actions

There are very few examples of specific laws/regulations for AIxBio. The majority of existing laws pertain to digital technologies, and sometimes to AI, or to some aspect of biological technology. From here, much of the needed oversight is developing. However, there is some activity taking place on this front, and a brief overview of current actions in the global space is presented to demonstrate not only the challenges due to the paucity of developing regulations and policies, but also that development is slow and takes time to work its way through the system. There have already been AI standards and regulations proposed and promulgated in many jurisdictions (Table 8.2). However, none of these focus on AI convergence or any aspect of AI mixing with biotechnology. Little specific regulatory action exists for AIxBio, and most of what exists is governed by existing legislation or recently introduced AI laws, standards, and policies; however, several guidelines and rules that could potentially apply more towards regulating AIxBio are currently under development.

Canada has several efforts underway to bolster responsible use and development of AI that could bridge the gap with converging biotechnologies. The Pan-Canadian AI for Health (AI4H) Guiding Principles "outline shared values intended to guide Federal, Provincial, and Territorial (FPT) governments' efforts to support the responsible and ethical adoption and use of AI technologies across Canada's health systems

Table 8.2 Examples of existing and emerging AI specific regulatory approaches in select countries [20]

Country	Legislation and regulation	Standards	Principles
Brazil	• Report and proposed substitute text for draft bills 5051/2019, 21/2020 and 872/2021 (2022) • Proposed Bill 705 on the compatibility of AI use in the public sector with ESG practices (2022)	• Incorporation of international standards National standards by the Brazilian Association of Technical Norms (ABNT)	• Proposed Art. 3 of the proposed substitute text for draft bills 5051/2019, 21/2020 and 872/2021 (2022)
Canada	• Directive on Automated Decision-Making (2019) • Proposed Bill C-27, Digital Charter Implementation Act, including AI and Data Act (AIDA) (2022 1)	• Proposed CAN-ASC-6.2: Accessible and Equitable AI Systems (2023)	• Canada's Digital Charter (2019)
China	• Chinese Internet Information Service Algorithmic Recommendation Provisions (2021) • Opinion on Strengthening the Ethics and Governance of Science and Technology (2022)	• National Standards for Autonomous Vehicle Testing (2018)	• New Generation AI Ethics Specifications (2019) • New Generation AI Code of Ethics (2021) • White Paper on Trustworthy AI (2021) • Internet Information Service Algorithmic Recommendation Management Provisions (2021)
European Union	• Proposed EU AI Act (2021) • Proposed updates to the EU Product Liability Directive (2022) • Proposed AI Liability Directive (2022) • EU's Digital Services Act (2022)	• CEN/CENELEC standards for AI and related data (forthcoming)	• Ethics guidelines on AI (2018)

(continued)

Table 8.2 (continued)

Country	Legislation and regulation	Standards	Principles
Intergovernmental organizations	• Proposed Council of Europe Convention on AI, Human Rights, Democracy and the Rule of Law (2023)	• ISO 31000 Risk management (2009, 2018) • ISO/IEC 23053:2022 Framework for AI Systems Using Machine Learning (ML) (2022)	• OECD Recommendation of the Council on AI (2019) • UNESCO Recommendation on the Ethics of AI (2021)
United Kingdom	• Proposed Online Safety Bill (2022) • Proposed Data Protection and Digital Information Bill (2023)	• Algorithmic Transparency Standard (Central Digital Data Office, 2021)	• A pro-innovation approach to AI regulation (2023)
United States	• Federal Trade Commission Act, for deceptive practices from deepfakes or chatbots (1914) • Proposed Algorithmic Accountability Act (US AAA) (2022) • Executive Order 13960: Promoting the Use of Trustworthy Artificial Intelligence in the Federal Government (2020)	• National Institute of Standards and Technology (NIST) AI Risk Management Framework (2023)	• Blueprint for an AI Bill of Rights (2023)

[21]. The Artificial Intelligence and Data Acta (AIDA) was introduced into the Canadian Legislature in 2022. While the Act has not been approved, yet, several areas could potentially pertain to the convergence of AI and biotech [22].

The US Food and Drug Administration (FDA) has taken a leading approach in developing methods for developing regulatory guidance for using AI in pharmaceutical manufacturing. While nor just specific to AI, the Framework for Regulatory Advanced Manufacturing Evaluation (FRAME) Initiative focuses on four key priorities in regulating advanced technologies [23]:

1. Seek and analyze input
2. Address risks
3. Clarify expectations
4. Harmonize

The FDA has also published a series of papers on AI in pharmaceutical development and medical device integration, including:

- Artificial Intelligence and Medical Products: How CBER, CDER, CDRH, and OCP are Working Together, rev. 2025 [24]
- Using Artificial Intelligence and Machine Learning in the Development of Drug and Biological Products, 2023 [25]
- Artificial Intelligence in Drug Manufacturing, 2023 [26]
- Clinical Decision Support Software: Guidance for Industry and Food and Drug Administration Staff, 2022 [27]

In July 2023, two bills were introduced in the US Congress that focused more specifically on AI and biological threats. H.R. 4704 and S. 2399, "Artificial Intelligence and Biosecurity Risk Assessment Act," would require the Department of Health and Human Services (HHS) Assistant Secretary for Preparedness and Response to "conduct risk assessments and implement strategic initiatives or activities to address whether technical advancements in artificial intelligence, such as open-source artificial intelligence models and large language models, can be used intentionally or unintentionally to develop novel pathogens, viruses, bioweapons, or chemical weapons" [28]. S. 2346, "Strategy for Public Health Preparedness and Response to Artificial Intelligence Threats," would require the Secretary of HHS to develop a strategy for public health preparedness and response to AI threats, and for other purposes [29]. In 2021, the Defense Innovation Unit introduced "Responsible AI Guidelines in Practice," which might apply to any sort of technology involving AIxBio [30].

The *United States Government Policy for Oversight of Dual Use Research of Concern and Pathogens with Enhanced Pandemic Potential* was released in 2024 with an expected effective date of May 2025 [31]. According to the Policy, it "recognizes the rapidly evolving nature of computational biology and the increasing use of computational models and approaches, including the use of artificial intelligence, that potentially contributes to the production of dual-use biological knowledge, information, technologies, and products."

Several US states have either enacted to introduced bills related to AI in the past few years (Fig. 8.6). While many of these are more general in nature or focus more on data and privacy, two states have recognized the capacity for AI and biotech to converge and have sought out to address this issue in legislation. *The Safe and Secure Innovation for Frontier Artificial Intelligence Systems Act*, SB 1047, was introduced in California in early 2024. If passed, the bill would have included several proteins discussed earlier, including ensuring models cannot create a "chemical, biological, radiological, or nuclear weapon in a manner that results in mass casualties" and requiring developers to meet specified requirements. However, the bill was vetoed by Governor Newson. *The Ethical Artificial Intelligence Act* was also introduced in 2024 in Oklahoma but failed to pass the legislature. An amended bill, SB 611, was introduced in early 2025 and would "mandate that government, or businesses cannot use AI and biotech applications to determine who will live or die in any situation, receive medical care, or receive insurance coverage/determine the amount of the coverage," if passed [32].

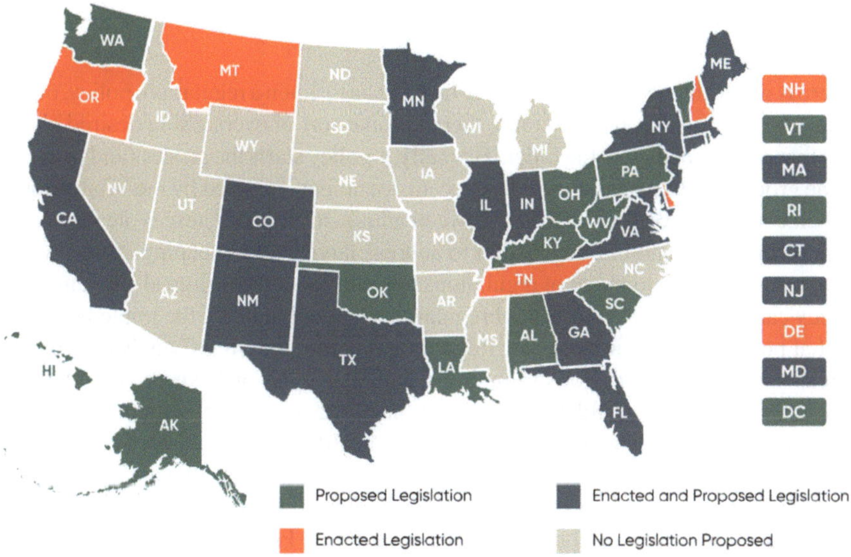

Fig. 8.6 State-by-state AI legislation (as of June 2024) [32]

The governance of AIxBio is extremely complex and still very much unfolding, and several resources provide overviews for ideas on proposed policy guidance:

- The Convergence of Artificial Intelligence and the Life Sciences, 2023—NTI [33]
- Machine Learning and gene editing at the helm of a societal evolution, 2023—RAND [34]
- Artificial Intelligence in the Biological Sciences: Uses, Safety, Security, and Oversight, 2023—Congressional Research Service [35]
- Chemical and Biological Weapons and Artificial Intelligence: Problem Analysis and US Policy Recommendations, 2024—Future of Life Institute [36]
- Developing Guardrails for AI Biodesign Tools, 2024—NTI [37]

Additionally, the OECD Artificial Intelligence Policy Observatory is an excellent resource for ongoing regulatory and policy developments in AI in each of the OECD nations [38].

Tools under development for the governance of AI, biotechnology, and AIxBio are still very much in flux, especially in a competitive global environment. However, these initiatives provide an excellent point to begin thinking about the gaps, needs, and challenges for overseeing the safe, secure, and realistic implementation of this new convergence. The global environment is extremely challenging to regulate as each country views its laws and regulations differently based on cultural norms, historical context, and current geopolitical situation. As more information is acquired, updated aspects of how and what to govern will likely be implemented into this new calculus.

8.8 Summary and Conclusions

There is generally no specific hard-and-fast rule as to when a regulation, policy, or law would be appropriate, but generally, there is some sort of overarching desired effect, and that effect can dictate the order. Typically, a policy starts the process and provides some ground rules, hence some of the recent documents released by the White House (albeit conflicting policies at times between administrations). These typically set the tone and may expand into legislation to address large-scale sweeping issues, which can prompt the development of regulations to handle specific technical challenges, or policy can drive the development of regulations that will enforce the implementation of that policy by specific agencies. Regarding AIxBio, the SIPOC analysis provides a reasonable framework for evaluating what may work and where. Regulations, especially in the US, would be most appropriate for agency-level requirements, such as what would be expected from the FDA for health and medical applications. For example, point 1 in the SIPOC analysis (Regulate / control data that are available for use) could be a good entry point for an agency regulation if it is overseeing data access or data use for what would end up being used in the Process step of SIPOC (e.g., Centers for Disease Control and Prevention or Centers for Medicare and Medicaid Services). Then again, a sweeping law could also require any and all users of data that will specifically be fed into an AI model to register with their state or some existing government agency: it depends on the ultimate intent—usually targeted action versus more societal effect.

In many cases with governance, policy, and regulation, the past can be the best predictor of future outcomes. In this case, new technologies that seemingly were once open-ended and unfettered eventually became guided by practical advice, and there is no reason to suspect this time will be any different. Biotechnology and artificial intelligence are not new technologies, but the ways in which they are being used and are planned to be used challenge the limits to our very understanding of not just science and nature but also governance, risk, and ethics. The added complexity of the convergence of these technologies brings another added wrinkle, so to speak, in terms of their regulation and predicted effects on society. While not a completely new frontier, as existing guidelines can put some limitations and guardrails on the practice of AIxBio and related technologies, there is certainly much more that needs to be done with oversight. As technology continues to push forward, the law lags behind, as is often the case. In a seemingly constant push and pull for the ages, regulatory guidance and governance seem to constantly remain in pursuit of ensuring the safe, responsible, and effective research, development, and use of emerging technologies. However, technological governance, while rooted in science and societal values, often takes the form of art much more than it does science. There is a fine line between over- and underregulating a technology. The balance lies in ensuring public health, safety, and security while also not stifling innovation, economic development, and societal benefit.

References

1. Shortliffe EH (1977) Annual meeting of the Society for Computer Medicine. In: Proceedings of the annual symposium for computer applications in medical care, vol 5, pp 66–69, Las Vegas, NV
2. Pray L. Discovery of DNA structure and function: Watson and Crick. Scitable by Nature Education. Last accessed 21 April 2025
3. University of Washington. What is Rosetta@home? What is rosetta@home? Last accessed 21 April 2025
4. Google DeepMind. AlphaFold protein structure database. Last accessed 21 April 2025
5. Rahman-Jones I (2025) Ex-Google boss Eric Schmidt fears for "bin Laden" AI scenario. BBC News. Last accessed 21 April 2025
6. Meta AI. ESM Metagenomic Atlas by Meta AI. Last accessed 21 April 2025
7. Luo R, Sun L, Xia Y, Qin T, Zhang S, Poon H, Liu T-Y (2022) BioGPT: generative pre-trained transformer for biomedical text generation and mining. Brief Bioinform 23(6)
8. EvolutionaryScale. ESM: Frontier AI for the life sciences. Last accessed 21 April 2025
9. Hayes T, Rao R, Akin H, Sofroniew NJ, Oktay D, Lin Z, Verkuil R, Tran VQ, Deaton J, Wiggert M, Badkundri R, Shafkat I, Gong J, Derry A, Molina RS, Thomas N, Khan Y, Mishra C, Kim C, Rives A et al (2025) Simulating 500 million years of evolution with a language model. Science 387(6736):850–858
10. Google DeepMind: AlphaProteo generates novel proteins for biology and health research. Last accessed 21 April 2025
11. Ali M, Shabbir K, Ali S, Mohsin M, Kumar A, Aziz M, Zubair M, Sultan HM (2024) A new era of discovery: how artificial intelligence has revolutionized the biotechnology. Nepal J Biotechnol 12(1)
12. Nobel Prize Outreach (2025) Nobel Prizes 2024. The Nobel Prize. Last accessed 21 April 2025
13. Helbing D, Ienca M (2024) Why converging technologies need converging international regulation. Ethics Inf Technol 26(15)
14. Dixon TA, Freemont PS, Johnson RA, Pretorius IS (2022) A global forum on synthetic biology: the need for international engagement. Nat Commun 13(1)
15. Berg P, Baltimore D, Brenner S, Roblin RO, Singer MF (1975) Summary statement of the Asilomar conference on recombinant DNA molecules. Proc Natl Acad Sci USA 72(6):1981–1984
16. Future of Life Institute (2017) Asilomar AI principles. Future of Life Institute. Last accessed 21 April 2025
17. Existential2021 (2017) Group photo taken at the Asilomar conference on beneficial AI in January 2017. Photograph, Pacific Grove, CA
18. Scott-Fordsmand JJ, Amorim MJ, de Garidel-Thoron C, Castranova V, Hardy B, Linkov I, Feitshans I, Nichols G, Petersen EJ, Spurgeon D, Tinkle S, Vogel U, Westerhoff P, Wiesner MR, Hendren CO (2021) Bridging international approaches on nanoEHS. Nat Nanotechnol 16(6):608–611
19. Schulte PA, Leso V, Iavicoli I (2024) Responsible development of emerging technologies. J Occup Environ Med 66(7):528–535
20. OECD. The state of implementation of the OECD AI principles four years. Last accessed 28 March 2025
21. Government of Canada (2025) Pan-Canadian AI for Health (AI4H) guiding principles. Health Canada. Last accessed 21 April 2025
22. Government of Canada (2025) The Artificial Intelligence and Data Act (AIDA)—companion document. Innovation for a Better Canada. Last accessed 21 April 2025
23. U.S. Food and Drug Administration (2025) CDER's framework for regulatory advanced manufacturing evaluation (FRAME) initiative. Last accessed 21 April 2025
24. U.S. Food and Drug Administration (2025) Artificial intelligence and medical products: how CBER, CDER, CDRH, and OCP are working together. Report

25. U.S. Food and Drug Administration (2023) Using artificial intelligence & machine learning in the development of drug & biological products. Last accessed 21 April 2025
26. U.S. Food and Drug Administration (2023) Artificial intelligence in drug manufacturing. Last accessed 21 April 2025
27. U.S. Food and Drug Administration (2022) Clinical decision support software: guidance for industry and food and drug administration staff. Clinical Decision Support Software. Last accessed 21 April 2025
28. Artificial Intelligence and Biosecurity Risk Assessment Act. Bill (2023)
29. Strategy for Public health preparedness and response to artificial intelligence threats. Bill (2023)
30. Defense Innovation Unit (2023) Responsible AI guidelines. Artificial Intelligence Portfolio. Last accessed 21 April 2025
31. National Science and Technology Council (2024) United States government policy for oversight of dual use research of concern and pathogens with enhanced pandemic potential. Washington, DC
32. BCLP (2024) US state-by-state AI legislation snapshot. Insights. Last accessed 21 April 2025
33. Carter SR, Wheeler NE, Chwalek S, Isaac CR, Yassif J (2023) The convergence of artificial intelligence and the life sciences. Last accessed 21 April 2025
34. Zakaria S, Marler T, Cabling M, Genc S, Honich A, Virdee M, Stockwell S (2023) Machine learning and gene editing at the helm of a societal evolution. Last accessed 21 April 2025
35. Kuiken T (2023) Artificial intelligence in the biological sciences: uses, safety, security, and oversight. Congressional Research Service, Washington, DC
36. Chaudhry H, Klein L (2024) Chemical and biological weapons and artificial intelligence: problem analysis and US policy recommendations. Report. Future of Life Institute. Last accessed 28 March 2025
37. Carter SR, Wheeler NE, Isaac C, Yassif JM (2024) Developing guardrails for AI biodesign tools. Report. Future of Life Institute. Last accessed 28 March 2025
38. OECD Artificial Intelligence Policy Observatory (n.d.) Policies, data and analysis for trustworthy artificial intelligence. Last accessed 21 April 2025

Chapter 9
Non-genetic Optoelectronic Biointerfaces

Pengju Li and Bozhi Tian

Abstract Non-genetic optoelectronic biointerfaces are transforming biomedical engineering in exciting new ways. These devices harness light to produce finely tuned electrical or chemical signals, allowing us to modulate and monitor biological systems without genetic alteration. This chapter explores three innovative modalities—nanostructured phototransducers, micropatterned photoelectrode arrays, and thin-film flexible photoelectrodes—that together form the foundation for next-generation therapies in neurology, cardiology, oncology, regenerative medicine, and beyond. Emphasizing their minimally invasive nature and high spatiotemporal precision, we discuss both the engineering principles and the translational potential of these biointerfaces. Finally, we examine the broader societal implications, including ethical considerations and the need for robust regulatory frameworks to guide their clinical implementation.

9.1 Introduction

The merging of nanotechnology, optoelectronics, and biomedicine is paving the way for innovative therapeutic platforms that change how clinicians engage with the human body [1–4]. Traditional methods of neuromodulation or tissue stimulation often rely on invasive electrical implants [5–7] or genetic modification (e.g., optogenetics [8–13]), which pose challenges in terms of safety, ethical concerns, and regulatory hurdles. Non-genetic optoelectronic biointerfaces, by contrast, harness light to generate localized stimuli at both the cellular and tissue levels without altering genetic makeup [1, 14–29]. This intrinsic advantage not only circumvents the limitations associated with gene therapies but also paves the way for more direct clinical translation into human healthcare applications.

P. Li · B. Tian (✉)
The University of Chicago, Chicago, IL, USA
e-mail: btian@uchicago.edu

C. L. Cummings et al. (eds.), *Biotechnology and AI: Technological Convergence and Information Hazards*, NATO Science for Peace and Security Series A: Chemistry and Biology, https://doi.org/10.1007/978-3-032-05246-9_9

This chapter provides an exploration of these non-genetic platforms [1, 14–18]. We first outline the underlying principles that allow light to be converted into biologically active signals. We then describe three important device configurations—nanostructured phototransducers, micropatterned photoelectrode arrays, and thin-film flexible photoelectrodes—detailing their design, functionality, and demonstrated applications. Next, we explore additional engineering considerations for these optoelectronic systems and their current and future clinical applications. Finally, we consider the societal, regulatory, and ethical dimensions of deploying these transformative technologies in clinical settings.

9.2 Overview of Non-genetic Optoelectronic Biointerfaces

At the heart of non-genetic optoelectronic biointerfaces lies the capacity to convert optical energy into localized electrical or electrochemical signals capable of modulating biological function [1, 16]. Unlike genetic interventions [8–12], these devices do not require the expression of exogenous proteins or the use of viral vectors, thereby eliminating associated safety and ethical concerns. Instead, they rely on the intrinsic physical properties of materials engineered at the nano- to microscale.

9.2.1 Basic Working Principles

The fundamental operating mechanism behind these biointerfaces is primarily based on photoelectrochemical or photovoltaic energy conversion [1, 16, 30], with only a limited number of systems relying on photothermal processes [14, 30–32] (Fig. 9.1). More specifically:

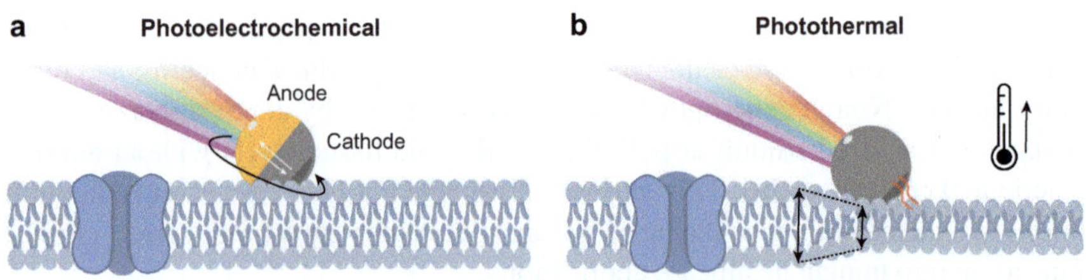

Fig. 9.1 Key optical techniques used in photostimulation applications: **a** photoelectrochemical stimulation involves the use of semiconductors to inject current into cellular membranes through a closed electrical circuit, enabling precise control of stimulation by either capacitive or redox processes. **b** Photothermal activation utilizes materials to convert light energy into heat, locally increasing temperatures and altering membrane properties. Figure panels are reproduced with permission from *Springer Nature*, 2024 [30]

Photoelectrochemical Effect. The photoelectrochemical effect (Fig. 9.1a) plays a critical role in non-genetic optoelectronic biointerfaces. When light of a specific wavelength strikes the semiconductor material within the device, it is absorbed by the material, causing electrons in the valence band to gain sufficient energy to overcome their binding and transition into the conduction band. This transition results in the formation of electron–hole pairs—electrons that are free to move and holes that act as positive charge carriers. The design of these biointerfaces is optimized so that these charges are separated efficiently; the electrons and holes are guided to different regions of the device by built-in electric fields or engineered material interfaces. This localized collection of charges generates an electrical current, which in turn can influence the activity of adjacent cells by altering their membrane potential. The precise control over charge generation and collection allows for targeted stimulation of cells, making it possible to modulate cellular functions with a high degree of spatial resolution 17–20. Ultimately, the photoelectrochemical effect enables these devices to convert optical signals into electrical stimuli, providing a powerful tool for precise bio-interfacing without the need for genetic modifications. A closely related process involves the photovoltaic mechanism, where photovoltages are generated and harnessed to stimulate cells and tissues [21, 22, 24, 26, 27, 33, 34]. The ability to directly stimulate biological systems with light-driven electrical signals provides a powerful and non-invasive approach to manipulate cellular functions with high spatial and temporal resolution.

Photothermal Effect. The photothermal effect (Fig. 9.1b) is another mechanism underpinning the operation of these innovative biointerfaces [14, 31, 32]. In this process, when electrons within the material are excited by light, not all the absorbed energy is used to generate free charge carriers; a portion of the energy is dissipated as heat through nonradiative relaxation processes. This localized heat generation occurs because excited electrons return to their ground state by releasing energy in the form of phonons, which are quantized vibrations of the lattice structure. The resulting increase in temperature, although confined to a very small region, can have significant biological implications. Localized heating can alter the physical properties of cell membranes, such as electrical capacitance, fluidity, and permeability, thereby influencing the behavior of ion channels and receptors. This thermal modulation can facilitate cellular stimulation by creating transient changes in membrane potential or by triggering heat-sensitive signaling pathways. Consequently, the photothermal effect complements the photoelectrochemical effect by offering an alternative route to modulate cell activity through controlled, localized temperature changes, thereby expanding the versatility of optoelectronic biointerfaces in biomedical applications.

9.2.2 Primary Materials

The development of optoelectronic modulation interfaces has advanced significantly, driven by innovations in materials, enabling the integration of diverse semiconductor technologies for more precise and effective therapeutic applications in medicine.

Inorganic Semiconductors. Inorganic semiconductors, particularly silicon photodiodes, are widely employed in optoelectronic biointerfaces due to their exceptional efficiency and long-term stability [1, 14–20]. Silicon-based devices benefit from decades of advancements in microfabrication technology, which have honed their performance and reliability for a myriad of applications. These photodiodes efficiently convert incident light into electrical signals with minimal energy loss, a critical factor when precise modulation of biological tissues is required. Moreover, silicon's well-characterized electronic properties and robustness under physiological conditions make it an ideal candidate for implantable devices. Its intrinsic stability ensures that once integrated into a biological system, the device can consistently perform over extended periods, reducing the need for frequent replacements or adjustments. This reliability is particularly crucial for clinical applications where long-term performance and patient safety are paramount, such as in retinal prostheses or neural stimulation devices.

Organic Semiconductors. Organic semiconductors, on the other hand, offer a complementary set of advantages that make them particularly attractive for applications where flexibility and biocompatibility are key considerations. Materials such as P3HT:PCBM not only exhibit favorable optoelectronic properties, such as efficient light absorption and charge transport, but also possess mechanical flexibility that allows the devices to conform to the complex, curved surfaces of biological tissues [35, 36]. This inherent flexibility significantly reduces mechanical mismatch between the device and the surrounding tissue, thereby minimizing irritation and inflammatory responses that could compromise device performance. Organic semiconductors have been successfully integrated into innovative devices such as the foldable wide-field epiretinal prosthesis, known as *POLYRETINA* [37]. *POLYRETINA* is designed to cover a large visual field with minimal surgical invasiveness, leveraging the pliability of organic materials to achieve a seamless interface with retinal tissue. The combination of biocompatibility, low-cost processing, and adaptability makes organic semiconductors an indispensable component in the next generation of optoelectronic biointerfaces [21, 22, 33], especially for applications that demand both high-performance and a gentle, patient-friendly approach.

9.2.3 *The Key Advantages*

Several intrinsic advantages present opportunities for non-genetic biological modulation to address a variety of medical conditions, including neurodegenerative diseases, cardiac dysrhythmias, sensory impairments, and tissue regeneration.

Precision and Control. The non-genetic approach offers unprecedented precision and control [19, 20], enabling stimulation to be localized with remarkable accuracy at the cellular or even subcellular level. By harnessing the unique properties of engineered nanomaterials and advanced optoelectronic devices, researchers can target specific cells or microdomains within tissues, minimizing the risk of activating non-targeted regions. This degree of specificity is essential for both therapeutic applications and basic biological research, as it allows for the modulation of discrete cellular circuits without causing widespread collateral effects. The precision inherent in these devices facilitates the study of complex biological processes in real time, ultimately advancing our understanding of cellular function and enabling the development of highly tailored treatments that address specific pathological conditions.

Minimally Invasive Application. Another significant advantage of this approach over traditional bulky implantable devices is its minimally invasive nature. Traditional methods often involve invasive procedures that can lead to substantial tissue damage and trigger chronic inflammatory responses. In contrast, many non-genetic optoelectronic biointerfaces are designed to be leadless and interface seamlessly with biological tissues, reducing mechanical stress and the risk of long-term adverse effects. The devices are typically fabricated to be injectable [14, 19, 20], flexible [15, 16] and biocompatible, allowing them to conform to the natural contours of organs and tissues without inducing significant trauma. This reduced invasiveness not only improves patient comfort and recovery times but also mitigates complications associated with long-term implantation, making these technologies highly attractive for chronic therapeutic applications.

Direct Clinical Translation. Perhaps most importantly, the direct clinical translation of non-genetic optoelectronic biointerfaces is greatly facilitated by the absence of genetic modifications [16]. Unlike optogenetics, which often encounter stringent regulatory hurdles and ethical debates, these approaches leverage physical and chemical mechanisms to achieve biological modulation. The reduced regulatory burden may accelerate the development process and ensure that patients could benefit from these innovative therapies sooner, without the delays typically associated with the approval and monitoring of gene-based interventions such as optogenetics.

9.3 Device Configurations of Non-genetic Optoelectronic Biointerfaces

Recent research has largely concentrated on three distinct configurations of non-genetic biointerfaces (Fig. 9.2) [38]: nanostructured phototransducers, micropatterned photoelectrode arrays, and thin-film flexible photoelectrodes, each designed to meet specific therapeutic needs and scale requirements.

9.3.1 Nanostructured Phototransducers for Cellular-Level Stimulation

Design and Materials. Nanostructured phototransducers (Figs. 9.2a and 9.3a) are engineered to interact with individual cells. These devices are often constructed from semiconductor materials—such as silicon nanowires [14, 17], gold nanoparticle-coated titania arrays [39], or polymeric nanoparticles [23, 40–42]—that exhibit strong photoelectrochemical properties. When exposed to specific wavelengths of light (e.g., near-infrared for deeper tissue penetration), these nanostructures generate highly localized electrical currents. This current can induce membrane depolarization, trigger action potentials, and ultimately modulate cellular function.

Applications for Nanostructured Phototransducers. Nanostructured phototransducers have been widely investigated for their potential in neuromodulation. For example, coaxial silicon nanowires have been shown to modulate neuronal excitability with single-cell precision. By functionalizing these nanowires with biomolecules (e.g., antibodies against specific synaptic receptors), researchers can target particular neuronal subpopulations and precisely control the spatiotemporal pattern of stimulation (Fig. 9.3b, c) [20]. This precision is invaluable for studying complex neural circuits and could lead to novel therapies for disorders such as epilepsy or Parkinson's disease.

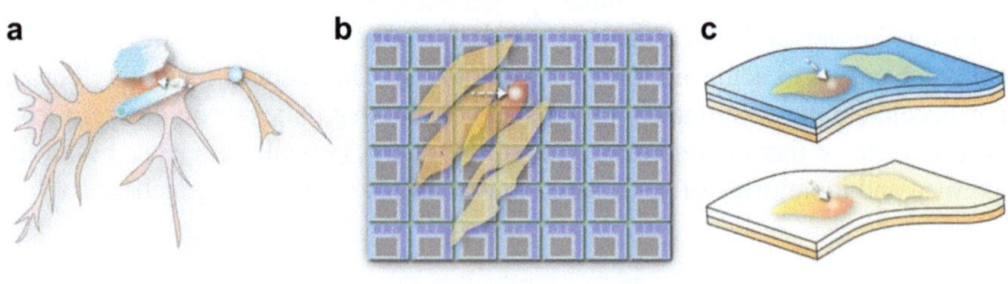

Fig. 9.2 **Three types of optoelectronic modulation biointerfaces**: **a** nanostructured phototransducers, **b** micropatterned arrays, and **c** flexible thin-film electrodes. Figure panels are reproduced with permission from *Springer Nature*, 2025 [38]

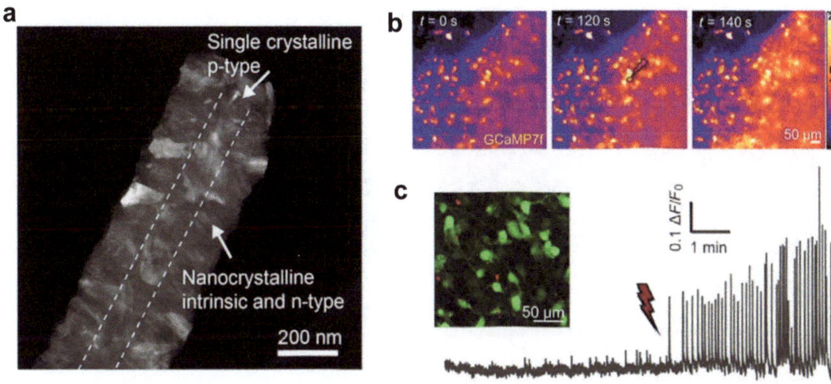

Fig. 9.3 **Nanowire-based transducers for optoelectronic modulation.** **a** Scanning transmission electron microscopy image of a Si nanowire used for neuromodulation, featuring a p-type core and intrinsic and n-type shells. **b–c** Si nanowires for spinal cord dorsal horn neuromodulation. **b** NIR light stimulation of nanowires on dorsal OSC neurons enhances network activity, with GCaMP7f fluorescence showing baseline, during stimulation, and post-stimulation changes. **c** GluA2 antibody functionalization targets nanowires to DH excitatory pathways, with confocal images of OSCs expressing GCaMP7f and functionalized nanowires, showing increased network activity upon light stimulation of nanophototransducers. (**a**) Adapted from [17], with permission from Springer Nature, 2018. (**b**) and (**c**) Adapted from [20], with permission from the American Association for the Advancement of Science, 2022

In addition, polymeric nanoparticles that mimic natural photoreceptors have demonstrated promising results in animal models of retinal degeneration [23, 40–42]. These nanoparticles convert light into electrical signals that stimulate retinal neurons, thereby restoring visual function without the need for genetic modification—a significant step toward clinical translation for patients with advanced retinitis pigmentosa.

9.3.2 *Micropatterned Photoelectrode Arrays for Tissue-Specific Interventions*

Design and Materials. Micropatterned photoelectrode arrays (Figs. 9.2b and 9.4a) are designed to stimulate tissues at a larger scale compared to nanoscale devices. These arrays are fabricated with precise geometric patterns that match the architecture of the target tissue. For instance, retinal prostheses based on silicon photodiodes [26, 27, 43–45] convert pulsed near-infrared light into electrical signals, directly stimulating retinal neurons. The micropatterned design enables high-resolution spatial control over the stimulation area, which is critical for achieving functional vision restoration.

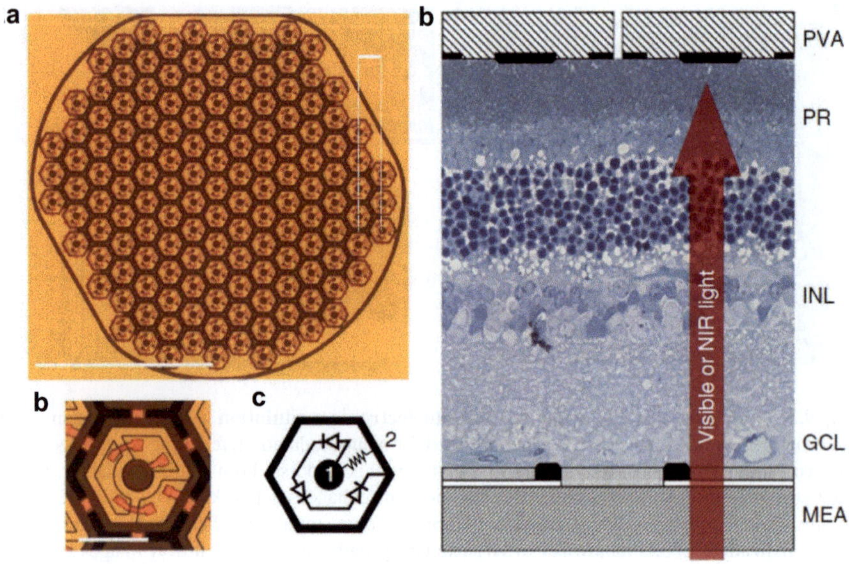

Fig. 9.4 **Micropatterned optoelectronic systems for modulation interfaces.** **a** Image of a single photovoltaic prosthesis module with 70-µm-wide pixels and 5-µm trenches in a 1-mm-wide hexagonal pattern. Scale bars: 65 µm (top right), 500 µm (bottom left). **b** Close-up of a 70-µm-wide pixel. Scale bar, 50 µm. **c** Wiring diagram for each pixel, with photodiodes in series between active and return electrodes. **d** Schematic of a rat retina with a multielectrode array (MEA) and photovoltaic array (PVA), where NIR light generates current pulses in the photovoltaic pixels to stimulate the inner nuclear layer (INL). (**a–d**) Adapted from [26], with permission from Springer Nature, 2015

Applications for Micropatterned Photoelectrode Arrays. One of the most compelling applications of micropatterned photoelectrode arrays is in the restoration of vision. In patients with degenerative retinal diseases, the loss of photoreceptor function leads to blindness. By implanting an array that replicates the function of natural photoreceptors [21, 22, 26, 27, 29, 33, 37, 43–46], it may be possible to reintroduce patterned electrical stimulation to the retina, thereby eliciting visual percepts. Preclinical studies have demonstrated that these arrays could deliver visual acuity close to that of natural vision and maintain stable functionality over extended periods (Fig. 9.4) [26].

Beyond vision restoration, these arrays have potential applications in other tissue-specific therapies. For example, patterned arrays could be designed to target specific regions of the heart or brain [15, 16], offering new avenues for treating arrhythmias or localized neurological disorders.

9.3.3 *Thin-Film Flexible Photoelectrodes for Multiscale Applications*

Design and Materials. Thin-film flexible photoelectrodes (Figs. 9.2c and 9.5a, b) combine the advantageous properties of high light absorption, biocompatibility, and mechanical flexibility. Fabricated from materials such as silicon-based membranes [16] or organic semiconductor films [47–49], these devices can conform to complex tissue geometries, making them suitable for a wide range of applications. Their flexibility is particularly critical for interfacing with soft tissues like the brain, heart, or peripheral nerves, where rigid devices might cause mechanical stress or chronic inflammation.

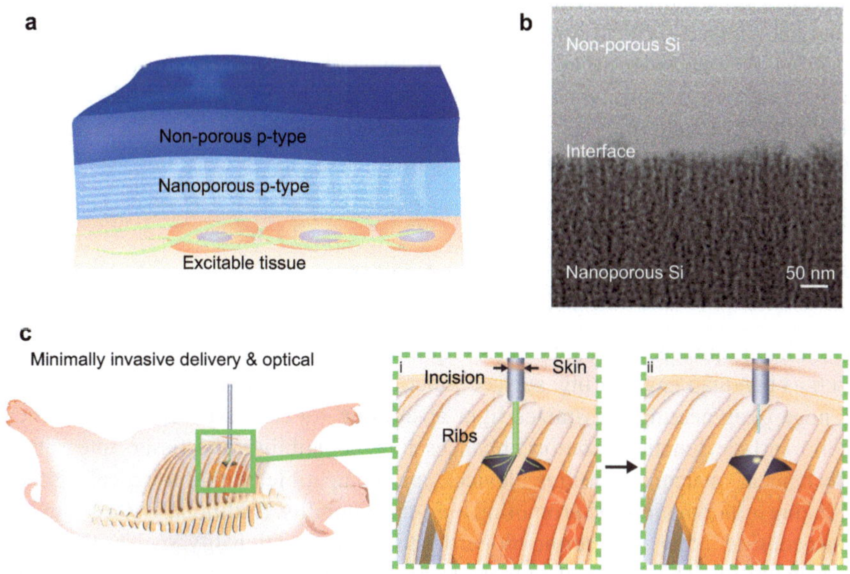

Fig. 9.5 Monolithic semiconductor membrane for optoelectronic stimulation of cells and tissues. a Schematic diagram illustrating the biointerfaces of the semiconductor membrane. The membrane is composed of a porosity-based heterojunction, which generates diode-like behavior and a photoelectrochemical response when exposed to saline environments. This configuration allows for efficient light-to-electrical conversion, enabling precise control of cellular and tissue stimulation. **b** Cross-sectional transmission electron microscopy (TEM) image of the membrane, clearly showing the heterojunction structure and highlighting the integration of the semiconductor layers. **c** Schematic diagram depicting a minimally invasive surgical procedure for optical pacing of a pig heart using the silicon-based optoelectronic membrane. This diagram illustrates how the membrane can be implanted with minimal disruption to the heart's natural function, allowing for targeted, non-invasive pacing and stimulation through optical methods. This approach showcases the potential of the semiconductor membrane in clinical applications, including cardiac rhythm management and other therapeutic interventions. (**a–b**) Adapted from [18], with permission from Springer Nature, 2022. (**c**) Adapted from [16], with permission from Springer Nature, 2024

Advances in thin-film fabrication techniques have enabled the production of devices that are not only flexible but also scalable. By engineering p-n or heterojunction configurations within these films, researchers can enhance photocurrent generation and optimize the efficiency of tissue-level stimulation [16, 47–49].

Applications for Thin-Film Flexible Photoelectrodes. Thin-film photoelectrodes have shown tremendous promise in multiscale stimulation. In cardiac applications, for instance, these devices have been used to pace heart tissue in both small animal models and large mammals (Fig. 9.5c) [15, 16]. Their ability to deliver low-intensity, highly localized light pulses ensures that stimulation is both effective and minimally invasive. In one study, monolithic devices enabled multiscale and "random-access" stimulation of cardiac tissue [16]—precisely modulating the activity of individual cells as well as larger tissue regions (Fig. 9.5c). This capability is particularly important for treating arrhythmias, where precise timing and spatial distribution of stimulation can be critical for restoring normal rhythm.

Similarly, in neuromodulation, thin-film devices have been employed to stimulate cortical regions or peripheral nerves [15, 18]. Their high biocompatibility and the possibility of integrating them with wireless power and control systems make them strong candidates for chronic implantable therapies.

9.4 Other Device Engineering Aspects

Optimizing optoelectronic interfaces requires careful attention to other key factors such as bandgap engineering, surface functionalization, and mechanical matching. Bandgap engineering maximizes light absorption for improved device efficiency, while surface functionalization enhances biocompatibility and tissue integration. Mechanical matching ensures devices are compatible with biological tissues, reducing inflammation and promoting long-term stability. Additionally, these devices must consider other engineering properties to ensure functionality, such as flexibility and durability. The future of optoelectronic bioelectronics also lies in integrating multiple therapeutic modalities, such as photostimulation and drug delivery, into multifunctional platforms for personalized, adaptive treatments.

9.4.1 Considerations for Optimal Biointerface Performance

Bandgap Engineering. Bandgap engineering is a fundamental strategy used to optimize the performance of semiconductor materials in optoelectronic devices. By carefully adjusting the bandgap—the energy difference between the valence and conduction bands—researchers can tailor the material's light absorption properties to match specific wavelengths of interest. This adjustment can be achieved through various techniques, such as alloying different semiconductor compounds

[50], doping with specific impurities, or manipulating the size and shape of nanostructures to take advantage of quantum confinement effects. In practice, a precisely engineered bandgap ensures that the semiconductor absorbs light with maximum efficiency, thereby increasing the generation of electron–hole pairs and enhancing the overall photoresponse of the device. The ability to fine-tune the bandgap is especially critical in applications like retinal prostheses or neural stimulation, where capturing low-intensity light over a specific spectral range can directly influence therapeutic outcomes. This level of control not only maximizes device efficiency but also contributes to reduced energy consumption and improved signal-to-noise ratios, ultimately paving the way for more effective and reliable bioelectronic interfaces.

Surface Functionalization. Surface functionalization plays a crucial role in bridging the gap between inorganic or organic nanostructures and the complex biological environment they are designed to interact with. By modifying the surface of these nanostructures with specific biomolecules—such as peptides, antibodies, or aptamers—researchers can enhance cellular targeting and improve the integration of the device with surrounding tissues. This process involves the deliberate attachment of functional groups that not only increase biocompatibility but also provide a means for selective binding to cell membranes or extracellular matrix components. Such targeted functionalization [20] can direct the nanostructures to specific cell types or tissue regions, thereby increasing the efficacy of stimulation or sensing. Moreover, surface modifications can help reduce non-specific interactions and minimize immune responses, ensuring that the device remains effective over long periods. This tailored approach is essential for achieving precise bio-interfacing, as it allows for more localized and controlled interactions with cells, ultimately enhancing both the performance and safety of the device in clinical applications.

Mechanical Matching. Mechanical matching is an essential design consideration when fabricating bioelectronic devices intended for long-term implantation. Achieving a mechanical match between the device and the soft tissues of the human body minimizes the risk of chronic inflammation and other adverse reactions that can arise from a mismatch in stiffness or flexibility. This involves selecting materials and designing device architectures [3, 4] that mimic the mechanical properties of biological tissues, such as their elasticity and compliance. By reducing mechanical stress at the interface, these devices can adapt more naturally to the dynamic movements of tissues, thereby promoting stable integration and reducing the potential for fibrosis or scar tissue formation. Techniques such as using ultrathin flexible membranes, incorporating soft polymers, or employing advanced fabrication methods like porosification [18] help create devices that are not only biocompatible but also durable over extended periods. In this way, mechanical matching not only improves patient comfort but also enhances the long-term functionality and reliability of optoelectronic biointerfaces, ultimately supporting better clinical outcomes. For example, the use of ultrathin flexible membranes reduces the mechanical mismatch between the implant and the host tissue, thereby limiting fibrosis and improving device longevity [3, 4]. In addition, porosification techniques can not only enhance light absorption and charge separation but also reduce the effective stiffness of the device.

9.4.2 Combining Multiple Modalities for Multifunctional Therapies

One of the most exciting prospects in the field is the integration of multiple modalities within a single platform. By combining photostimulation with additional therapeutic functions [51, 52]—such as drug delivery or real-time sensing—it is possible to create multifunctional biointerfaces that address complex clinical challenges. For instance, a device might simultaneously modulate neuronal activity while releasing anti-inflammatory agents, thereby achieving both symptomatic relief and tissue regeneration.

Advances in microfabrication and materials science are paving the way for such hybrid systems. The integration of advanced imaging techniques and machine learning algorithms for feedback control is also likely to enhance the precision and adaptability of these devices, ensuring that therapies can be tailored to individual patient needs in real time.

9.5 Potential Clinical Implications

The potential clinical applications of non-genetic optoelectronic biointerfaces could span multiple fields, offering transformative solutions in neurology, cardiology, oncology, and regenerative medicine (Fig. 9.6).

Neurology and Neuroscience. The ability to modulate neural circuits with single-cell precision offers new opportunities for treating neurological disorders [15, 20]. Conditions such as epilepsy, Parkinson's disease, and chronic pain could be

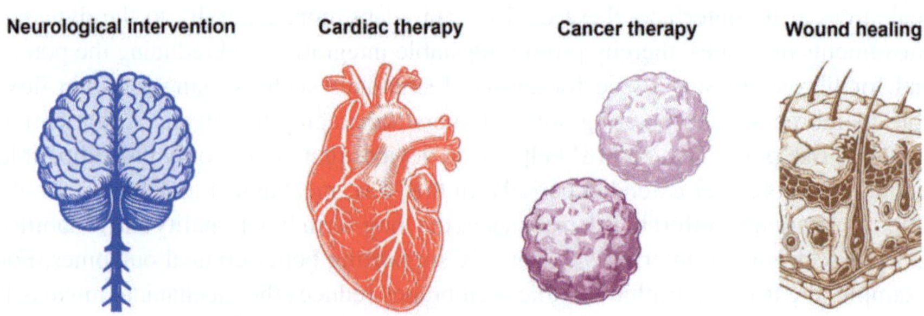

Fig. 9.6 Non-genetic optoelectronic biointerfaces could revolutionize clinical applications across neurology, cardiology, oncology, and regenerative medicine

addressed by selectively targeting aberrant neural pathways. Moreover, these devices provide a powerful research tool for mapping neural networks and understanding the fundamental mechanisms of brain function.

Cardiology. In cardiology, optoelectronic biointerfaces have been shown to provide precise control over cardiac pacing [16, 18]. The ability to stimulate specific regions of heart tissue with minimal invasiveness could revolutionize the management of arrhythmias and other cardiac dysrhythmias. Furthermore, the scalability of thin-film devices makes them well suited for integration into implantable systems that operate without bulky batteries or leads.

Oncology. Optoelectronic devices also hold potential in oncology, particularly in enhancing the precision of photodynamic therapy (PDT). By generating reactive oxygen species (ROS) in a controlled manner, these devices could selectively target cancerous cells while sparing healthy tissue. In combination with imaging modalities, they may also enable real-time monitoring of treatment efficacy.

Regenerative Medicine and Wound Healing. The capacity to deliver targeted stimulation to promote tissue regeneration is another promising application. For example, localized photostimulation may accelerate wound healing by promoting angiogenesis and collagen synthesis. Optoelectronic biointerfaces could also be used to stimulate stem cells or modulate the microenvironment of damaged tissues, thereby enhancing the repair and regeneration processes.

9.6 Societal Implications and Ethical Considerations

One of the most compelling advantages of non-genetic optoelectronic biointerfaces is their direct applicability to human subjects. Because they do not require genetic modification, these devices bypass many of the ethical and regulatory concerns that have traditionally hampered the clinical translation of gene-based therapies. This advantage is particularly significant in areas such as neuromodulation and retinal prostheses, where the risks associated with viral vectors or exogenous gene expression can be substantial.

As with any emerging technology, the widespread adoption of non-genetic optoelectronic biointerfaces comes with important societal and ethical implications.

9.6.1 Risks of Misuse and Information Hazards

The very attributes that make these devices powerful—precise control and remote modulation capabilities—also present potential risks if misused. There is a need for strict guidelines and oversight to prevent unauthorized or unethical applications. For example, the possibility of remotely altering neural activity raises questions about privacy, autonomy, and security. Addressing these issues will require not only technical safeguards but also comprehensive ethical frameworks.

9.6.2 Establishing Robust Regulatory Frameworks

Given the rapid pace of technological innovation in the field of non-genetic optoelectronic biointerfaces, establishing robust and adaptive regulatory frameworks is essential [1, 14–18]. These frameworks must address not only the technical and clinical aspects of device deployment but also the broader ethical and social implications. Regulatory agencies and stakeholders should collaborate to develop guidelines that ensure:

Safety and Efficacy. Ensuring the safety and efficacy of non-genetic optoelectronic biointerfaces is paramount and requires rigorous preclinical and clinical testing. In preclinical studies, these devices are subjected to extensive in vitro and in vivo experiments to confirm that they function as intended, without eliciting harmful side effects. Researchers systematically evaluate parameters such as biocompatibility, device stability, and long-term performance under simulated physiological conditions. Animal models are often used to study the device's integration with biological tissues, its potential for causing adverse immune responses, and its overall therapeutic benefits. Following successful preclinical assessments, clinical trials are designed to verify both safety and efficacy in human subjects. These trials include multiple phases—ranging from initial safety studies with a small cohort to larger-scale trials that assess therapeutic outcomes across diverse patient populations. Regulatory agencies, such as the FDA or EMA, require comprehensive data demonstrating that the devices not only meet stringent safety standards but also provide measurable clinical benefits. This rigorous testing process ensures that when these devices are finally deployed in a clinical setting, they are both effective and pose minimal risks to patients, thereby building trust among clinicians and patients alike.

Privacy and Security. Privacy and security represent critical pillars in the development and implementation of these advanced biointerfaces [53, 54]. As these devices often interface directly with the human body and collect sensitive biological data, robust measures must be in place to protect patient information from unauthorized access or manipulation. The design of these systems incorporates advanced encryption techniques to safeguard data during transmission and storage, ensuring that personal health information remains confidential. Additionally, secure authentication protocols are implemented to restrict access to authorized personnel only, thereby preventing potential cyberattacks or unauthorized control over the implanted devices. Manufacturers and regulatory bodies work together to develop standards and guidelines that address cybersecurity risks, continuously updating them in response to emerging threats. The integration of secure data management systems is crucial not only for protecting patient privacy but also for ensuring the reliable operation of the devices. By preventing unauthorized access, these security measures also help to maintain the integrity of the therapeutic interventions, thereby preserving both patient safety and the overall trust in bioelectronic medical technologies.

9.6.3 Mitigating Risks of Misuse and Information Hazards

The potential for misuse—whether intentional or inadvertent—is a critical consideration when it comes to non-genetic optoelectronic biointerfaces. While these technologies offer significant therapeutic advantages, such as remotely modulating neural or cardiac activity, they also pose risks if used improperly. To address these concerns and ensure the safe deployment of optoelectronic biointerfaces, it is essential to implement several protective measures:

Develop Robust Security Protocols. To safeguard these advanced biointerfaces, it is essential to develop robust security protocols that ensure the integrity and confidentiality of communications between the implanted device and its control systems. This involves implementing sophisticated encryption methods that render transmitted data unreadable to unauthorized parties, effectively mitigating risks associated with data breaches and cyberattacks. In addition to encryption, multi-factor authentication and secure key management systems should be integrated to further bolster the device's defenses against hacking attempts. The development process must also incorporate regular vulnerability assessments and security audits to identify potential weaknesses and update protocols accordingly. By designing security measures that are both resilient and adaptive, developers can ensure that the system remains secure even in the face of evolving threats, thereby protecting sensitive patient information and maintaining the reliability of therapeutic interventions.

Establish Ethical Use Guidelines. Alongside technical safeguards, establishing clear ethical use guidelines is critical to ensure that these bioelectronic technologies are applied in a manner that respects individual rights and societal values. This requires creating comprehensive policies that define the acceptable use of such devices, detailing both the permissible applications and the limitations to prevent misuse. Ethical guidelines should be developed through collaborative efforts involving ethicists, legal experts, clinicians, and patient advocacy groups, ensuring that diverse perspectives are considered. These guidelines should also outline strict penalties and accountability measures for any instances of abuse or unauthorized manipulation, serving as both a deterrent and a framework for corrective action.

Implement Continuous Monitoring. Continuous monitoring is an important strategy in maintaining the safe and effective operation of these devices throughout their lifecycle. By incorporating real-time monitoring systems, developers can establish a dynamic feedback loop that continuously tracks the optoelectronic device performance and detects any deviations from expected operational parameters. This involves integrating advanced sensors and telemetry systems that provide constant data on key metrics such as the optoelectronic output, temperature, and interface integrity. When anomalies are detected—whether due to device malfunction, environmental interference, or potential security breaches—automated alert mechanisms can trigger immediate countermeasures or initiate remote diagnostics. This proactive approach ensures that any issues are identified and addressed promptly, safeguarding both the device's functionality and the wellbeing of the patient.

While current systems often rely on external control elements—such as telemetry and diagnostic interfaces—to monitor key parameters, there is a strong drive toward developing integrated, stand-alone solutions that reduce external vulnerabilities and mitigate risks of misuse. In the case of an inorganic retinal prosthesis, future advancements in materials science, sensor miniaturization, and on-board processing could enable these devices to incorporate internal control systems. This integration would allow the prosthesis to self-regulate and operate autonomously, maintaining the essential safety and reliability functions without relying on external monitoring. However, achieving a fully autonomous, implanted system is a formidable challenge. It necessitates that all critical parameters—such as light-to-electrical conversion efficiency, temperature stability, and interface integrity—be continuously and reliably monitored internally. Although progress in this field is promising, substantial research and development will be required before such self-contained prostheses can effectively mitigate the risks of misuse and information hazards.

By implementing these safeguards, the non-genetic optoelectronic biointerface community can ensure that these technologies are used responsibly, securely, and ethically, maximizing their therapeutic potential while minimizing associated risks.

9.7 Concluding Remarks

The convergence of nanotechnology, optoelectronics, and biomedicine is ushering in a new era of medical practice—one in which the modulation and monitoring of biological systems can be achieved with unparalleled precision and minimal invasiveness. Non-genetic optoelectronic biointerfaces offer a transformative alternative to traditional genetic approaches, enabling direct application in human subjects while bypassing many of the ethical, regulatory, and technical challenges associated with gene-based therapies [1, 14–18].

By harnessing the power of light to generate localized electrical or electrochemical stimuli, these devices open up new possibilities for treating a wide range of conditions—from neurological disorders and cardiac arrhythmias to retinal degeneration and chronic wounds. The three device configurations discussed in this chapter—nanostructured phototransducers, micropatterned photoelectrode arrays, and thin-film flexible photoelectrodes—each contribute unique capabilities that, when integrated, offer a comprehensive toolkit for precision medicine.

The potential applications of these technologies are vast. In neurology, they promise to enable precise modulation of neural circuits with high spatiotemporal resolution. In cardiology, they offer a novel means of pacing and rhythm control that minimizes the invasiveness of current electrical implants. In sensory restoration, particularly for vision, these devices can replicate or even surpass the function of natural photoreceptors, restoring meaningful visual function in patients with degenerative retinal diseases. Beyond these clinical applications, the integration of multifunctional devices that combine photostimulation with drug delivery or real-time monitoring promises to further expand the therapeutic potential of optoelectronic biointerfaces.

However, as with any rapidly advancing technology, the path forward is not without challenges. Key issues remain in ensuring long-term biocompatibility, optimizing the efficiency of energy conversion processes, and addressing the ethical and societal implications of deploying these devices. In particular, the risks associated with remote modulation of biological systems must be carefully managed through rigorous regulatory oversight and ethical stewardship.

Looking to the future, the continued integration of advanced imaging, machine learning, and wireless power technologies will undoubtedly enhance the performance and adaptability of non-genetic optoelectronic biointerfaces. Innovations such as trainable and adaptive interfaces may soon lead to systems that not only stimulate and monitor biological tissues but also learn and evolve in response to their environment, further personalizing and optimizing therapeutic outcomes [2, 55].

In conclusion, non-genetic optoelectronic biointerfaces stand at the forefront of a biotechnological revolution. Their ability to deliver precise, minimally invasive, and clinically translatable solutions positions them as key enablers of the next generation of medical therapies. With thoughtful governance, interdisciplinary collaboration, and a commitment to ethical innovation, these technologies have the potential to transform medical practice—delivering improved outcomes, enhancing patient quality of life, and fundamentally redefining the interface between technology and biology.

References

1. Jiang YW, Tian BZ (2018) Inorganic semiconductor biointerfaces. Nat Rev Mater 3:473–490. https://doi.org/10.1038/s41578-018-0062-3
2. Li PJ, Kim S, Tian BZ (2024) Beyond 25 years of biomedical innovation in nano-bioelectronics. Device 2. https://doi.org/10.1016/j.device.2024.100401
3. Rogers JA, Lagally MG, Nuzzo RG (2011) Synthesis, assembly and applications of semiconductor nanomembranes. Nature 477:45–53. https://doi.org/10.1038/nature10381
4. Song EM, Li JH, Won SM, Bai WB, Rogers JA (2020) Materials for flexible bioelectronic systems as chronic neural interfaces. Nat Mater 19:590–603. https://doi.org/10.1038/s41563-020-0679-7
5. Bensmaia SJ, Miller LE (2014) Restoring sensorimotor function through intracortical interfaces: progress and looming challenges. Nat Rev Neurosci 15:313–325. https://doi.org/10.1038/nrn3724
6. Cingolani E, Goldhaber JI, Marbán E (2018) Next-generation pacemakers: from small devices to biological pacemakers. Nat Rev Cardiol 15:139–150. https://doi.org/10.1038/nrcardio.2017.165

7. Shen K, Chen O, Edmunds JL, Piech DK, Maharbiz MM (2023) Translational opportunities and challenges of invasive electrodes for neural interfaces. Nat Biomed Eng. https://doi.org/10.1038/s41551-023-01021-5
8. Bansal A, Shikha S, Zhang Y (2023) Towards translational optogenetics. Nat Biomed Eng 7:349–369. https://doi.org/10.1038/s41551-021-00829-3
9. Emiliani V, Entcheva E, Hedrich R, Hegemann P, Konrad KR, Lüscher C, Mahn M, Pan ZH, Sims RR, Vierock J, Yizhar O (2022) Optogenetics for light control of biological systems. Nat Rev Methods Primers 2. https://doi.org/10.1038/s43586-022-00136-4
10. Entcheva E, Kay MW (2021) Cardiac optogenetics: a decade of enlightenment. Nat Rev Cardiol 18:349–367. https://doi.org/10.1038/s41569-020-00478-0
11. Kim CK, Adhikari A, Deisseroth K (2017) Integration of optogenetics with complementary methodologies in systems neuroscience. Nat Rev Neurosci 18:222–235. https://doi.org/10.1038/nrn.2017.15
12. Song CC, Knöpfel T (2016) Optogenetics enlightens neuroscience drug discovery. Nat Rev Drug Discov 15:97–109. https://doi.org/10.1038/nrd.2015.15
13. Berryessa CM, Cho MK (2013) In: Chakravarti A, Green E (eds) Annual review of genomics and human genetics, vol 14, pp 515–534
14. Jiang YW, Carvalho-de-Souza JL, Wong RCS, Luo ZQ, Isheim D, Zuo XB, Nicholls AW, Jung IW, Yue JP, Liu DJ, Wang YC, De Andrade V, Xiao XH, Navrazhnykh L, Weiss DE, Wu XY, Seidman DN, Bezanilla F, Tian BZ (2016) Heterogeneous silicon mesostructures for lipid-supported bioelectric interfaces. Nat Mater 15:1023–1030. https://doi.org/10.1038/nmat4673
15. Jiang YW, Li XJ, Liu B, Yi J, Fang Y, Shi FY, Gao X, Sudzilovsky E, Parameswaran R, Koehler K, Nair V, Yue JP, Guo KH, Tsai HM, Freyermuth G, Wong RCS, Kao CM, Chen CT, Nicholls AW, Wu XY, Shepherd GMG, Tian BZ (2018) Rational design of silicon structures for optically controlled multiscale biointerfaces. Nat Biomed Eng 2:508–521. https://doi.org/10.1038/s41551-018-0230-1
16. Li PJ, Zhang J, Hayashi H, Yue JP, Li W, Yang CW, Sun CX, Shi JY, Huberman-Shlaes J, Hibino N, Tian BZ (2024) Monolithic silicon for high spatiotemporal translational photostimulation. Nature. https://doi.org/10.1038/s41586-41024-07016-41589
17. Parameswaran R, Carvalho-de-Souza JL, Jiang Y, Burke MJ, Zimmerman JF, Koehler K, Phillips AW, Yi J, Adams EJ, Bezanilla F, Tian B (2018) Photoelectrochemical modulation of neuronal activity with free-standing coaxial silicon nanowires. Nat Nanotechnol 13:260–266. https://doi.org/10.1038/s41565-017-0041-7
18. Prominski A, Shi JY, Li PJ, Yue JP, Lin YL, Park J, Tian BZ, Rotenberg MY (2022) Porosity-based heterojunctions enable leadless optoelectronic modulation of tissues. Nat Mater 21:647–+. https://doi.org/10.1038/s41563-022-01249-7
19. Rotenberg MY, Yamamoto N, Schaumann EN, Matino L, Santoro F, Tian BZ (2019) Living myofibroblast-silicon composites for probing electrical coupling in cardiac systems. Proc Natl Acad Sci USA 116:22531–22539. https://doi.org/10.1073/pnas.1913651116
20. Thalhammer A, Fontanini M, Shi J, Scaini D, Recupero L, Evtushenko A, Fu Y, Pavagada S, Bistrovic-Popov A, Fruk L, Tian B, Ballerini L (2022) Distributed interfacing by nanoscale photodiodes enables single-neuron light activation and sensory enhancement in 3D spinal explants. Sci Adv 8:eabp9257. https://doi.org/10.1126/sciadv.abp9257
21. Chenais NAL, Leccardi M, Ghezzi D (2019) Capacitive-like photovoltaic epiretinal stimulation enhances and narrows the network-mediated activity of retinal ganglion cells by recruiting the lateral inhibitory network. J Neural Eng 16. https://doi.org/10.1088/1741-2552/ab3913
22. Chenais NAL, Leccardi M, Ghezzi D (2021) Photovoltaic retinal prosthesis restores high-resolution responses to single-pixel stimulation in blind retinas. Commun Mater 2. https://doi.org/10.1038/s43246-021-00133-2
23. Francia S, Shmal D, Di Marco S, Chiaravalli G, Maya-Vetencourt JF, Mantero G, Michetti C, Cupini S, Manfredi G, DiFrancesco ML, Rocchi A, Perotto S, Attanasio M, Sacco R, Bisti S, Mete M, Pertile G, Lanzani G, Colombo E, Benfenati F (2022) Light-induced charge generation in polymeric nanoparticles restores vision in advanced-stage retinitis pigmentosa rats. Nat Commun 13:3677. https://doi.org/10.1038/s41467-022-31368-3

24. Han M, Jalali HB, Yildiz E, Qureshi MH, Sahin A, Nizamoglu S (2021) Photovoltaic neurointerface based on aluminum antimonide nanocrystals. Commun Mater 2. https://doi.org/10.1038/s43246-021-00123-4
25. Han M, Karatum O, Nizamoglu S (2022) Optoelectronic neural interfaces based on quantum dots. ACS Appl Mater Interfaces 14:20468–20490. https://doi.org/10.1021/acsami.1c25009
26. Lorach H, Goetz G, Smith R, Lei X, Mandel Y, Kamins T, Mathieson K, Huie P, Harris J, Sher A, Palanker D (2015) Photovoltaic restoration of sight with high visual acuity. Nat Med 21:476–482. https://doi.org/10.1038/nm.3851
27. Mathieson K, Loudin J, Goetz G, Huie P, Wang L, Kamins TI, Galambos L, Smith R, Harris JS, Sher A, Palanker D (2012) Photovoltaic retinal prosthesis with high pixel density. Nat Photonics 6:391–397. https://doi.org/10.1038/nphoton.2012.104
28. Savchenko A, Cherkas V, Liu C, Braun GB, Kleschevnikov A, Miller YI, Molokanova E (2018) Graphene biointerfaces for optical stimulation of cells. Sci Adv 4. https://doi.org/10.1126/sciadv.aat0351
29. Vijjapu MT, Fouda ME, Agambayev A, Kang CH, Lin CH, Ooi BS, He JH, Eltawil AM, Salama KN (2022) A flexible capacitive photoreceptor for the biomimetic retina. Light Sci Appl 11:3. https://doi.org/10.1038/s41377-021-00686-4
30. Lanzani G, Chiaravalli G, Colombo E, Manfredi G, Di Marco S, Vurro V, Benfenati F (2024) Nanotechnology for vision restoration. Nat Rev Bioeng 2:829–848. https://doi.org/10.1038/s44222-024-00210-4
31. Bassetto CAZ Jr, Pfeffermann J, Yadav R, Strassgschwandtner S, Glasnov T, Bezanilla F, Pohl P (2024) Photolipid excitation triggers depolarizing optocapacitive currents and action potentials. Nat Commun 15. https://doi.org/10.1038/s41467-024-45403-y
32. Carvalho-de-Souza JL, Treger JS, Dang B, Kent SBH, Pepperberg DR, Bezanilla F (2015) Photosensitivity of neurons enabled by cell-targeted gold nanoparticles. Neuron 86:207–217. https://doi.org/10.1016/j.neuron.2015.02.033
33. Ferlauto L, Airaghi Leccardi MJI, Chenais NAL, Gilliéron SCA, Vagni P, Bevilacqua M, Wolfensberger TJ, Sivula K, Ghezzi D (2018) Design and validation of a foldable and photovoltaic wide-field epiretinal prosthesis. Nat Commun 9:992. https://doi.org/10.1038/s41467-018-03386-7
34. Wang BY, Chen ZC, Bhuckory M, Huang T, Shin A, Zuckerman V, Ho E, Rosenfeld E, Galambos L, Kamins T, Mathieson K, Palanker D (2022) Electronic photoreceptors enable prosthetic visual acuity matching the natural resolution in rats. Nat Commun 13:6627. https://doi.org/10.1038/s41467-022-34353-y
35. Kroon R, Lenes M, Hummelen JC, Blom PWM, De Boer B (2008) Small bandgap polymers for organic solar cells (polymer material development in the last 5 years). Polym Rev 48:531–582. https://doi.org/10.1080/15583720802231833
36. Laquai F, Andrienko D, Mauer R, Blom PWM (2015) Charge carrier transport and photogeneration in P3HT:PCBM photovoltaic blends. Macromol Rapid Commun 36:1001–1025. https://doi.org/10.1002/marc.201500047
37. Vagni P, Leccardi M, Vila CH, Zollinger EG, Sherafatipour G, Wolfensberger TJ, Ghezzi D (2022) POLYRETINA restores light responses in vivo in blind Gottingen minipigs. Nat Commun 13. https://doi.org/10.1038/s41467-022-31180-z
38. Majmudar A, Kim S-H, Li PJ, Tian BZ (2024) Perspectives on non-genetic optoelectronic modulation biointerfaces for advancing healthcare. Med-X 2. https://doi.org/10.1007/s44258-024-00030-6
39. Yang R, Zhao P, Wang L, Feng C, Peng C, Wang Z, Zhang Y, Shen M, Shi K, Weng S, Dong C, Zeng F, Zhang T, Chen X, Wang S, Wang Y, Luo Y, Chen Q, Chen Y, Jiang C, Jia S, Yu Z, Liu J, Wang F, Jiang S, Xu W, Li L, Wang G, Mo X, Zheng G, Chen A, Zhou X, Jiang C, Yuan Y, Yan B, Zhang J (2023) Assessment of visual function in blind mice and monkeys with subretinally implanted nanowire arrays as artificial photoreceptors. Nat Biomed Eng. https://doi.org/10.1038/s41551-023-01137-8
40. Francia S, Shmal D, Di Marco S, Chiaravalli G, Maya-Vetencourt JF, Mantero G, Michetti C, Cupini S, Manfredi G, DiFrancesco ML, Rocchi A, Perotto S, Attanasio M, Sacco R,

Bisti S, Mete M, Pertile G, Lanzani G, Colombo E, Benfenati F (2022) Light-induced charge generation in polymeric nanoparticles restores vision in advanced-stage retinitis pigmentosa rats. Nat Commun 13. https://doi.org/10.1038/s41467-022-31368-3
41. Benfenati F, Lanzani G (2021) Clinical translation of nanoparticles for neural stimulation. Nat Rev Mater 6:1–4. https://doi.org/10.1038/s41578-020-00267-8
42. Maya-Vetencourt JF, Manfredi G, Mete M, Colombo E, Bramini M, Di Marco S, Shmal D, Mantero G, Dipalo M, Rocchi A, DiFrancesco ML, Papaleo ED, Russo A, Barsotti J, Eleftheriou C, Di Maria F, Cossu V, Piazza F, Emionite L, Ticconi F, Marini C, Sambuceti G, Pertile G, Lanzani G, Benfenati F (2020) Subretinally injected semiconducting polymer nanoparticles rescue vision in a rat model of retinal dystrophy. Nat Nanotechnol 15:698–+. https://doi.org/10.1038/s41565-020-0696-3
43. Palanker D, Le Mer Y, Mohand-Said S, Sahel JA (2022) Simultaneous perception of prosthetic and natural vision in AMD patients. Nat Commun 13. https://doi.org/10.1038/s41467-022-28125-x
44. Tan BY, Li HK, Zhuo YM, Han L, Mupparapu R, Nanni D, Barathi VA, Palanker D, Schmetterer L, Ling T (2024) Light-evoked deformations in rod photoreceptors, pigment epithelium and subretinal space revealed by prolonged and multilayered optoretinography. Nat Commun 15. https://doi.org/10.1038/s41467-024-49014-5
45. Wang BY, Chen ZC, Bhuckory M, Huang T, Shin A, Zuckerman V, Ho E, Rosenfeld E, Galambos L, Kamins T, Mathieson K, Palanker D (2022) Electronic photoreceptors enable prosthetic visual acuity matching the natural resolution in rats. Nat Commun 13. https://doi.org/10.1038/s41467-022-34353-y
46. Long Z, Qiu X, Chan CLJ, Sun Z, Yuan Z, Poddar S, Zhang Y, Ding Y, Gu L, Zhou Y, Tang W, Srivastava AK, Yu C, Zou X, Shen G, Fan Z (2023) A neuromorphic bionic eye with filter-free color vision using hemispherical perovskite nanowire array retina. Nat Commun 14:1972. https://doi.org/10.1038/s41467-023-37581-y
47. Ghezzi D, Antognazza MR, Dal Maschio M, Lanzarini E, Benfenati F, Lanzani G (2011) A hybrid bioorganic interface for neuronal photoactivation. Nat Commun 2. https://doi.org/10.1038/ncomms1164
48. Ghezzi D, Antognazza MR, Maccarone R, Bellani S, Lanzarini E, Martino N, Mete M, Pertile G, Bisti S, Lanzani G, Benfenati F (2013) A polymer optoelectronic interface restores light sensitivity in blind rat retinas. Nat Photonics 7:400–406. https://doi.org/10.1038/nphoton.2013.34
49. Maya-Vetencourt JF, Ghezzi D, Antognazza MR, Colombo E, Mete M, Feyen P, Desii A, Buschiazzo A, Di Paolo M, Di Marco S, Ticconi F, Emionite L, Shmal D, Marini C, Donelli I, Freddi G, Maccarone R, Bisti S, Sambuceti G, Pertile G, Lanzani G, Benfenati F (2017) A fully organic retinal prosthesis restores vision in a rat model of degenerative blindness. Nat Mater 16:681–+. https://doi.org/10.1038/nmat4874
50. Ning CZ, Dou LT, Yang PD (2017) Bandgap engineering in semiconductor alloy nanomaterials with widely tunable compositions. Nat Rev Mater 2. https://doi.org/10.1038/natrevmats.2017.70
51. Elnathan R, Barbato MG, Guo XF, Mariano A, Wang ZX, Santoro F, Shi P, Voelcker NH, Xie X, Young JL, Zhao YL, Zhao WT, Chiappini C (2022) Biointerface design for vertical nanoprobes. Nat Rev Mater 7:953–973. https://doi.org/10.1038/s41578-022-00464-7
52. Sunwoo SH, Han SI, Park CS, Kim JH, Georgiou JS, Lee SP, Kim DH, Hyeon T (2024) Soft bioelectronics for the management of cardiovascular diseases. Nat Rev Bioeng 2:8–24. https://doi.org/10.1038/s44222-023-00102-z
53. Qiu H, Qiu MK, Liu MQ, Memmi G (2020) Secure health data sharing for medical cyber-physical systems for the Healthcare 4.0. IEEE J Biomed Health Inform 24:2499–2505. https://doi.org/10.1109/jbhi.2020.2973467

54. Zhang B, Rahmatullah B, Wang SL, Zaidan AA, Zaidan BB, Liu PH (2023) A review of research on medical image confidentiality related technology coherent taxonomy, motivations, open challenges and recommendations. Multimed Tools Appl 82:21867–21906. https://doi.org/10.1007/s11042-020-09629-4
55. Li PJ, Kim S, Tian BZ (2022) Nanoenabled trainable systems: from biointerfaces to biomimetics. ACS Nano 16:19651–19664. https://doi.org/10.1021/acsnano.2c08042

Chapter 10
AlphaFold and the Protein-Folding Problem

Hallam Stevens

Abstract In November 2020, organizers of a key scientific meeting on protein structure prediction announced that artificial intelligence (AI) had solved the protein-coding problem, a scientific challenge which had stood for over fifty years. The scientific and popular news media quickly lauded this breakthrough. Since 2020, however, a less visible (and less celebratory) set of debates has continued about what exactly AlphaFold has achieved and how important those achievements are for science and medicine. These discussions provide some important clues as to as how the usage of AI in science may play out. They suggest that AI will play an important role in redefining what counts as an interesting or important problem; as problems become tractable for AI, scientists are likely to reorient their work towards other questions and problems that they perceive may not be AI-solvable. Moreover, a shift towards AI would entail a further shift towards data-driven forms of science that have developed clear metrics for assessing solutions; it also may cause scientists to re-imagine and re-configure problems in these terms.

Keywords AlphaFold · Protein-folding problem · DeepMind · Google · AI · Science

10.1 Introduction

In November 2020, at the 14th iteration of the Community Wide Experiment on the Critical Assessment of Techniques for Protein Structure Prediction (CASP14), the organizers declared that artificial intelligence (AI) had triumphed. "Building on the work of hundreds of researchers across the globe," the CASP14 press release read, "an AI program called AlphaFold, created by London-based AI lab DeepMind, has proved capable of determining the shape of many proteins. It has done so to a level of accuracy comparable to that achieved with expensive and time-consuming lab

H. Stevens (✉)
James Cook University, Townsville, QLD, Australia
e-mail: hallam.stevens@jcu.edu.au

C. L. Cummings et al. (eds.), *Biotechnology and AI: Technological Convergence and Information Hazards*, NATO Science for Peace and Security Series A: Chemistry and Biology, https://doi.org/10.1007/978-3-032-05246-9_10

experiments" [1]. A scientific challenge which had stood for over fifty years appeared to have been definitively solved by AI. The scientific and popular news media quickly lauded this breakthrough. Noting the potential for protein structure prediction to be used in understanding disease and in drug design, scientists described the advance in hyperbolic terms: "It will change everything," the evolutionary biologist Andrei Lupas told Nature [2].

DeepMind attracted global attention again in 2024. First, in May, when it announced the release of AlphaFold 3, an updated and improved version of the software that predicted the structure of other biomolecules (apart from proteins) and also made progress in predicting interactions between biomolecules [3]. Second, in October, it was announced that the Nobel Prize in Chemistry had been awarded to John Jumper and Demis Hassabis for their work on developing AlphaFold [4]. Both these announcements were accompanied by bold pronouncements about the software's potential to speed up disease research, bring new treatments to market, combat plastic pollution, and increase the global food supply (amongst others) [5].

Since 2020, however, a less visible (and less celebratory) set of debates has continued about what exactly AlphaFold has achieved and how important those achievements are for science and medicine. Writing in a blog in December 2020, for example, Mohammed AlQuraishi, a CASP14 participant and a systems biologist now at Columbia University, argued that the word "solution" was potentially confusing or misleading: "Part of me wishes that DeepMind never used the term, that the CASP14 organizers didn't, and that I didn't, because all it has done is raise temperatures on all sides without advancing the discussion one bit." As AlQuraishi goes on to point out, the protein-folding problem can be posed (and solved) in different ways. And, even with some versions of these problems solved, there still may be a lot of scientific work that needs to be done before such solutions result in any meaningful applications [6].

My aim in this essay is not to adjudicate on the successes or shortcomings of AlphaFold or its competitors. Rather, my aim is to examine the debates around AlphaFold to better understand how scientists perceive, react to, and incorporate AI into their work. How does AI shift expectations around scientific discovery?

This short essay cannot answer these questions in general. However, the AlphaFold episode provides some important clues as to as how the usage of AI in science may play out. First, it suggests that AI will play an important role in redefining what counts as an interesting or important problem; as problems become tractable for AI, scientists are likely to reorient their work towards other questions and problems that they perceive may not be AI-solvable. Second, I argue here that a shift towards AI would entail a further shift towards data-driven forms of science that have developed clear metrics for assessing solutions; it also may cause scientists to re-imagine and re-configure problems in these terms.

The first part of this essay examines the history of the protein-folding problem and reviews attempts to address it computationally. Next, the essay reviews the ensuing debates about AlphaFold and its solution to the protein-folding problem. The final parts of the essay discuss what these debates reveal about the uses of AI in science and what they reveal about the uses and perceptions of AI amongst scientists.

10.2 The Protein-Folding Problem

Proteins—the critical molecular machinery of cells—consist of long, linear chains of amino acids. This was first demonstrated definitively in 1949 by Frederick Sanger who experimentally determined the amino acid sequence (or primary structure) of the insulin protein [7]. In the early 1950s, Linus Pauling and others suggested how these linear protein chains could fold into local sub-structures (secondary structures) and then into complex globular structures [8]. It is this "tertiary structure," or three-dimensional shape, that gives proteins their function. The first efforts to precisely determine these three-dimensional structures utilized the techniques of X-ray crystallography. This involved the painstaking work of purifying and crystallizing proteins, bombarding them with X-rays, and interpreting the resulting diffraction patterns. The first protein structures to be "solved" in this way were myoglobin (by John Kendrew) and hemoglobin (by Max Perutz), both in 1958 [9, 10].

Knowing both the primary and tertiary structure of a protein, however, did not reveal how a protein went from one state to the other: that is, it did not provide a mechanism for how a protein folded from a linear chain of amino acids into a globular three-dimensional shape. In 1973, the biochemist Christian Anfinsen proposed the "thermodynamic hypothesis," arguing that the tertiary structure of a protein is determined completely by the amino acid sequence [11]. In a given environment, Anfinsen argued, the linear sequence would spontaneously fold to reach a stable state with the minimum free energy. Importantly, this hypothesis suggested that it should be able to predict—using physical principles—the final folded shape of a protein from its linear sequence.

This prediction problem—known as the "protein-folding problem"—was taken up in the 1970s and 1980s by biochemists, physicists, and mathematicians who made a variety of attempts to attack the problem, often using computers. These approaches included advanced statistical mechanics and molecular dynamics simulations [12–15]. These efforts aimed to model—either in specific cases or in general—how forces acted on an amino acid chain to cause it to evolve through a series of conformations and into its final shape. De novo methods aimed to make such predictions entirely from first principles or by searching various possible conformations. Other so-called "template-based methods" utilized known structures in order to reduce the search space. Because of the large number of possible conformations that a protein can adopt, the computational cost of computing protein folding is extremely high. This meant that either such methods were limited to very short proteins or otherwise had to make a range of simplifying assumptions about the protein and its environment.

Alongside these more theoretical endeavours, experimental work continued to elucidate the three-dimensional structure of more proteins (and with greater resolution). Efforts to collect and make these structures available to researchers began in the late 1960s. In 1971, the Protein Data Bank (PDB) was announced as a joint effort between the Cambridge Crystallographic Data Centre and Brookhaven National Laboratory, providing online and offline access to protein structure data [16]. The number of known protein sequences in this database had only grown to 69 by 1980

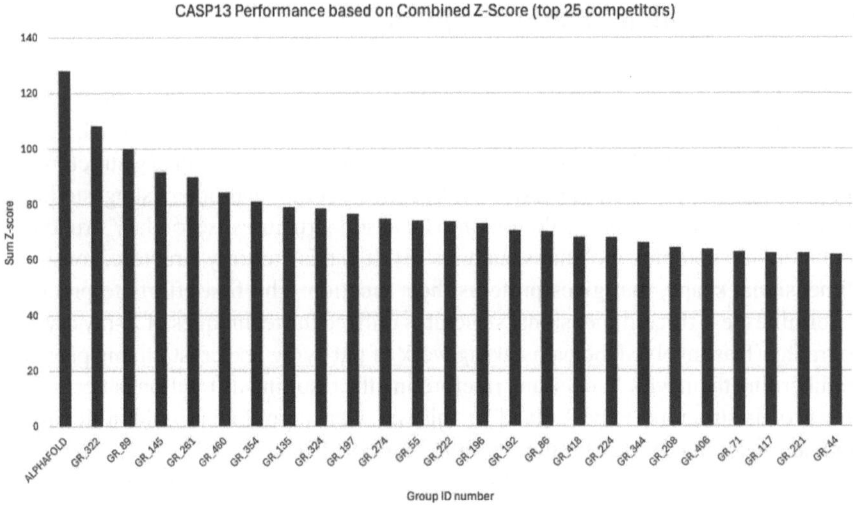

Fig. 10.1 Combined Z-scores of top twenty-five competitors at CASP13. *Image source* https://www.rcsb.org/stats/growth/growth-released-structures

and about 500 by 1990 (see Fig. 10.1). The emergence of new methods for finding protein structures—including nuclear magnetic resonance—helped to more rapidly grow this number from the 1990s onwards (to 13,500 by 2000; 70,000 in 2010; and 230,000 in 2025).[1]

Beginning in 1994, John Moult at the University of Maryland, along with other protein biochemists, began to organize a worldwide competition to test the success of protein structure prediction efforts [17]. CASP became a forum for protein researchers, often armed with different theoretical and software tools to compete with one another to correctly predict protein structures. Competitors would be provided with the primary structure of proteins for which the tertiary structure had recently been determined but had not yet been announced by the PDB. Evaluation methods would include visual inspection by the judges, but also the Global Distance Test Total Score (GDS_TS), a means for scoring the deviation between the predicted and the experimentally determined positions of the atoms in the structure.

The CASP competition fostered a flourishing and diverse set of approaches, pursued by different research groups, to attacking the protein folding problem. These included both de novo and template-based methods. Due to the significant computational resources required to simulate the molecular dynamics of protein-folding, several research groups pioneered distributed computing projects that farmed out computations across large numbers of individual machines. These included the Folding@home project (initiated in 2000 by Vijay Pande) and the Rosetta@home project (initiated in 2002 by David Baker), both of which utilized the idle CPU and

[1] Technically these figures are for the number of "structures" in the database, a category which is wider than merely proteins. The number of proteins in 2025 is approximately 170,000.

GPU time of home users' computers. These crowdsourced efforts were also taken up by the Human Proteome Folding Project (which used Rosetta software) and by FoldIt (released 2008, also initiated by David Baker), which used Rosetta as the basis for an online game that could be played by non-specialists [18, 19].

This abbreviated history of the protein-folding problem shows how, from the 1970s, scientific questions about protein folding became focused on the very specific problem of how to predict tertiary structure from amino acid sequence. Moreover, CASP framed this problem as a competition which could be evaluated with very specific criteria, namely the GDS_TS.

10.3 Enter AI

Many of the approaches mentioned in the previous section rely on Markov State Models (MSMs). MSMs simulate protein folding as a series of discrete states (each representing a different fold) and a set of probabilities for transitioning between these states. The discreteness of an MSM simulation allows it to be computationally tractable (since the model is working with a finite number of states), while the stochastic nature of the model allows it to explore a range of possible folds [20]. Such MSMs are built using data from molecular dynamics simulations. The model uses this to make predictions about the likelihood of transitioning from a given state to another. Although Markov models are statistical or stochastics models, they are a key tool of machine learning. Markov models have been used in pattern recognition, natural language processing, and other classic AI problem areas.

In the 2010s, a team at Google's DeepMind in London began to apply a different machine learning technique—the convolutional neural network rather than a Markov model—to the protein-folding problem. Although the resulting AlphaFold software was a new tool, many of the assumptions and methods utilized by the DeepMind team remained the same as previous approaches to the problem. In particular, the DeepMind team, like other approaches, utilized existing data from "solved" proteins in the PDB. Further, the basis of its techniques relied on covariance (that is, comparing variation in homologous sequences from different species—a well-established approach to structure prediction) to predict the distances between amino acids. Further, AlphaFold also borrowed the energy function from Rosetta in order to generate the final folded structure prediction [21, 22].

AlphaFold was first entered into the CASP13 competition in 2018, demonstrating remarkable success. Figure 10.2 representing the distance scores across the competition, illustrates the significance of its achievement compared to other competitors). As described in the introduction to this chapter, subsequent versions of AlphaFold—released for CASP14 in 2020, CASP15 in 2022, and CASP16 in 2024—have generated similarly impressive results.

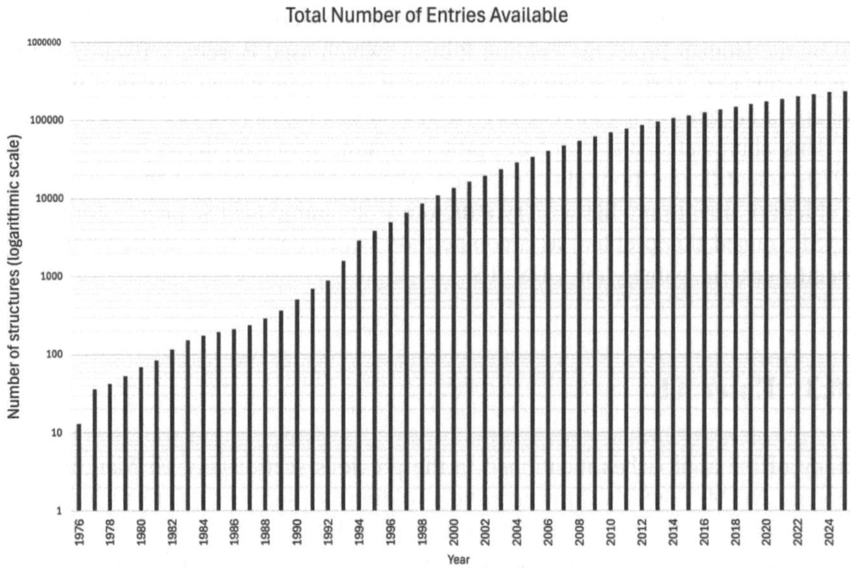

Fig. 10.2 Total number of entries in the Protein Data Base from 1976 to 2025. *Image Source* https://predictioncenter.org/casp13/zscores_final.cgi?formula=gdt_ts

10.4 Debates

The reaction of the scientific community to these achievements has been a mix of celebration and caution. This section draws on both published literature as well as blogs and other informal writings by scientists, describing the various responses to AlphaFold's successes. I focus here on the discussions at the end of 2020 when AlphaFold's achievements were being most widely celebrated. Some of the criticisms levelled against AlphaFold at that time were addressed by later versions (e.g. its ability to predict the structure of protein complexes). It is nevertheless important to include these critiques here in order to understand how scientists and commentators were reacting to the software. I divide the responses to AlphaFold into three categories: re-thinking and re-defining the "protein-folding problem"; questioning the use and applicability of training data; and questioning AlphaFold's ability to tackle the important problems.

10.4.1 What Is the Problem Anyway?

One response to the announcement that AlphaFold had "solved the protein-folding problem" was to draw into question what that problem consisted of in the first place. AlQuraishi, who in 2020 did believe the problem to be solved, argued that the protein-folding problem could be understood in at least five different ways. Did the problem consist of predicting the structure of a single protein domain, or the structure of a whole complex? Does this include giving insight into the folding pathway? Moreover, what information should be utilized to make such predictions? Just the protein sequence itself? Or are homologous sequences (or structures) allowed too? Does the solution have to solve the problem for every protein? And with what level of accuracy [6]?

Others argued that AlphaFold missed the point entirely. "The challenge," Philip Ball wrote in Chemistry World, "was to find those rules and predict the folding path." Here Ball is referring to the set of physical or thermodynamic principles that govern protein folding. Finding correlations between sequences and structures based on the PDB, Ball argues, "doesn't so much solve the protein-folding problem as evade it" [23]. This is analogous to the kinds of criticisms that Noam Chomsky has made of Large Language Models (LLMs)—namely, that they are merely generating correlations based on large amounts of data and do not provide anything that might even look like an understanding of human language [24].

Linked to this, some biologists argue that AlphaFold, because it is a deep neural network, is a "black box." Although it may have "learned" how proteins fold itself, it does nothing for our understanding of how this happens—in some sense, it cannot teach us what is has learned about protein folding. Nevertheless, the emergence of AlphaFold has caused structural biologists to carefully rethink what the "protein-folding problem" really is and to re-articulate it as a series of related but distinct problems.

10.4.2 The Training Data Problem

A second set of concerns about AlphaFold's predictions relates to the data on which the model was trained. Like other protein structure prediction programs, AlphaFold relied on the roughly 170,000 experimentally determined protein structures in the PDB. By some estimates, there are at least 20 million proteins across all species of life on earth; only 5% of human protein structures have been experimentally determined. Like other AI models, AlphaFold, some biologists argued, can only be as good as the data it is trained on. If that data is not representative of the larger universe of proteins, then it is likely to have serious shortcomings when predicting novel proteins. As Ball points out, proteins in the PDB are also likely to be "biased towards structures that are relatively easy to solve" or that are susceptible to existing methods of structure determination (like NMR or X-ray crystallography)—"there could be a systematic repertoire of protein structures that we just don't know about" [23].

Moreover, the success of AlphaFold will hardly obviate the need for experiments, commentators reminded their colleagues. At lot more experimental confirmation work would need to be done before biologists could be confident that any given predicted structures were actually correct. "While the majority of predicted structures were within experimental resolution, one can't tell a priori which are and which aren't—so you need experiments to check" [23]. Others note that very small changes in protein structure can be critical for determining a protein's function (or disfunction). This suggests that experimental determination of protein structure—not relying on AI prediction—is going to remain the gold standard for understanding function.

10.4.3 Structure Versus Function

Many structural biologists have also pushed back on the usefulness of AlphaFold for addressing important biological questions and problems. Yes, AlphaFold may be able to predict structures, they argue, but structures don't necessarily reveal the function of proteins (which is critical for applications in medicine and drug design). Downplaying the importance of structure in general, commentators argued that proteins may be modified and crosslinked after folding, meaning that identifying the initial static structure only solves a small part of the problem of understanding protein behaviour.

Others quickly pointed to more ambitious problems—what about protein complexes? Could AlphaFold solve them? What about proteins interacting with other molecules? (a problem that AlphaFold has now addressed in detail). As systems biologist Nazim Bouatta put it, "Structure is not the final story; we want to understand much more." This "much more" includes protein dynamics, how proteins behave in cellular environments, and why proteins fold incorrectly [25]. AlQuraishi argues that AlphaFold will lead to more work, not less, in subfields including variant prediction, protein design, comparative structuromics, synthetic biology, and drug discovery [6]. Although AlphaFold may ultimately help with (or even "solve") some of these problems, in the shorter term, its successes mainly serve to open up and accelerate new lines of research.

Of course, the importance of all these problems was known before AlphaFold's successes. This suggests that as AlphaFold's progress has become clearer, biologists have already moved towards thinking about and addressing other, related, problems that have not (or perhaps cannot) been attacked by AI. To some extent this reflects nature of scientific and academic competition. As AlQuraishi puts it, AlphaFold makes it "very risky for an academic group to swim in these waters" [6]. Significantly, though, this also reflects a more general phenomenon in the history of AI research in which humans have tended to identify the hard and important problems with things AI systems cannot do. When it turns out AI systems can do those things, what is hard and important gets redefined [26].

10.5 The Uses of AI in Science

The purpose here is not to adjudicate or add to these internal debates about AlphaFold and the protein-folding problem. Rather, my aim is to explore what this episode (the dramatic arrival of AI into this field) can tell us about the impacts AI is likely to have more generally and how scientists are likely to respond. It is notable here that the protein-folding problem has been one of the few instances to date in which AI has had such a dramatic effect—this, in itself, can also provide some valuable insight into the kinds of problems AI is likely to be able to address and what the consequences of this "problem selection" may be.

10.5.1 Continuities and Discontinuities

The presentation of AlphaFold in the media (and in some scientific coverage) suggested that its appearance marked a radical break with previous efforts and approaches. Although the application of a convolutional neural network to the problem was new, in many important respects, the AI approach built on and developed existing techniques and approaches within the field. Most notably, AlphaFold was trained on the same data from the PDB that other structural biologists had been using, built up through decades of experimental work. Even the application of AI techniques to the problem was not entirely new—as we have seen, Markov Models were an important part of protein folding prediction for decades before AlphaFold. And, of course, the AlphaFold team, while adding important insights of their own, used techniques for predicting protein structure that had been developed by others. The debates described above reveal these important continuities between before and after AlphaFold.

These continuities are important because they show that AI had pronounced success on a well-developed problem by building on past approaches. The team at DeepMind that built AlphaFold were well-versed in this science and designed their model accordingly. Coming in the wake of DeepMind's success in beating human Go champions (with AlphaGo), AlphaFold is often portrayed as "beating" humans to the solution of the protein-folding problem. But, with these continuities in mind, it may be more appropriate to understand AlphaFold's successes as collaborative and cooperative with humans. More speculatively, these continuities might also suggest that AI may find less success in tackling completely new or unexplored problems.

10.5.2 Data-Driven Science

It is also notable that AI has had its impact in a space that was already heavily "data-driven." Much of the work that has been done in the last two decades on the protein-folding problem has been based on the known sequences and structures in the PDB (and, to some extent, other databases). The emergence of the "big data" approaches in the first decade of the twenty-first century has already had a significant impact on science. Advocates of "data-driven" or "hypothesis free" science argued that discovery would be transformed by these approaches and more theoretical approaches rendered redundant [27].

The intercession of AI (which makes extensive use of training data) into this space suggests a further extension of data-driven approaches. Data-driven approaches already privilege correlation over causation and largely eschew questions about mechanism. Moreover, as I have argued elsewhere, the sheer scale of the data involved often makes it impractical or impossible for researchers to "see" patterns without extensive computational intervention [28]. In other words, pre-AI data-driven approaches already instantiated a kind of black box.

All this indicates another way in which AI approaches are continuous with, rather than an abrupt break from, existing science. Moreover, it suggests something about the type of problem where AI is likely to meet successful application—domains where significant data already exists or can be readily obtained.[2] Considering AI as a data-driven approach may also suggest something about the types of "solutions" we might expect from it—like data-driven science, it is likely to provide results that are correlative and agnostic with respect to underlying principles and mechanisms. This is consistent with some of the criticisms of AlphaFold.

10.5.3 Scoring and Measurement

AlphaFold's success came in a domain where an existing competition (CASP) had already established well-defined criteria for success. Specifically, the GDS_TS metric provided a clear and numerical criterion of success for the AI model. Such metrics are compatible with the way in which AI models are trained. Having an unambiguous scoring mechanism allows clear feedback to be provided to the model so that it can refine its neural network and maximize its score.

This is not to say that AI could not be successful in addressing other types of problems in the future. However, it may be necessary for humans to continue to provide clear metrics for judging solutions. Where criteria are fuzzier or where this is disagreement or ambiguity with respect to the aims, the application of AI may be less straightforward.

To put this another way, the development of AI approaches in science may necessitate (or go hand-in-hand with) the development of such metrics and competitions. Agreement on the "success" of AI, then, will also depend on building consensus with respect to such metrics.

---

[2] Since science is in the business of collecting data, in practice this may not appear very restrictive. However, it may indicate that AI will be limited to problems where particular quantities and qualities of data are available. For example, data that are highly representative across a domain.

10.5.4 Academic and Commercial Science

One aspect of the debates about AlphaFold not mentioned above involved concerns about where AlphaFold was developed. In particular, the fact that it was developed in a corporate lab (rather than in a strictly academic space) added to the "surprise" of AlphaFold's achievement. In addition, the fact that AlphaFold was developed by an Internet/search company, rather than, say, a pharmaceutical company (which might have a more direct interest in the problem of drug discovery) was also taken to be significant. Engaging in what AlQuraishi calls, "armchair sociology," he speculates that not only did DeepMind have significant computational resources not available to others, but also that:

> DeepMind is organized very differently from academic groups. There are minimal administrative requirements, freeing up time to do research. This research is done by professionals working at the same job for years and who have achieved mastery of at least one discipline. Contrast this with academic labs where there is constant turnover of students and postdocs [6].

These concerns were also paired with criticism of DeepMind for sharing insufficient details of their work in scientific talks and publications [29].

DeepMind (and subsequent work by others) has now made many of AlphaFold's insights and methods widely available. The commercialization of science is not particularly new.[3] Moreover, as AI develops, there may be a democratization of resources as the cost of training models comes down. Even with these caveats, there nevertheless may be important implications here for where AI-driven science takes place. Leading-edge work may continue to be driven by access to large computing resources and dedicated, mission-oriented scientific teams.

10.5.5 Science Versus Engineering

Related to the question of academic versus commercial science, AlphaFold also has provoked a set of questions about the boundaries of scientific work. Commentators have suggested that AlphaFold has made the protein-folding problem one of "engineering":

> [The protein-folding problem] is now an engineering rather than a scientific problem. That doesn't mean it's any less important or any less hard. I consider myself an engineer as much as I am a scientist, maybe even more so. But there is an important distinction between scientific and engineering problems that is pertinent to the discussion here: engineering problems can be exceedingly difficult, and require the marshaling of inordinate resources, but competent domain experts know the pieces that need to fall into place to solve them. Whether or not a problem can be solved is usually not a question in engineering tasks. It's a question whether it can be done given the resources available. In scientific problems on the other hand, we

[3] There are many historical examples of the influences of commercial enterprises in science, including the emergence of corporate laboratories such as Bell Labs and the biotechnology industry.

> don't know how to go from A to Z. It may turn out that, with a little bit of careful thought (as appears to be the case for AF2), one discovers the solution and it's easy to go from A to Z, easier than many engineering problems [6].

Although AlQuraishi states that he is not trying to diminish engineering here, he is certainly trying to draw a boundary around the kinds of problems he considers most interesting and important. The reaction tells us much about possible responses to uses of AI in science, as scientists attempt to redefine the kinds of problems and questions that are "scientific" away from what AI is capable of. In other words, at least in the perception of scientists, the kinds of problems that are amenable to AI will require different sorts of approaches (engineering), different ways of thinking, different sets of tools, and different kinds of institutions and organizations (large computing power, commercially driven, mission-oriented teams).

10.6 Conclusions

The computer has been a powerful tool that has shaped the kinds of questions that scientists can ask answer [30]. But the computer is not a single tool, but rather a range of software and hardware that have shaped scientific practice in various ways. In the long term, AI should likely be thought of in a similar way—as a multiple, rather than singular. AlphaFold—and even deep learning approaches more generally—are likely to be only one version of what AI looks like and can do. These different AIs will have different epistemological, institutional, and sociological effects.

We are now only in the early days of these developments. Nevertheless, it may still be possible to draw some preliminary conclusions from the AlphaFold episode. The hype around AI (and around AlphaFold in particular) has tended to suggest that AI approaches mark a radical break with past scientific practices and that AI will increasingly "replace" or make redundant human scientific thinking and working. A more careful look at what AlphaFold has actually done (and the responses to it) suggests quite the opposite. First, while AlphaFold made some breakthroughs that were shocking in the moment, they built on much existing science. A clear-eyed reflection on DeepMind's achievements would suggest that it made (important) contributions to an existing problem within an existing competition framework; it drew on knowledge and expertise already existing in the field and utilized some techniques that had already been applied to similar problems. This is not to diminish or undermine the novelty of AlphaFold's achievements; a similar characterization could be applied to almost any scientific breakthrough. However, highlighting the continuity of AlphaFold's approaches with existing science suggests that AI is likely to continue to build on existing problems and approaches in the same way. It is less likely AI will merely solve problems "out of the blue."

Second, human scientists also appeared to quickly adapt to these changes, focusing their work on problems that could not immediately be solved by AI. Although some reporting on AlphaFold's achievements suggested shock and surprise, the immediate community of scientists working on the protein-folding problem largely appeared to take the results in stride. They turned their attention rapidly to similar or related problems or sub-problems and to what had not been done by AlphaFold. If AI had solved one problem, it opened the door to several others, on which many in the field could now focus their attention.

Ultimately, the AlphaFold story, much like the longer story of scientific computing, suggests a picture in which humans and machines work together to solve problems and generate new ideas. This is the direction in which emerging companies such as Lila Sciences are headed: training AI systems to work with humans towards scientific discoveries [31]. This may result in shifts in the divisions of scientific labor (between machine and human and human and human) as various tasks become more routine. A similar dynamic has played out in other domains of automation. For example, in the 1970s, sequencing tens or hundreds of base pairs of a DNA molecular was sufficiently difficult to be the basis of a Ph.D. By the 1990s, such processes had been automated to the extent that DNA sequencing was no longer considered worthy of a doctorate [32]. Such shifts in the "units" of labor may have longer term effects on kinds (and amounts) of labor scientists are expected to perform.

This is not to say that AI just means an acceleration of science or problem solving. As we have seen here, software such as AlphaFold can exert a decisive influence on which kinds of problems are imagined to be fruitful and where scientists decide to expend their efforts. On the one hand, the rise of AI could lead to a doubling-down on data-driven and scoring-driven type problems that are tractable with deep learning. The protein-folding problem was already highly gamified and existed in a domain with lots of data ready to hand; these factors made the problem extremely amenable to reinforcement learning. There are others aspects of science however in which data is not readily available or is very scarce. Such fields are less likely to be susceptible to AI. One possible outcome here is that these fields will receive less attention and less funding. The excitement around big data approaches to science has already contributed to bolstering data-hungry fields; this process may accelerate with the advent of AI.

Indeed, in the longer term, some of the more pervasive effects of AI may be on the resourcing and organization of science. AlphaFold made clear the significant computing resources and money required to deploy AI. Such high-profile successes may lead to more interdisciplinary approaches or a greater need for cooperation between academic scientists and industry to obtain access to computational resources. Such arrangements would likely lead to new kinds of collaborations and teams—including collaborations between subject-matter experts and AI engineers—or even entirely new kinds of specializations and skills.

On the other hand, if data-heavy or AI-accessible problems come to be perceived as "engineering" rather than "science," then some scientists might make a conscious decision to define their work against such practices. "Science," at least for some, may become precisely what AI cannot (yet) do. Just as AI has the potential to shift the units of scientific labor, it also has potential to shift the definition of what counts as a valuable or interesting scientific contribution to knowledge.

References

1. Prediction Center (2020) Artificial intelligence solution to a 50-year-old science challenge could “revolutionise” medical research. Press release, 30 November 2020. Accessed 26 Feb 2025
2. Callaway E (2020) “It will change everything”: DeepMind’s AI makes gigantic leap in solving protein structures. Nature
3. Google DeepMind Team, Isomorphic Labs (2024) AlphaFold3 predicts the structure and interaction of life’s molecules. Blog post, 8 May 2024. Accessed 26 Feb 2025
4. Callaway E (2020) Chemistry Nobel goes to developers of AlphaFold AI that predicts protein structures. Nature
5. Google DeepMind (2025) AlphaFold. Accessed 25 Feb 2025
6. AlQuraishi M (2020) AlphaFold2 @ CASP14: “it feels like one child has left home.” Blog post, 8 December 2020. Accessed 25 Feb 2025
7. Sanger F (1949) The terminal peptides of insulin. Biochem J 45(5):563–574
8. Pauling L, Corey RB, Branson HR (1951) The structure of proteins; two hydrogen-bonded helical configurations of the polypeptide chain. Proc Natl Acad Sci USA 37(4):205–211
9. Kendrew JC, Bodo G, Dintzis HM, Parrish RG, Wyckoff H, Phillips DC (1958) A three-dimensional model of the myoglobin molecule obtained by X-ray analysis. Nature 181(4610):662–666
10. Muirhead H, Perutz MF (1963) Structure of haemoglobin. A three-dimensional Fourier synthesis of reduced human haemoglobin at 5.5 Å resolution. Nature 199(4894):633–638
11. Anfinsen CB (1973) Principles that govern the folding of protein chains. Science 181(4096):223–230
12. Chou PY, Fasman GD (1974) Prediction of protein conformation. Biochemistry 13:222–245
13. Karplus M, Weaver DL (1979) Diffusion-collision model for protein folding. Biopolymers 18:1421–1437
14. Levitt M (1983) Molecular dynamics of native protein. I. Computer simulation of trajectories. J Mol Biol 168:595–617
15. Li Z, Scheraga HA (1987) Monte Carlo–minimization approach to the multiple-minima problem in protein folding. Proc Natl Acad Sci USA 84:6611–6615
16. Protein Data Bank (1971) Nat New Biol 233:223–226
17. Moult J et al (1995) A large-scale experiment to assess protein structure prediction methods. Proteins Struct Funct Genet 23:ii–iv
18. Drew K et al (2011) The proteome folding project: proteome-scale prediction of structure and function. Genome Res 21(11):1981–1994
19. Cooper S et al (2010) Predicting protein structures with a multiplayer online game. Nature 466(7307):756–760
20. Pande VS, Beauchamp K, Bowman GR (2010) Everything you wanted to know about Markov state models but were too afraid to ask. Methods 52(1):99–105
21. Bkeop (2020) AlphaFold: machine learning for protein structure prediction. Blog post, 31 January 2020. Accessed 25 Feb 2025
22. Senior AW et al (2020) Improved protein structure prediction using potentials from deep learning. Nature 577:706
23. Ball P (2020) Behind the screens of AlphaFold. Chemistry World, 9 December 2020. Accessed 25 Feb 2025
24. Chomsky N (2023) The false promise of ChatGPT. The New York Times, 8 March 2023
25. McDonough M (2024) Did AI solve the protein-folding problem? Harvard Medicine. Accessed 25 Feb 2025
26. Hayles NK (1999) How we became posthuman: virtual bodies in cybernetics, literature, and informatics. University of Chicago Press
27. Anderson C (2008) The end of theory: the data deluge makes the scientific method obsolete. Wired, 23 June 2008

28. Stevens H (2017) A feeling for the algorithm: working knowledge and big data in biology. Osiris 32(1):151–174
29. Retraction Watch (2024) Nature earns ire over lack of code availability for Google DeepMind protein folding paper. Accessed 25 Feb 2025
30. Agar J (2006) What difference did computers make? Soc Stud Sci 36(6):869–907
31. Lohr S (2025) The quest for A.I. "scientific superintelligence". The New York Times, 10 March 2025
32. Stevens H (2013) Life out of sequence: a data-driven history of bioinformatics. University of Chicago Press

Chapter 11
Bridging the Biotechnology Valley of Death: Modernizing the TRL-MRL-IRL Acquisitions Process for Accelerated Technology Modernization

Benjamin D. Trump and Madison D. Horgan

Abstract This chapter explores the persistent challenge of the "Valley of Death" in biotechnology innovation—the gap between early-stage research and successful commercialization. It examines why many promising biotechnological advances, despite substantial scientific breakthroughs, fail to achieve widespread market adoption or societal impact. The chapter utilizes established readiness frameworks—Technology Readiness Levels (TRL), Manufacturing Readiness Levels (MRL), and Integration Readiness Levels (IRL)—to analyze this problem. It highlights the unique factors that make the Valley of Death particularly severe for biotechnology, including inherent biological variability and unpredictability during scale-up, the capital-intensive infrastructure required for biomanufacturing, significant regulatory uncertainty, fragmented funding ecosystems, risk-averse private capital, and the absence of a clear Regulatory Readiness Level (RRL) framework. The chapter then suggests ways to bridge this gap, such as developing biotech-specific readiness frameworks like RRLs, expanding pilot-scale infrastructure through biofoundries, reforming investment models, and leveraging government as an early customer to de-risk innovation. Addressing these systemic challenges is presented as crucial for realizing biotechnology's full economic, environmental, and societal benefits.

Keywords Biotechnology modernization · Readiness levels · Technology acquisition process

B. D. Trump
Tor Intelligence, LLC, Apex, NC, USA

M. D. Horgan (✉)
North Carolina State University, Raleigh, NC, USA
e-mail: mdhorgan@ncsu.edu

C. L. Cummings et al. (eds.), *Biotechnology and AI: Technological Convergence and Information Hazards*, NATO Science for Peace and Security Series A: Chemistry and Biology, https://doi.org/10.1007/978-3-032-05246-9_11

11.1 Introduction

Biotechnology modernization is constrained by the structural limitations of its translational pipeline, particularly in the transition from laboratory-scale research to industrial-scale production and regulatory integration. The stages of advancing biotechnology from concept to deployment is traditionally described using Technology Readiness Levels (TRL), Manufacturing Readiness Levels (MRL), and Integration Readiness Levels (IRL) [1]. As depicted in Fig. 11.1, these frameworks together assess the technical maturity, manufacturability, and operational deployment of a technology [2]. Like historically transformative technologies (railroads, semiconductors, etc.), biotechnology modernization necessitates substantial upfront public investment to bridge the "Valley of Death" in innovation. Unlike other technological domains, however, biotechnology modernization is also subject to unique constraints arising from biological variability, complex regulatory environments, and capital-intensive infrastructure requirements [3]. These factors contribute to the sector's disproportionately severe Valley of Death, where promising innovations fail to progress from validated research findings to scalable, market-ready applications.

The Valley of Death in biotechnology is defined by the difficulty of securing funding, achieving scalable production processes, and navigating uncertain regulatory pathways between TRL 4–7 and MRL 4–8 [4]. Unlike hardware-driven industries, where prototypes can be iteratively improved with predictable performance enhancements, biological systems introduce an inherent degree of stochasticity that complicates scale-up [5]. The transition from laboratory-scale experiments to pilot-scale biomanufacturing often exposes metabolic inefficiencies, instability in engineered organisms, and unpredictable process failures that cannot be resolved through standard engineering approaches. This problem is exacerbated by the high capital

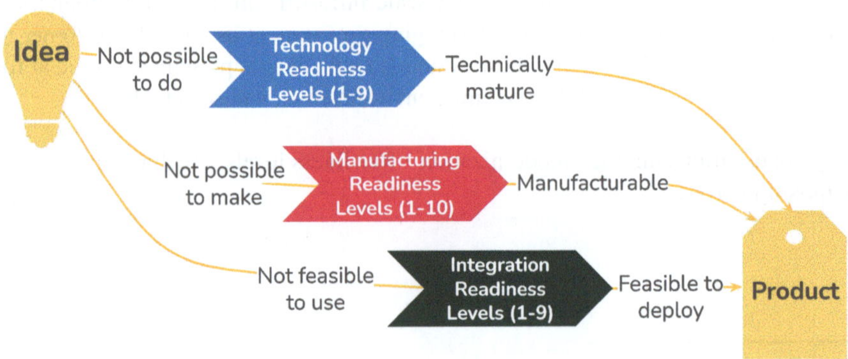

Fig. 11.1 Representation of the parallel and overlapping Technology Readiness Levels (TRL), Manufacturing Readiness Levels (MRL), and Integration Readiness Levels (IRL), which together track the progress from an idea for a technology to a product

requirements of biotechnology infrastructure, which disincentivize private investment in early-stage commercialization. The absence of robust pilot-scale manufacturing ecosystems limits the ability to test and refine biotechnology applications before full-scale production, leading to premature market exits of otherwise viable innovations.

Regulatory uncertainty further compounds the difficulty of crossing the Valley of Death—developers rely upon regulators to provide clear signals regarding whether and under which conditions biotechnology products would be permitted to reach the marketplace [6]. Unlike sectors where compliance requirements are well-defined and evolve incrementally, biotechnology regulations are frequently retroactive, fragmented across jurisdictions, and subject to evolving scientific and ethical considerations. These challenges are intensified by the inherently high stakes of biotechnology, where unanticipated consequences—ranging from ecological disruption to off-target genetic effects—can carry significant risks for public health and the environment [7]. The absence of a clear Regulatory Readiness Level (RRL) framework forces biotechnology developers to engage in costly, iterative regulatory negotiations that further delay commercialization. In some cases, technologies that have achieved high TRLs remain non-deployable due to the absence of corresponding regulatory infrastructure [8]. For instance, synthetic biology applications in agriculture, gene-editing therapies, and lab-grown meats have all encountered extended commercialization delays due to shifting safety and ethical guidelines, as well as confusion regarding the governance of associated information hazards [9–11]. Even well-validated products face the risk of societal resistance or policy reversals, which can abruptly shift the regulatory landscape and undermine long-term development strategies [12]. The result is an innovation ecosystem where biotechnology advancements frequently stall at the transition between fundamental research and applied deployment, limiting their impact on healthcare, industrial biomanufacturing, and environmental sustainability [13].

This chapter examines the specific mechanisms by which biotechnology modernization is impeded, distinguishing its translational challenges from those of other technology sectors. It provides a structured analysis of how TRL, MRL, and IRL frameworks intersect in biotechnology, why the Valley of Death is particularly severe in this domain, and what policy interventions are necessary to overcome these constraints. The discussion will focus on key failure points within biotechnology's development pipeline, such as inadequate biomanufacturing infrastructure, risk-averse private capital markets, and regulatory misalignment. By mapping these barriers to concrete policy and investment solutions, this chapter outlines an approach to biotechnology modernization that accounts for its unique scientific and economic constraints.

11.2 Background: About the TRL, MRL, and IRL Acquisitions Process for Technology Modernization

The Department of Defense (DoD) employs structured readiness frameworks to evaluate the maturity of technologies under development, ensuring that innovations can be transitioned effectively from research and development (R&D) to operational deployment [14]. TRLs provide a standardized methodology for assessing the scientific and technical maturity of a given technology, ranging from TRL 1 (basic scientific principles observed) to TRL 9 (actual system proven in operational conditions). This framework is widely applied across multiple industries, including aerospace, defense, and biotechnology, to benchmark innovation progression and guide investment decisions, with applications beyond the United States into the EU and beyond [15]. However, achieving high TRL does not guarantee successful deployment, particularly in biotechnology, where laboratory-scale achievements do not necessarily translate to viable industrial-scale processes.

MRLs complement TRLs by measuring the scalability and production feasibility of a given technology. While TRLs focuses on technical validation, MRL assesses the ability to manufacture a technology at cost-effective and consistent quality levels [16]. The MRL scale ranges from MRL 1 (basic manufacturing concepts) to MRL 10 (full-rate production). For biotechnology, MRL is particularly relevant because bio-based systems often experience process unpredictability when transitioning from laboratory conditions to pilot-scale and commercial-scale manufacturing. Biomanufacturing requires rigorous process validation, control of biological variability, and extensive infrastructure that is capital-intensive to develop. Unlike hardware-based industries, where production scalability is largely an engineering challenge, biotechnology manufacturing involves inherent biological complexity that introduces greater uncertainty into the transition from prototype to full-scale production.

IRLs address the challenge of incorporating a new technology into a larger system. In the defense context, IRL evaluates how well a given innovation can be integrated into existing operational frameworks, including infrastructure, logistics, and mission-critical applications [17]. IRL 1 represents an unintegrated concept, while IRL 9 signifies fully integrated operational use. The biotechnology sector often encounters IRL challenges due to the difficulty of integrating biological products into regulatory frameworks, supply chains, and end-user applications. For instance, synthetic biology-based materials, advanced biologics, and engineered microbial systems must be compatible with existing industrial and healthcare infrastructures. Misalignment between TRL, MRL, and IRL results in stalled innovation pipelines, particularly when technologies achieve laboratory validation (TRL 6+) but lack scalable production methods (MRL < 6) or regulatory pathways (IRL < 5).

TRL, MRL, and IRL are integrated into the acquisitions process to ensure that new technologies advance through a structured development pipeline before being adopted for operational use [17]. In defense and commercial biotechnology, these readiness levels provide decision-making frameworks for investment, procurement, and risk assessment. TRL establishes whether a technology has been validated at a fundamental scientific level, ensuring that only mature technologies progress to later stages of acquisition. MRL determines whether a given technology can be produced at scale with sufficient quality control and cost efficiency, reducing the risk of manufacturing bottlenecks post-procurement. IRL assesses whether a technology can be integrated into existing operational systems, ensuring that new capabilities do not create logistical incompatibilities or require additional infrastructure investments. By aligning TRL, MRL, and IRL, acquisition officers and policymakers make strategic funding and deployment decisions that minimize risk while ensuring that technologies meet mission and operational requirements.

These readiness levels shape strategic decisions by defining gateways for transition between research, pilot-scale testing, and full-scale production. Technologies that meet high TRL but lack corresponding MRL and IRL levels often require additional investment in manufacturing infrastructure before advancing in the acquisitions process. For instance, a synthetic biology-based material may achieve TRL 6–7 in a laboratory setting but remain at MRL 4–5 due to the absence of an established production pipeline. In such cases, acquisitions strategy may involve funding pilot-scale production through biofoundries or establishing public–private partnerships with industry to accelerate MRL progression. Similarly, IRL misalignment can delay adoption if a biotechnology innovation cannot be seamlessly integrated into existing supply chains, regulatory frameworks, or operational environments. Decision-makers use TRL-MRL-IRL assessments to prioritize technologies that demonstrate maturity across all three dimensions, ensuring that acquisitions investments yield deployable and scalable capabilities.

In defense acquisitions, TRL, MRL, and IRL serve as risk management tools to determine whether a given technology should advance to full procurement or require further validation. DoD acquisition programs use milestone-based decision points where technologies must meet specific TRL, MRL, and IRL criteria before progressing to the next phase. For example, the Defense Innovation Unit (DIU) or the Defense Advanced Research Projects Agency (DARPA) may fund a biotechnology project through TRL 5–6, but transitioning to a Program of Record requires demonstrating MRL 7–8 and IRL 7+ to ensure deployability [18, 19]. By structuring acquisitions around readiness levels, DoD ensures that biotechnology capabilities are cost effective and capable of delivering desired performance before mass procurement and deployment. This approach minimizes cost overruns, prevents premature fielding of immature technologies, and ensures that investments result in functional, scalable, and mission-ready solutions [20].

However, this process is fraught with chokepoints where otherwise desirable technologies become bogged down at key 'go/no-go' decision points. The Valley of Death describes the failure of technologies to transition from laboratory-scale research to full-scale production and operational deployment. In biotechnology, this occurs most frequently between TRL 4–7 and MRL 4–8, where scientific validation has been achieved but technical, regulatory, and financial barriers prevent commercialization. Unlike in sectors such as aerospace, where DoD funding provides sustained support through TRL and MRL progression, biotechnology companies face a fragmented funding landscape that leaves gaps in mid-stage development.

Several structural factors contribute to the formation of the Valley of Death in biotechnology. First, funding asymmetry creates a disconnect between early-stage research support and late-stage commercialization investment. Agencies such as the US National Institute of Health (NIH), National Science Foundation (NSF), and DARPA provide grants for fundamental research (TRL 1–4) but do not support manufacturing scale-up or regulatory approval costs. At the same time, private capital is risk-averse toward biotechnology projects that require capital-intensive infrastructure, long regulatory timelines, and uncertain returns on investment. The result is a funding gap for companies attempting to bridge TRL 5–8 and MRL 4–7, where scale-up costs increase exponentially but revenue generation remains uncertain.

Second, biomanufacturing scalability is highly constrained compared to other industries. While semiconductor and aerospace manufacturing follow predictable engineering models, biotechnology requires precise control over living systems, bioprocess optimization, and contamination risk management. Many TRL 4–6 biotechnologies cannot advance to MRL 7+ due to instability in metabolic pathways, batch-to-batch variability, and difficulty in maintaining process reproducibility. Unlike traditional engineering solutions, where system performance improves predictably with refinement, biotechnology scale-up often introduces new failure points that are not evident at small-scale testing. This limits investment in early-stage commercialization efforts, as process failures at MRL 4–6 introduce high financial risks.

Third, regulatory uncertainty slows IRL progression, making biotechnology's Valley of Death particularly severe. While hardware-based defense technologies follow standardized testing and certification pathways, biotechnology products must undergo extensive safety, efficacy, and environmental impact assessments before deployment. US Food and Drug Administration (FDA), Environmental Protection Agency (EPA), and Department of Agriculture (USDA) regulations impose lengthy approval timelines, often requiring years of additional testing beyond achieving TRL 9 [21]. Even when a technology has reached high TRL and MRL, failure to secure regulatory approval can prevent integration into existing operational frameworks, leaving promising innovations stalled at IRL 4–6.

Traditional approaches to overcoming the Valley of Death focus on increasing public-sector funding, establishing procurement programs, and forming public–private partnerships. Programs funded by US federal agencies such as Small Business Innovation Research (SBIR) and Small Business Technology Transfer (STTR) grants provide translational research support, but their impact is limited to TRL 3–6 and does not address MRL 5–8 scale-up challenges. Similarly, DoD procurement mechanisms such as Other Transaction Authority (OTA) agreements facilitate early-stage technology adoption but do not resolve manufacturing and regulatory readiness bottlenecks [22]. In the pharmaceutical industry, advanced market commitments have successfully accelerated vaccine development by guaranteeing a market for successful products. However, this strategy has limited applicability for synthetic biology, industrial biomanufacturing, and bio-based materials, where long-term market demand is uncertain.

Efforts to address MRL challenges have included the establishment of biofoundries and pilot-scale production infrastructure, supported by agencies such as the Department of Energy (DOE), DoD, and Advanced Research Projects Agency for Health (ARPA-H). These facilities provide shared resources for biotechnology scale-up, reducing financial risk for early-stage companies. However, most existing biofoundries focus on TRL 4–6 research rather than MRL 6–9 manufacturing, leaving a gap in transitioning from pilot-scale production to full-scale biomanufacturing. Further, the lack of standardized facility design compounds this issue, as most biomanufacturing facilities in the U.S. are built for single-product processes, making it difficult to repurpose them for emerging technologies [23]. Without coordinated investment in MRL and IRL progression, many biotechnology innovations remain confined to academic or laboratory settings without reaching commercial viability.

Traditional solutions remain insufficient due to the structural misalignment between TRL, MRL, and IRL in biotechnology. Funding mechanisms are fragmented across different readiness levels, leaving biotechnology startups without sustained financial support through scale-up. Regulatory frameworks remain unpredictable, preventing companies from achieving the necessary IRL for deployment. Biomanufacturing constraints introduce process instability and cost escalation, increasing financial risks and discouraging investment. These challenges underscore the need for new policy and funding mechanisms that specifically address TRL-MRL-IRL gaps in biotechnology, ensuring that promising innovations can successfully transition from research to scalable, regulated deployment.

11.3 Why the Biotechnology Valley of Death Is More Severe than in Other Sectors

Unlike software or hardware, biotechnology innovations rely on living systems that introduce high variability, stochastic behavior, and environmental sensitivity into the development process. Unlike deterministic engineering disciplines, where system performance improves through iterative refinement, biotechnology faces inherent biological unpredictability that complicates the transition from early-stage research to industrial-scale application. Even when a biotechnology reaches TRL 4–7, scaling up from laboratory validation to applied deployment presents unique challenges that are not present in sectors like aerospace, semiconductors, or software development. Many synthetic biology applications, such as engineered microbes for biofuel production or recombinant protein synthesis, demonstrate promising results in controlled laboratory conditions but fail to maintain metabolic stability, process efficiency, or yield consistency at pilot scale.

One major constraint in biotechnology modernization is non-deterministic biological behavior, where engineered systems do not perform identically across different production environments. Even well-characterized cell lines, microbial strains, or biochemical pathways can exhibit variability due to resource limitations, genetic instability, or unintended interactions with environmental factors. This problem is particularly pronounced when transitioning from bench-scale (TRL 3–4) to pilot-scale bioreactors (TRL 5–6), where nutrient flow, oxygenation, and metabolic waste accumulation introduce new stressors that were not present in small-scale validation studies. Unlike semiconductor fabrication, where scaling laws follow predictable physics-based principles, biotechnology must contend with evolutionary drift, mutational instability, and pathway crosstalk, all of which can degrade system performance in unpredictable ways [24].

Synthetic biology startups illustrate these challenges. Many achieve TRL 5–6 by demonstrating proof-of-concept in well-controlled laboratory conditions, yet fail to transition to commercial-scale deployment due to metabolic bottlenecks, loss of engineered traits, and process inefficiencies. Companies developing microbial strains for biofuel synthesis, for example, often encounter unexpected yield drops when transitioning from small fermenters (10–100 L) to pilot-scale reactors (1,000–10,000 L) due to substrate limitation, toxic byproduct accumulation, or shifts in microbial community dynamics [24]. These scale-up failures are often not evident at TRL 4–5, meaning that companies enter the Valley of Death unaware of critical barriers to manufacturing feasibility. Unlike software or electronics, where optimization cycles are rapid and incremental, biotechnology solutions require multi-year refinement cycles with no guarantee of successful scale-up, making the TRL progression far more complex and capital-intensive.

The challenges of biological complexity at TRL 4–7 are compounded by the manufacturing bottleneck at MRL 4–6, where process scalability remains an unresolved constraint for many biotechnology innovations. Unlike hardware-based industries, where TRL progression is closely linked to MRL advancement, biotechnology experiences a disconnect between technical validation and production feasibility. Even when a process is validated at TRL 6, there is no guarantee that it can be manufactured at an economically viable scale, particularly in industries that require precise bioprocess control, batch reproducibility, and strict regulatory oversight. The transition from TRL 6 to MRL 7 requires overcoming key manufacturing challenges, including fermentation dynamics, purification bottlenecks, and supply chain constraints that are not present in early-stage research. Without access to pilot-scale biomanufacturing infrastructure, many biotechnology firms stall in the Valley of Death, unable to generate the data required for investment, regulatory approval, or commercial adoption.

Progressing beyond MRL 4–6 requires resolving several fundamental barriers in bioprocess engineering, quality control, and compliance with Good Manufacturing Practice (GMP) regulations. Unlike semiconductor or aerospace manufacturing, where production processes can be precisely engineered to minimize variability, biomanufacturing processes are highly sensitive to raw material quality, microbial contamination, and environmental fluctuations. This creates scalability constraints that extend beyond engineering optimization, requiring extensive process validation, batch-to-batch consistency testing, and regulatory documentation. GMP compliance further increases manufacturing costs, requiring dedicated infrastructure for sterile production, contamination mitigation, and process standardization. Companies operating at MRL 4–6 often struggle to secure the capital required to transition to MRL 7–8, where biomanufacturing costs escalate due to the need for large-scale fermentation, downstream purification, and quality assurance measures. The inability to navigate this transition results in a high failure rate for biotechnology startups, where innovations that have achieved TRL 6–7 ultimately fail due to unresolved manufacturing constraints.

Biotechnology modernization is constrained not only by technical and manufacturing barriers but also by regulatory uncertainty, which serves as an invisible yet significant obstacle to IRL progression [25]. Unlike sectors such as defense, aerospace, or information technology, where integration into existing systems is primarily a logistical or engineering challenge, biotechnology products must navigate complex, evolving, and often inconsistent regulatory landscapes before they can be deployed [26]. Even when a biotechnology innovation reaches TRL 9 and demonstrates efficacy and safety in controlled studies, it may fail to progress beyond IRL 6–7 due to unresolved regulatory issues [27, 28]. The regulatory frameworks governing synthetic biology, gene editing, and cell therapies are still in development across different jurisdictions, leading to inconsistent approval pathways, prolonged review cycles, and variable compliance requirements [29, 30]. These uncertainties extend the Valley of Death by delaying product deployment, increasing costs, and discouraging investment in high-risk biotechnology innovations.

Regulatory readiness is a hidden component of IRL, affecting how seamlessly a biotechnology product can transition from technical validation to real-world use. Unlike MRL, which is constrained by process scalability and cost feasibility, IRL is limited by regulatory approval, compliance requirements, and market integration barriers. In pharmaceuticals and therapeutics, for example, achieving TRL 9 (fully developed technology) does not guarantee that a product will reach IRL 7+, as regulatory bodies such as the FDA, European Medicines Agency (EMA), and National Medical Products Administration (NMPA) impose varying clinical trial requirements, safety evaluations, and risk assessment procedures. Gene-editing technologies such as CRISPR-based therapies illustrate this challenge. While CRISPR applications have successfully demonstrated TRL 8–9 in preclinical and clinical studies, their regulatory approval pathways remain unclear and inconsistent across different regions. The FDA and EMA apply different risk classification models, requiring separate and often contradictory validation steps, which prolongs commercialization and increases financial risk for biotechnology firms seeking global market entry [31]. Moreover, unpredictability is an inherent feature of biotechnology development, particularly in the transition from technical maturity to real-world implementation. Even when a technology performs reliably in controlled settings, its path to deployment may be disrupted by shifting regulatory standards, emerging safety concerns, or unforeseen socio-political dynamics, often at timelines much faster than other emerging and enabling technologies like nanotechnology [32]. This uncertainty complicates planning and investment decisions, as timelines, approval requirements, and market conditions may change without warning. Developers must therefore navigate not only scientific and engineering hurdles but also evolving external expectations that can upend established strategies [33].

Comparing IRL progression in biotechnology versus aerospace and defense industries highlights why regulatory uncertainty is a greater barrier in biotech. In aerospace, once a technology reaches TRL 7+, it follows a structured certification and testing pathway, leading to predictable IRL progression. Defense acquisitions operate under defined procurement cycles, where TRL 7+ technologies are integrated through established evaluation protocols. In contrast, biotechnology products that reach TRL 9 may still face years of regulatory delays and shifting compliance requirements, preventing IRL advancement. Lab-grown meat and mRNA vaccine technologies before COVID-19 illustrate this issue. Prior to the pandemic, mRNA vaccines were stuck in regulatory limbo despite demonstrating TRL 8–9 for years in preclinical studies. The lack of an established regulatory framework prevented rapid deployment until emergency-use authorizations were enacted. The COVID-19 response (Operation Warp Speed) mirrored historical wartime technological accelerations, where government intervention, urgent demand, and rapid regulatory adjustments pushed critical innovations—similar to how military procurement historically advanced aviation, computing, and semiconductor industries [34]. Meanwhile, lab-based agriculture products such as lab-grown meat have reached TRL 9 but remain stalled at IRL 5–6 due to regulatory uncertainty regarding food safety, labeling, and public acceptance criteria. Without a massive shift in demand, regulatory alignment, or a concerted policy push, these technologies are likely to remain in limbo, as the lack of clear commercialization pathways discourages investment and delays widespread adoption.

The financial burden of biotechnology modernization is significantly greater than in other technology sectors, creating a capital intensity bottleneck that reinforces the Valley of Death. Advancing from TRL 4 to TRL 8 in biotechnology requires exponentially increasing investment, particularly due to the costs associated with clinical trials, biomanufacturing scale-up, and regulatory compliance. Unlike aerospace and defense, where the DoD, DARPA, and National Aeronautics and Space Administration (NASA) provide long-term contracts to fund TRL 4–8 development, or AI/software, where venture capital (VC) funding scales predictably as startups iterate and expand, biotechnology firms face a high-risk, high-cost funding gap that most private investors are unwilling to support. The transition from TRL 4–8 and MRL 4–8 requires tens to hundreds of millions of dollars before a product reaches commercialization, with no guarantee of return on investment (ROI) due to technical, regulatory, and market risks. The absence of patient capital in biotechnology contributes to prolonged time-to-market, higher attrition rates, and an overreliance on government grants, corporate partnerships, or public–private funding mechanisms.

Investor risk aversion is particularly pronounced in biotechnology because the commercialization timeline is long, failure rates are high, and liquidity events are unpredictable. Unlike SaaS (software as a service) startups, where iterative development, rapid scaling, and low capital expenditure allow for frequent pivots, biotechnology firms operate under a fundamentally different risk model. Clinical trial attrition rates exceed 90% for new drugs, meaning that even promising TRL 6–7 innovations may never progress to market deployment due to unforeseen toxicity, efficacy issues, or regulatory rejections [35]. Time horizons for biotechnology commercialization extend beyond 10–15 years, further discouraging traditional VC investors who seek shorter exit cycles through acquisitions or public offerings. Additionally, while MRL 5–8 funding is critical for scaling production and optimizing cost efficiency, this phase is particularly underfunded due to the high capital requirements, regulatory uncertainty, and long ROI timelines. Most private investors are reluctant to support biomanufacturing infrastructure development, leaving companies unable to progress beyond pilot-scale production.

The structural misalignment between biotechnology's capital needs and the investment landscape contributes to low survival rates for startups attempting to bridge the Valley—or, in the case of biotechnology, Valleys—of Death. Figure 11.2 illustrates this point using a hypothetical example of a microbial-derived biopolymer intended to replace conventional plastic packaging. In the biopolymer's imagined journey from idea to product, various challenges arise that require significant investments to overcome technical, scale-up, regulatory, and market integration challenges that would otherwise prevent the product from making it to market.

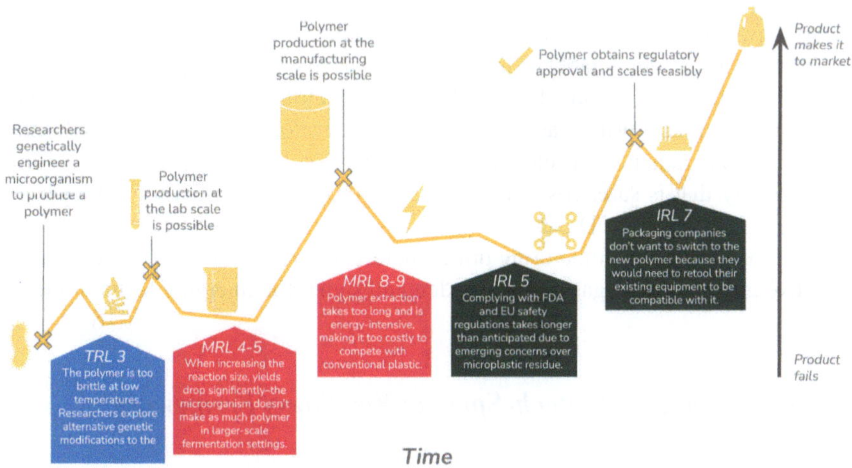

Fig. 11.2 Hypothetical development pathway and associated "Valleys of Death" for a biopolymer produced by a genetically engineered microbe. The figure illustrates key milestones and common pain points across the Technology Readiness Level (TRL), Manufacturing Readiness Level (MRL), and Integration Readiness Level (IRL) frameworks. This example illustrates a subset of potential challenges and transition points and is not intended to represent the full range of obstacles that may arise during biotechnology development. Further, the labeled stages of TRL, MRL, and IRL challenges are approximate and may vary depending on the specific technology, application context, and development pathway

A critical barrier in biotechnology development is the persistent funding gap between early-stage research and late-stage commercialization, which leaves mid-stage innovations financially stranded. Early-stage funding is available through NIH, NSF, and DARPA grants, but these sources do not fund MRL 5–8 scale-up costs or regulatory compliance efforts. Large pharmaceutical companies and industrial biotechnology firms only invest in TRL 7+ assets with demonstrated market potential, meaning that mid-stage innovations struggle to secure funding for clinical trials, pilot-scale production, or regulatory approval processes. The result is a high rate of technology abandonment at TRL 6–7, where promising discoveries fail to transition into commercial viability due to insufficient capital, unresolved manufacturing challenges, and prolonged regulatory review cycles. Overcoming these financial barriers requires new funding mechanisms, risk-sharing investment models, and public–private initiatives that bridge the capital gap between TRL validation and full-scale deployment. Without these interventions, biotechnology innovation will continue to stagnate in the Valley of Death, unable to realize its full potential.

11.4 Early Ideas to Bridge the Biotechnology Valley of Death

The biotechnology sector requires targeted interventions to mitigate the systemic challenges of TRL-MRL-IRL misalignment, capital constraints, and regulatory uncertainty. Unlike software or defense technologies, biotechnology cannot rely on rapid iteration, predictable scaling, or existing regulatory precedents to ensure efficient commercialization. The goal is not to eliminate all risks—biological complexity, manufacturing variability, and long development timelines make that infeasible—but rather to accelerate the biotechnology research acquisition timeline in a way that is safe, cost-efficient, and appropriately scaled for deployment. Addressing the Valley of Death in biotechnology will require new readiness frameworks, expanded biomanufacturing infrastructure, reformed investment models, and proactive government engagement as a demand signal for emerging technologies.

11.4.1 Creating a Biotech-Specific Readiness Framework

The existing TRL-MRL-IRL framework does not account for regulatory uncertainty, which remains a critical constraint in biotechnology commercialization. Unlike hardware-based technologies, where achieving TRL 7+ and MRL 6+ typically ensures that a product is ready for adoption, biotechnology must undergo extensive regulatory evaluation before integration into clinical, industrial, or agricultural systems.

To address regulatory gaps in pharmaceutical commercialization—a significant but relatively historically precedented arm of biotechnology—researchers proposed new regulatory readiness frameworks that supplement the existing TRL-MRL-IRL structure. The Biomanufacturing Readiness Level (BRL) framework, developed by the National Institute for Innovation in Manufacturing Biopharmaceuticals (NIIMBL), provides a structured approach to assessing the maturity of biopharmaceutical manufacturing technologies. Unlike traditional MRLs, BRLs incorporate quality and regulatory considerations, recognizing that manufacturing scalability in biopharmaceuticals must meet stringent process validation and quality management requirements before a product can advance to full commercialization [36]. While this is a good start, the proposed BRLs do not fully address the complexity of scaling live biological systems (which are increasingly at the heart of biomanufacturing), where process stability and reproducibility can vary significantly between products. Complementing NIIMBL's approach, other academics have constructed a Regulatory Readiness Level (RRL) framework to address the critical role of regulatory science in pharmaceutical scaleup. Designed specifically for academic researchers developing biopharmaceuticals, RRLs define nine levels of regulatory maturity, helping researchers align product development with regulatory expectations from an early stage [35]. While these RRLs offer a structured approach to regulatory awareness, they do not yet resolve deeper challenges such as navigating international regulatory divergence or securing sustained investment for compliance-related development. These two proposed pharmaceutical frameworks provide a decent starting point, but more work must be done to expand a biotech-specific readiness framework beyond pharmaceuticals, particularly since other areas of biotechnology—such as gene editing, synthetic biology, and agricultural biotech—face even greater uncertainty due to evolving regulatory landscapes, the lack of standardized manufacturing pathways, and greater variability in real-world deployment conditions.

Introducing Regulatory Readiness Levels (RRLs) alongside the TRL-MRL-IRL framework for biotechnology would provide a structured assessment of a technology's compliance trajectory, offering a standardized framework for quantifying regulatory risk, approval likelihood, and remaining compliance requirements. An RRL framework could define key stages of regulatory maturity, including preclinical validation, clinical trial progression, market authorization, and post-market surveillance. Each level would correspond to specific data requirements, compliance milestones, and risk mitigation steps, allowing researchers and investors to anticipate which aspects of regulatory clearance remain unresolved. This would improve decision-making on resource allocation, ensuring that funding is directed towards the most critical barriers to approval rather than areas of already established technical feasibility.

Beyond RRLs, regulatory processes must shift towards adaptive licensing and real-time oversight mechanisms to reduce approval delays without compromising safety [37]. Traditional regulatory pathways rely on sequential data submission and rigid validation phases, which can prolong review cycles even when safety and efficacy data are already sufficient for early deployment. A model based on rolling submission of preclinical and clinical data, continuous risk monitoring, and real-time safety assessments would facilitate earlier market access while maintaining long-term compliance accountability. This approach would be particularly useful for gene-editing therapeutics, engineered biologicals, and cell-based manufacturing systems, where iterative process optimization is necessary even after initial approval.

Agile regulatory and approval pathways through the Singapore Food Agency have already attracted a swath of companies and investors to commercialize food biotechnologies like cultured meat in Singapore [38]. However, cultured meat has highlighted the inadequacy of existing regulatory frameworks to fully address consumer safety concerns. While Singapore has emerged as a leader by establishing regulations for cultured meat and allowing its retail sale, consumer hesitancy raises questions about whether initial regulations comprehensively address what truly worries consumers [39]. In alignment with the principles of Responsible Research and Innovation (RRI), it is essential that regulatory bodies demonstrate adaptability and responsiveness to evolving public concerns regarding novel biotechnologies, ensuring that regulations are not static but can evolve to maintain public trust in the safety and oversight of these innovative products.

This need for regulatory adaptability reinforces the limitations of the TRL-MRL-IRL framework in biotechnology, where regulatory uncertainty remains one of the most significant barriers to commercialization. While the introduction of Regulatory Readiness Levels (RRLs) would provide a clearer structure for evaluating regulatory progress, they must be integrated with adaptive licensing models that allow for continuous regulatory engagement rather than binary approval decisions. Without such a shift, biotechnology products—even those that have achieved technical feasibility—will continue to face lengthy and unpredictable approval processes that delay market entry and hinder investment. A biotech-specific readiness framework must therefore go beyond static regulatory milestones to account for real-time oversight, risk-adjusted approval pathways, and the evolving nature of biological manufacturing processes, ensuring that innovations in synthetic biology, gene editing, and cellular agriculture can transition from research breakthroughs to real-world impact more efficiently.

11.4.2 Expanding Biofoundries and Pilot-Scale Manufacturing Infrastructure

The progression from MRL 4–6 to MRL 7–9 is constrained by the lack of scalable bioprocessing platforms capable of handling variable biological inputs, continuous fermentation, and high-throughput production environments. Unlike industries such as semiconductor manufacturing, where scaling is largely an issue of equipment precision and automation, biotechnology manufacturing introduces dynamic system variability, requiring infrastructure capable of iterative process refinement, adaptive bioreactor optimization, and real-time metabolic monitoring. The absence of pilot-scale biomanufacturing infrastructure means that most biotechnology innovations fail to generate the necessary process data to inform large-scale industrial deployment.

Expanding government-backed biofoundries would provide shared-access facilities for process validation, scale-up optimization, and compliance standardization, allowing biotechnology firms to refine manufacturing protocols without incurring prohibitive capital expenditures. These biofoundries would need to support high-density cell culture, microbial fermentation, cell-free synthesis platforms, and modular bioprocessing units, ensuring that different classes of biotechnology applications—from pharmaceutical biologics to industrial enzyme production—can undergo structured process maturation. Public–private investment into biofoundries and pilot-scale manufacturing hubs in Singapore (e.g., Biopolis and Tuas Biomedical Park) has helped lower the cost of scale-up and attracted international talent.

In the process of expanding biofoundries and pilot-scale manufacturing infrastructure, there is an opportunity to minimize future shortages of suitable biomanufacturing facilities through coordinated, standardized design. In manufacturing sectors, is common practice to construct plants tailored for a single product with an emphasis on speed and cost-efficiency, leading to designs that are difficult to repurpose. When market conditions change or products fail, these specialized facilities are often shut down and dismantled, losing the potential for future use. Additionally, in 2023 it was estimated that almost a third of existing biomanufacturing facilities have at least some idle capacity that theoretically could be used to test and scale up new products, though constraints like lower steel quality make the plants incompatible with many fermentation processes that use high-salt or high-chloride broths [23]. To avoid this problem in the future, new biomanufacturing sites should be designed for flexibility. This would allow easier adaptation to different bioprocesses and products, addressing the current bottleneck and fostering a more resilient bioeconomy.

Additionally, advanced process analytics and real-time quality control systems should be integrated into these facilities to enable in-situ monitoring of metabolic flux, biosynthetic pathway fidelity, and contaminant detection. These capabilities are necessary for stabilizing process yields across batch production cycles, ensuring that early-stage pilot-scale manufacturing accurately reflects full-scale production conditions. Without these infrastructures, companies remain stuck at MRL 4–6, unable to generate the necessary process reproducibility and scale-up validation to secure regulatory and market adoption.

11.4.3 Reforming Biotechnology Investment Models

Biotechnology investment structures must be reformed to align funding incentives with the financial and technical realities of biomanufacturing scale-up. The transition from TRL 4–7 to MRL 6–9 requires substantially higher capital investment than comparable technology sectors, yet biotechnology firms struggle to secure funding due to extended commercialization timelines, complex regulatory dependencies, and the high cost of establishing production-scale infrastructure. Unlike industries such as artificial intelligence (AI), software, or electronics, where capital investment follows a predictable revenue and scaling model, biotechnology requires multi-year R&D cycles, risk-intensive process validation, and compliance-driven approval pathways before monetization occurs.

To address this, new financial mechanisms must be developed to de-risk investment in biotechnology scale-up while maintaining incentives for long-term return on investment. One approach is the implementation of milestone-based funding models, where capital allocation is tied to demonstrated advancements in MRL and IRL readiness metrics. Rather than providing lump-sum investments that require biotechnology firms to navigate scale-up barriers independently, this structure would allow for incremental capital deployment based on process reproducibility, manufacturability improvements, and validated regulatory benchmarks.

Additionally, hybrid public–private investment models could be introduced to support biotechnology scale-up beyond MRL 5–6, where traditional venture capital is insufficient, but commercial revenue is not yet achievable. For example, Singapore directly funds biotechnology scale-up through agencies like the Economic Development Board (EDB) and the Agency for Science, Technology and Research (A*STAR) [40]. By integrating low-interest government-backed financing, tax credits for biomanufacturing infrastructure, and conditional grant structures that mitigate investor downside risk, biotechnology firms could secure the long-term capital necessary to sustain process development and validation. This would enable faster MRL progression while reducing the risk of early-stage technology abandonment due to funding shortfalls.

11.4.4 Government as a First Customer for Emerging Biotech

To reap the full benefits of the biotechnology, which many believe will uphold the next long-term economic cycle [41–43], governments may need to strategically reprise their historical role as a first customer. This approach proved instrumental in de-risking previous technological revolutions—including aircraft, semiconductors, computers, and the internet—by providing essential early demand signals and patient capital where market incentives alone were insufficient [34]. Such intervention is particularly vital for biotechnology due to significant market failures; the private sector often shies away from the high-risk, long-term, and often public-good-oriented investment required for foundational bio-innovations, especially those aligned with societal priorities like health, biodefense, and environmental sustainability. Federal R&D funding and targeted procurement can uniquely address these gaps, guiding innovation towards national needs, and drawing in subsequent private investment [44]. However, simply replicating past procurement models is insufficient; strategies must be carefully adapted to navigate biotechnology's distinct biological complexities, extended development timelines, unique regulatory landscapes, and the need for sustained public support.

Government procurement strategies can play a critical role in stabilizing biotechnology commercialization by serving as an initial demand signal for high-readiness innovations. Unlike consumer-driven technology sectors, where market pull is strong for high-TRL products, biotechnology innovations often require pre-committed purchase agreements, industrial-scale validation trials, and regulatory coordination before commercial deployment is feasible. Without an established first-customer framework, biotechnology firms struggle to transition from TRL 8–9 to sustained industrial-scale production, as investors are unwilling to back technologies without a clear market entry pathway.

Government entities could serve as a guaranteed early-stage buyer for biotechnology products that demonstrate high TRL and MRL progression but lack immediate commercial marketability. This would be particularly effective in biodefense, bio-based industrial materials, and next-generation medical countermeasures, where government procurement contracts could sustain production viability until commercial adoption scales. The structuring of these procurement agreements could include performance-based contracting, risk-sharing mechanisms, and long-term demand commitments, ensuring that early-stage biotechnology products can generate revenue during the final stages of regulatory and production scale-up.

Additionally, procurement policies should incorporate domestic biomanufacturing capacity requirements, ensuring that government-backed biotechnology deployments contribute to national-scale supply chain resilience. This would support not only TRL-MRL-IRL progression but also industrial stability by preventing the offshoring of critical biomanufacturing capabilities. Such strategies would enable a more predictable market entry for high-readiness biotechnology solutions, ensuring that innovations that clear TRL, MRL, and IRL bottlenecks do not remain stalled due to market access constraints.

While government procurement can effectively de-risk early-stage biotechnology and provide a vital initial demand signal, it must be recognized as a temporary catalyst rather than a perpetual market substitute. Long-term success hinges on a deliberate transition plan enabling these innovations to compete independently in the commercial marketplace. This necessitates proactive engagement with the broader ecosystem to address barriers beyond initial production scale-up. As highlighted in studies on novel bio-products like cultured meat, achieving market viability involves navigating complex regulatory approvals, establishing robust safety assessments agreeable to authorities like the European Food Safety Authority (EFSA), the FDA, and/or the USDA, and managing consumer perception through transparency and clear communication [40]. Therefore, parallel to procurement efforts, ongoing dialogue between developers, regulators, scientists, and the public is crucial to clarify regulatory pathways, build trust through comprehensive safety data, and foster consumer acceptance, ensuring that government-nurtured biotechnologies ultimately achieve sustainable commercial integration.

11.5 Conclusions

Successfully harnessing transformative technologies has historically required substantial, decades-long national investment, frequently catalyzed by government action. Previous technological revolutions in areas like aerospace, semiconductors, and the internet were often heavily supported by government investment that leveraged the urgency and directed funding associated with defense needs and wartime conditions to overcome initial market failures and development. We may now stand at the cusp of the next great economic wave—one driven by biotechnology modernization. However, fully realizing this potential demands a departure from past strategies. While foundational public investment remains critical, the unique complexities inherent in biotechnology mean we cannot simply rely on basic research funding to propel innovations across the translational Valleys of Death encompassing Technology, Manufacturing, and Integration Readiness Levels (TRL, MRL, IRL). Instead, navigating biotechnology's intricate path from lab to market requires a more deliberate, targeted approach specifically designed to address its distinct biological, manufacturing, and regulatory hurdles throughout the entire development lifecycle.

Biotechnology modernization progresses more slowly than other technology sectors due to biological complexity, manufacturing constraints, regulatory uncertainty, and capital-intensive scale-up requirements. Unlike fields such as aerospace, semiconductors, or artificial intelligence, where TRL advancement typically aligns with MRL and IRL progression, biotechnology faces distinct scaling and integration challenges that make TRL 4–7 a persistent barrier to commercialization. The transition from proof-of-concept validation to full-scale deployment is hindered by biomanufacturing bottlenecks, fragmented regulatory pathways, and a lack of patient capital for long-term development. Without structural interventions, biotechnology's potential in medicine, sustainability, and industrial applications will remain unrealized, as technologies that demonstrate early success fail to reach commercial viability.

Bridging the biotechnology Valley of Death requires targeted interventions at critical TRL-MRL-IRL junctions to enable efficient scale-up, regulatory approval, and market integration. Expanding pilot-scale manufacturing infrastructure through biofoundries and process validation hubs would provide companies with the resources needed to transition from laboratory-scale production to commercial-scale deployment. Establishing Regulatory Readiness Levels (RRLs) as a parallel framework to TRL-MRL-IRL would provide greater clarity on the approval process, aligning regulatory expectations with the pace of technological innovation. Additionally, public–private investment models and government procurement programs could mitigate financial risk, ensuring that high-readiness biotechnology solutions do not remain stalled due to market-entry barriers. These approaches would create a structured biotechnology development pipeline, reducing the economic inefficiencies that prevent TRL 6–8 technologies from reaching industrial or clinical use.

Not all countries experience the same biotechnology Valleys of Death, as structural, regulatory, and financial environments vary significantly across global markets.

In some regions, including Singapore, state-backed biotechnology investments, vertically integrated regulatory frameworks, and sovereign biofoundries enable faster MRL 5–9 progression, while in others, fragmented approval pathways and limited domestic manufacturing capacity create persistent bottlenecks. This creates asymmetries in economic and defense competition, as nations that successfully navigate TRL-MRL-IRL transitions gain strategic advantages in synthetic biology, biopharmaceuticals, and bio-industrial applications. Addressing biotechnology modernization through systems-based policy interventions is therefore not only an economic imperative but also a matter of technological sovereignty and global competitiveness. If left unaddressed, nations that fail to resolve biomanufacturing, regulatory, and investment constraints will fall behind in the global bioeconomy, becoming dependent on foreign biotechnological infrastructure, supply chains, and innovation ecosystems.

References

1. Sauser B, Ramirez-Marquez JE, Magnaye R, Tan W (2008) A system approach to expanding the technology readiness level within defense acquisition. Acquisition research program
2. Azizian N, Sarkani S, Mazzuchi T (2009) A comprehensive review and analysis of maturity assessment approaches for improved decision support to achieve efficient defense acquisition. In: Proceedings of the world congress on engineering and computer science, vol 2, pp 20–22, San Francisco
3. Trump B, Cummings C, Klasa K, Galaitsi S, Linkov I (2023) Governing biotechnology to provide safety and security and address ethical, legal, and social implications. Front Genet 13:1052371
4. Ellwood P, Williams C, Egan J (2022) Crossing the valley of death: five underlying innovation processes. Technovation 109:102162
5. Kampers LF, Asin-Garcia E, Schaap PJ, Wagemakers A, Dos Santos VAM (2021) From innovation to application: bridging the valley of death in industrial biotechnology. Trends Biotechnol 39(12):1240–1242
6. Roca JB, O'Sullivan E (2022) The role of regulators in mitigating uncertainty within the Valley of Death. Technovation 109:102157
7. National Academies of Sciences, Engineering, and Medicine (NASEM) (2017) Understanding risks related to future biotechnology products. In: Preparing for future products of biotechnology. National Academies Press, pp 107–136
8. Salazar G, Russi-Vigoya MN (2021) Technology readiness level as the foundation of human readiness level. Ergon Des 29(4):25–29
9. Sheahan T, Wieden HJ (2021) Emerging regulatory challenges of next-generation synthetic biology. Biochem Cell Biol 99(6):766–771
10. Nieuwenweg AC, Trump BD, Klasa K, Bleijs DA, Oye KA (2021) Emerging biotechnology and information hazards. In: Emerging threats of synthetic biology and biotechnology: addressing security and resilience issues, pp 131–140
11. Berube DM (2023) Oxitec. In: Pandemics and resilience: lessons we should have learned from Zika. Springer International Publishing, Cham, pp 413–479
12. Tait J, Levidow L (1992) Proactive and reactive approaches to risk regulation: the case of biotechnology. Futures 24(3):219–231
13. Linton JD, Xu W (2021) Understanding and managing the biotechnology valley of death. Trends Biotechnol 39(2):107–110

14. Linkov I, Trump BD, Pabon N, Collier ZA, Keisler JM, Scriven J (2012) A decision analytic approach for Department of Defense acquisition risk management. Mil Oper Res 57–70
15. Héder M (2017) From NASA to EU: the evolution of the TRL scale in public sector innovation. Innov J 22(2):1–23
16. Fernandez JA (2010) Contextual role of TRLs and MRLs in technology management (No. SAND2010-7595). Sandia National Laboratories (SNL), Albuquerque, NM, and Livermore, CA (United States)
17. US Department of Defense (2023) Technology readiness assessment guidebook. Office of the Under Secretary of Defense for Research and Engineering. Published June 2023
18. Calhoun P (2016) DARPA emerging technologies. Strateg Stud Q 10(3):91–113
19. Predd JB, Blickstein I, Birkler JL, Ochmanek DA, Ohlandt CJ (2021) Is the defense acquisition system postured to solve national defense strategy operational problems? RAND
20. Housel T, Mun J, Jones R, Carlton MB (2019) A comparative analysis of advanced methodologies to improve the acquisition of information technology in the Department of Defense for optimal risk mitigation and decision support systems to avoid cost and schedule overruns. Acquisition research program
21. Maxon ME (2023) Racing to be first to be second: a coordinated regulatory framework for the bioeconomy. Iss Sci Technol 39(3):64–66
22. Schwartz M, Peters HM (2019) Department of Defense use of other transaction authority: background, analysis, and issues for congress. Congressional Research Service
23. Stower C (2023) Strengthening the US biomanufacturing sector through standardization. Ind Biotechnol 19(4):201–203
24. Takors R (2012) Scale-up of microbial processes: impacts, tools and open questions. J Biotechnol 160:3–9
25. Titus AJ, van Opstal E, Rozo M (2020) Biotechnology in defense of economic and national security. Health Secur 18(4):310–312
26. Trump BD, Cummings CL, Kuzma J, Linkov I (eds) (2020) Synthetic biology 2020: frontiers in risk analysis and governance. Springer, Cham
27. Greer SL, Trump B (2019) Regulation and regime: the comparative politics of adaptive regulation in synthetic biology. Policy Sci 52(4):505–524
28. Roberts JP, Stauffer S, Cummings C, Kuzma J (2015) Synthetic biology governance: Delphi study workshop report (No. 2015.2). GES Center report
29. Carter SR, Rodemeyer M, Garfinkel MS, Friedman RM (2014) Synthetic biology and the US biotechnology regulatory system: challenges and options (No. DOE-JCVI-SC0004872). J. Craig Venter Institute, Rockville, MD
30. National Academies of Sciences, Engineering, and Medicine (NASEM) (2017) Preparing for future products of biotechnology. The National Academies Press
31. Lis Y, Roberts MH, Kamble S, Guo JJ, Raisch DW (2012) Comparisons of food and drug administration and European Medicines Agency risk management implementation for recent pharmaceutical approvals. Value Health 15(8):1108–1118
32. Trump BD, Cegan JC, Wells E, Keisler J, Linkov I (2018) A critical juncture for synthetic biology: lessons from nanotechnology could inform public discourse and further development of synthetic biology. EMBO Rep 19(7):e46153
33. DiEuliis D (2019) Key national security questions for the future of synthetic biology. Fletcher Forum World Affairs 43:127
34. Ruttan VW (2006) Is war necessary for economic growth? Hist Speak 7(6):17–19
35. McGowran E, Harris E (2020) Regulatory readiness level: a tool to enhance early regulatory adoption in academic discovery. Level 3 15(2):6
36. Kedia SB, Baker JC, Carbonell RG, Lee KH, Roberts CJ, Erickson J, Schiel JE, Rogers K, Schaefer G, Pluschell S (2022) Biomanufacturing readiness levels (BRL): a shared vocabulary for biopharmaceutical technology development and commercialization. Biotechnol Bioeng 119:3526–3536
37. Trump BD, Keisler JM, Galaitsi SE, Palma-Oliveira JM, Linkov I (2020) Safety-by-design as a governance problem. Nano Today 35:100989

38. Ong S (2024) Singapore doubles down on lab-grown meat as Silicon Valley backs off. Rest of World
39. Ketelings L, Kremers S, de Boer A (2021) The barriers and drivers of a safe market introduction of cultured meat: a qualitative study. Food Control 130:108299
40. International Trade Administration (2024) Singapore biotechnology. Market intelligence report
41. Linstone HA (2006) The information and molecular ages: will K-waves persist? In: Devezas TC (ed) Kondratieff waves, warfare and world security. NATO Security through Science series. IOS Press
42. Naumer HJ, Nacken D, Scheurer S (2010) The sixth Kondratieff—long waves of prosperity. Allianz Global Investors
43. Vega-González LR (2017) Using patents and innovation strings to anticipate the next Kondratieff long waves. Int J Technol Intell Plan 11(3):212–229
44. Mandt R, Seetharam K, Cheng CHM (2020) Federal R&D funding: the bedrock of national innovation. MIT Sci Policy Rev 1:44–54

Chapter 12
Synergizing Minds and Machines: Harnessing Generative AI for Transformative Biotechnology Education

Juan C. Cruz and Luis H. Reyes

Abstract Integrating Generative Artificial Intelligence (GenAI) into biotechnology education presents a transformative opportunity to revolutionize the preparation of future biotechnology innovators. This chapter examines the potential of GenAI to enhance biotechnology education through the Cognitive-AI Synergy Framework (CASF), which aligns AI capabilities with the cognitive developmental stages of students, progressing from foundational understanding to advanced research skills. The framework seeks to optimize the utilization of GenAI tools while cultivating students' critical thinking, problem-solving abilities, and ethical awareness. By leveraging GenAI, educators can create personalized learning experiences, simulate complex biological systems, and provide real-time feedback on laboratory techniques, thereby accelerating the learning process and equipping students to thrive in the rapidly evolving biotechnology industry. Moreover, ethical considerations and governance challenges associated with integrating GenAI are also addressed, highlighting the necessity for adaptive governance frameworks to ensure responsible innovation. Furthermore, we outline our ongoing efforts at Universidad de los Andes to implement and evaluate this innovative approach within our biotechnology curriculum, presenting practical insights into the challenges and opportunities encountered. A more dynamic, engaging, and effective biotechnology education system can be established by adopting a balanced approach that synergizes the capabilities of GenAI with human expertise. This dynamic interplay between human cognition and artificial intelligence possesses the potential to accelerate scientific discovery, enhance global health solutions, and drive sustainable biotechnological innovations. As demonstrated by our initiatives at Universidad de los Andes, we remain steadfast in our commitment to promoting ethical awareness, critical thinking,

J. C. Cruz (✉)
Department of Biomedical Engineering, Universidad de los Andes, Bogotá, Colombia
e-mail: jc.cruz@uniandes.edu.co

L. H. Reyes
Department of Chemical and Food Engineering, Universidad de los Andes, Bogotá, Colombia
e-mail: lh.reyes@uniandes.edu.co

C. L. Cummings et al. (eds.), *Biotechnology and AI: Technological Convergence and Information Hazards*, NATO Science for Peace and Security Series A: Chemistry and Biology, https://doi.org/10.1007/978-3-032-05246-9_12

and interdisciplinary collaboration to realize the transformative potential of GenAI in biotechnology education fully.

Keywords Generative AI (GenAI) · Biotechnology education · Ethical AI integration · Cognitive-AI synergy framework (CASF) · Educational transformation

12.1 Introduction

12.1.1 The Need for an Evolving Biotechnology Curriculum

Education in biotechnology must adapt to a complex and rapidly evolving environment, necessitating a curriculum that aligns with scientific advancements and incorporates interdisciplinary approaches to equip graduates with the requisite skills for contemporary problem-solving. The integration of STEM disciplines with arts and humanities (STEAM) and industry partnerships creates versatile professionals equipped for today's challenges. Research by Madden et al. highlights how this multidisciplinary approach fosters creative thinking essential for biotechnology, while partnerships with industry ensure education aligns with practical workforce needs [1]. These interdisciplinary collaborations enhance critical communication skills for articulating complex scientific concepts [2]. Moreover, the integration of advanced instructional methodologies, including the utilization of digital technologies and interactive educational games, presents significant opportunities for the domain of biotechnology education. In their 2016 publication, Tsalapatas et al. explored how digital skill-building tools enhance employment prospects and facilitate social integration, particularly in sectors reliant on innovation [3]. Research conducted by Nietbaeva elucidates the capacity of digital tools to bolster students' digital literacy and engagement with biological concepts [4]. These digital tools, such as educational applications and virtual laboratories, are pivotal in fostering an interactive and practical comprehension of biology. Furthermore, the creation and execution of Open Educational Resources (OER) for hands-on instruction in molecular biology, as articulated by Rojas et al., demonstrate how digital resources can facilitate remote and distance learning, providing flexible and readily accessible educational opportunities [5]. Progress in pedagogical approaches is paramount for preparing students with the requisite skills to address the demands of contemporary industry and comprehend the broader societal implications of biotechnology, including immersive platforms like virtual laboratories and data visualization software that create interactive learning environments [6, 7].

The rapidly evolving biotechnology industry demands innovative educational approaches that prepare graduates to address complex challenges in healthcare, sustainability, and technology. This requires integrating diverse perspectives and digital technologies while emphasizing career readiness. Moralista et al. and Ningrum et al. suggest that students particularly benefit from exposure to the complementary perspectives of biology, chemistry, engineering, and data science when tackling complex biotechnology problems [8, 9]. Moreover, by highlighting the myriad professional pathways available within the biotechnology sector, encompassing pharmaceutical research and agricultural innovation, educators can assist students in identifying their strengths and passions, thus guiding them toward rewarding and meaningful career trajectories. According to Ningrum et al., bioinformatics methodologies provide a powerful platform for interdisciplinary education, enabling students to develop practical expertise in data analysis, computational modeling, and biological data interpretation while exploring diverse applications in pharmacy, medicine, agriculture, and forensics [9]. This approach prepares students for emerging areas such as genomics, proteomics, and systems biology, fostering the data-driven decision-making skills essential for addressing complex global challenges.

12.1.2 The Role of GenAI in Shaping Biotechnology Education

The convergence of AI and biotechnology marks the inception of a groundbreaking era in scientific exploration and educational frameworks. GenAI is Central to this transformation, a specialized subset of AI characterized by its ability to produce original content by analyzing extensive datasets. In biotechnology education, GenAI serves not merely as a tool but as a potent and transformative force that is fundamentally altering our approaches to teaching, knowledge acquisition, and research within this rapidly evolving field. The impact of GenAI on biotechnology education is intricate and foundational. Additionally, it can democratize access to cutting-edge information, tailor learning experiences, and accelerate the pace of scientific advancements.

A significant achievement of GenAI lies in its ability to furnish intricate visual representations and simulations of biological processes. The research conducted by Jumper et al. on AlphaFold exemplifies the extraordinary potential of artificial intelligence (AI) in precisely predicting protein structures [10]. The pedagogical implications of this accomplishment are profound, as it allows students to interact with and manipulate three-dimensional protein models, thereby fostering a deeper understanding of molecular interactions and processes. Moreover, GenAI is enabling the creation of adaptive learning systems that can tailor educational materials to address the unique needs of each learner. In their study, Holstein et al. explored the capacity of AI-driven tutoring systems to facilitate personalized instruction [11]. These systems can assess a student's performance and learning patterns to generate customized exercises, explanations, and challenges. Such a level of customization can significantly enhance student engagement and understanding, particularly in complex disciplines such as genomics, proteomics, and metabolomics. Another area of considerable promise is the integration of GenAI into laboratory education. GenAI-powered virtual and augmented reality simulations provide students with immersive and enriching experiences in biotechnology procedures and scientific research. These simulations can range from fundamental laboratory techniques to complex bioprocessing tasks. Potkonjak et al. have underscored that virtual laboratories offer numerous advantages, including cost-effectiveness, safety, and the opportunity to conduct experiments without constraints repeatedly [12]. In biotechnology education, this could lead to students developing practical skills in expensive or potentially hazardous methodologies within a secure and controlled environment.

The influence of GenAI extends beyond the confines of educational environments. It permeates the fields of biotechnology research and scientific exploration. GenAI algorithms exhibit their effectiveness as substantial collaborators in the development of hypotheses, the design of experiments, and the analysis of data. A noteworthy instance illustrating how artificial intelligence may accelerate drug development is the study conducted by Stokes et al., which employed machine learning models to predict novel antibacterial compounds. By incorporating these research methodologies, facilitated by artificial intelligence, into biotechnology curricula, educators can prepare students for a future in which human expertise and AI capabilities function symbiotically [13]. Furthermore, GenAI is assuming a fundamental role in genomics and bioinformatics education. The ability of artificial intelligence to analyze and interpret vast genomic datasets is revolutionizing our understanding of genetic information. Zou et al. demonstrated that deep learning models can predict the effects of genetic variations on gene regulation and the subsequent susceptibility to diseases [14]. Integrating these AI-based methodologies into bioinformatics courses may give students the essential skills to adeptly navigate the data-intensive landscape of contemporary biotechnology. The potential of GenAI in synthetic biology education is exceedingly promising. Artificial intelligence algorithms can assist in developing synthetic genetic circuits and metabolic pathways, thereby allowing students to engage with the leading edges of biological engineering. The review conducted by Lawson et al. exemplifies the use of machine learning in metabolic engineering, particularly in optimizing biological systems to achieve designated outcomes [15]. Incorporating these artificial intelligence techniques into synthetic biology curricula empowers educators to nurture a new generation of bioengineers capable of addressing complex biological design challenges.

12.1.3 Ethical Considerations in the Integration of AI

While the potential of GenAI in biotechnology education is extensive, it concurrently raises substantial ethical concerns that require resolution. Among these concerns are data privacy, algorithmic bias, and the potential exploitation of content generated by artificial intelligence. The research conducted by Jobin et al. rigorously explored principles of AI ethics, underscoring the growing imperative for ethical frameworks to guide the development and implementation of artificial intelligence in sensitive fields such as biotechnology [16]. Therefore, educators are tasked with imparting knowledge regarding the technological aspects of GenAI and fostering a deep sense of ethical responsibility among students. This aligns with the principles articulated by Makarova et al., who advocate for integrating ethics education within STEM disciplines to prepare students for the complex decision-making processes they will encounter in their professional careers [17]. This may include discussions on the ethical implications of genetic engineering driven by artificial intelligence, the potential societal impacts of accelerated drug development enabled by artificial intelligence, and the responsible use of artificial intelligence in analyzing sensitive biological data. Another challenge remains in ensuring that the incorporation of GenAI into biotechnology education enhances rather than diminishes human creativity and analytical skills. An overreliance on AI technologies might lead to a degradation of problem-solving capabilities and scientific intuition among students. Educators must design curricula combining AI-assisted learning with traditional problem-solving strategies to address this concern. Stolpe and Hallström underscore the importance of AI literacy in scientific education, suggesting that students should receive instruction not only on the functionalities of AI tools but also on critically evaluating their outcomes and limitations [18].

12.1.4 Future Directions in Biotechnology Education with AI

As GenAI continues to advance, we anticipate a significant deepening and expansion of its integration into biotechnology education. We foresee several significant developments soon: AI-assisted laboratory simulations are expected to become increasingly sophisticated, enabling students to conduct virtual experiments that accurately emulate real-world scenarios. These simulations will not only reduce costs but also enhance accessibility. Still, they will also facilitate the exploration of experimental conditions that are impractical or unattainable in traditional laboratory settings. Collaborative AI systems will emerge as virtual research assistants, partnering with students on research initiatives. These AI collaborators will offer valuable insights and experimental methodologies while assisting in analyzing complex data sets. Such systems are being developed under the framework of human-AI collaboration, as Karpatne et al. articulated in their research concerning AI-supported scientific exploration [19].

Assessment systems that AI drives will provide immediate feedback regarding student performance and comprehension. In addition to basic grading, these tools will deliver individualized feedback and accurately identify areas where students may require further assistance. This paper by Holstein et al. examines the capabilities of such systems within the context of AI-powered instructional support [20]. As GenAI gradually transcends the boundaries of conventional academic disciplines, the integration of interdisciplinary approaches will increase significantly. Biotechnology education must incorporate computer science, data analytics, and cognitive science elements to adequately prepare students for an increasingly interconnected future in biological and computational thinking.

Incorporating GenAI into biotechnology education represents a transformative shift with the potential to enhance learning outcomes and prepare students for future advancements. This integration requires a balanced approach: leveraging AI capabilities while fostering critical thinking, creativity, and ethical reasoning. The goal is to develop graduates who can utilize AI technologies, innovate beyond them, and navigate related ethical complexities. Collaboration among educators, academics, ethicists, and industry experts will guide this early-stage transformation. This ensures that AI enhances human capabilities rather than displaces them, ultimately empowering the next generation to address challenges in personalized medicine, sustainable bioengineering, and related fields (Fig. 12.1).

At the forefront of this educational transformation is the need for innovative frameworks that align GenAI capabilities with students' cognitive developmental stages. The ethical considerations previously discussed must be systematically addressed through structured implementation approaches. The following section introduces the Cognitive-AI Synergy Framework (CASF), a comprehensive methodology designed

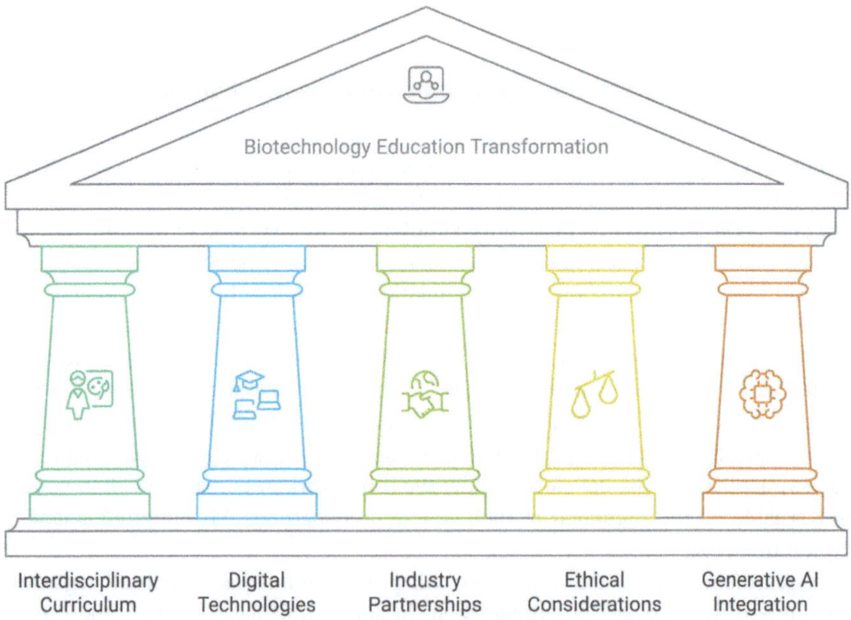

Fig. 12.1 Biotechnology education transformation pillars. This figure illustrates the foundational elements of transformative biotechnology education, emphasizing the critical pillars that support an adaptive and interdisciplinary curriculum. These include integrating digital technologies, fostering partnerships with industry, addressing ethical considerations, and embedding generative AI to optimize educational practices. The synergy between these pillars ensures a robust, future-oriented biotechnology curriculum to prepare students for the evolving challenges in biotechnology research and applications

to incorporate AI tools into biotechnology education in a staged manner that respects ethical boundaries while ensuring students build solid foundations before progressing to advanced applications. This framework provides educators with a practical roadmap for responsible AI integration that maximizes learning outcomes while minimizing potential ethical pitfalls.

12.2 The Cognitive-AI Synergy Framework (CASF)

The Cognitive-AI Synergy Framework (CASF) offers an innovative framework for integrating GenAI into biotechnology education that is aligned with the cognitive development levels of students. Grounded in Julian De Zubiría's dialogic pedagogical model and informed by the Artificial Intelligence Assessment Scale (AIAS), CASF introduces AI tools progressively, enhancing the learning process without disruption [21, 22]. Its tripartite structure—including Foundational AI Integration, Intermediate AI Collaboration, and Advanced AI Synergy—provides a balanced strategy corresponding to students' evolving cognitive capacities (Fig. 12.2).

At the Foundational AI Integration stage, CASF introduces GenAI tools as supplementary resources to reinforce students' understanding of fundamental bioscience principles. Students enhance their foundational knowledge by integrating basic AI applications—such as simulations to elucidate cellular processes or interactive models to explore genetic inheritance patterns. This approach fosters essential cognitive skills and solidifies core concepts, aligning with the research of Tsalapatas et al., which demonstrates the effectiveness of digital skills development in enhancing educational outcomes [3].

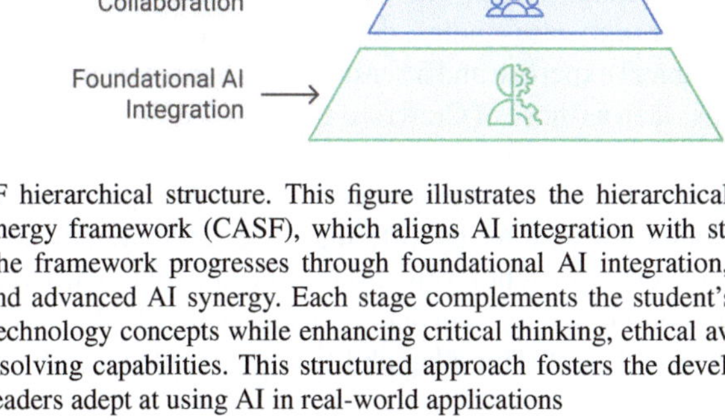

Fig. 12.2 CASF hierarchical structure. This figure illustrates the hierarchical structure of the cognitive-AI synergy framework (CASF), which aligns AI integration with students' cognitive development. The framework progresses through foundational AI integration, intermediate AI collaboration, and advanced AI synergy. Each stage complements the student's evolving understanding of biotechnology concepts while enhancing critical thinking, ethical awareness, and AI-driven problem-solving capabilities. This structured approach fosters the development of future biotechnology leaders adept at using AI in real-world applications

The Intermediate AI Collaboration phase represents a progression to more sophisticated AI applications. Here, students use machine learning techniques to analyze extensive genomic datasets or employ AI-supported design tools in synthetic biology. This gradual introduction of advanced tools accelerates the learning process by enhancing analytical capabilities while building upon the firm foundation established earlier. This approach compels students to critically evaluate the role of AI in biotechnology, deepening their comprehension. It is consistent with the findings of Nietbaeva, who highlighted the importance of digital technologies in enhancing students' literacy and conceptual understanding of biology [4].

At the Advanced AI Synergy level, CASF facilitates the comprehensive integration of AI into biotechnology education. Students engage with cutting-edge AI systems for applications such as in silico drug design, predictive modeling of complex biological systems, and optimization of large-scale bioprocesses. This advanced integration significantly enhances learning and research capabilities, allowing students to fully harness AI technologies' potential. This aligns with Byrne's emphasis on the necessity of a multidisciplinary approach to address current challenges in the field [23].

CASF's tiered methodology effectively addresses essential requirements in biotechnology education that conventional curricula have not adequately fulfilled. By aligning AI tool integration with students' cognitive development, CASF enhances interdisciplinary integration—a critical component of modern education. Moreover, CASF substantially bolsters students' research and innovation competencies in a sector characterized by rapid technological advancement. The framework emphasizes employing AI for literature analysis, hypothesis formulation, and experimental design, equipping students with essential skills for navigating the continuously evolving domain of biotechnology (Fig. 12.2). This systematic approach aligns with the research conducted by Moralista, which underscores the necessity of incorporating diverse viewpoints and digital technologies to prepare students for forthcoming scientific challenges [8].

CASF is distinguished by its strong emphasis on ethical considerations when using AI in biotechnology. The framework integrates analytical reasoning concerning AI ethics throughout all educational levels, focusing on issues such as data privacy in genomic research, societal ramifications of AI-enhanced genetic engineering, and potential biases in AI algorithms. This strategy promotes active engagement with ethical dilemmas through case studies, debates, and practical projects, empowering students with technical expertise and a sense of ethical responsibility. This addresses experts' concerns, such as those of Crawford et al., regarding the imperative of ethical leadership in AI education [24].

Furthermore, CASF navigates the challenges of keeping pace with swift advancements in biotechnology and AI by offering a versatile framework adaptable to students' cognitive development levels. Educators can continually revise and adapt instructional methodologies and curricula, ensuring that biotechnology education remains current and relevant. This adaptability equips students with the cutting-edge technologies they are likely to encounter in their professional endeavors, consistent with the conclusions drawn by Srivastava and Duda et al. on the importance of immersive digital platforms [6, 7].

In addition, CASF promotes the development of critical interpersonal communication skills, which are often overlooked in traditional STEM education. The framework's collaborative model encourages teamwork, effective communication, and leadership abilities. Students learn to articulate complex scientific concepts, collaborate on AI-directed projects, and effectively convey their findings to diverse audiences. These skills are exceedingly valuable in the contemporary biotechnology sector, where interdisciplinary collaboration and clear communication are paramount.

The Cognitive-AI Synergy Framework (CASF) thus equips biotechnology students with the necessary tools, knowledge, and ethical grounding to navigate the intersection of biological sciences and AI. By progressively integrating AI tools that are aligned with students' cognitive development, CASF enhances the educational experience without overwhelming students. Beyond enriching current education, this framework prepares students to address global challenges with innovative, ethically sound solutions. Considering the rapid evolution of biotechnology, CASF serves as a robust foundation for training the next generation of biotechnologists. The synergy between human insight and AI capabilities holds the potential to drive the field forward in ways currently only beginning to be envisioned.

12.3 Enhancing Learning Through GenAI

GenAI has emerged as a catalyst for change in education, transforming personalized learning by adapting to each student's specific needs with precision and flexibility. Its ability to analyze vast datasets and complex patterns provides tailored educational content that aligns with individual learning paces, styles, and cognitive preferences, creating a dynamic and responsive educational environment [25, 26]. Customized learning platforms powered by GenAI employ adaptive algorithms to assess student input, performance metrics, and engagement trends. These systems adjust the complexity of the material in real-time and optimize presentation formats, learning sequences, and examples to clarify concepts. Maintaining a state of "flow" is critical for effective learning, ensuring students remain engaged without facing excessive challenges or insufficient stimulation from overly simplistic content [25, 26].

Advancements in adaptive learning algorithms, such as Bayesian Knowledge Tracing (BKT) and Deep Knowledge Tracing (DKT), have enhanced GenAI's accuracy in predicting and modeling students' knowledge levels. These models incorporate response times, error histories, and behavioral indicators to deliver comprehensive student understanding profiles. Real-time adaptation of instructional materials and teaching methodologies facilitates targeted interventions precisely when needed, improving learning efficacy, retention rates, and overall academic success [27, 28]. By providing immediate forecasts of student progress and cognitive states, these systems enable educators to offer personalized support on a large scale, democratizing access to tailored instruction.

Real-time feedback facilitates the development of metacognitive skills, allowing learners to adjust their study strategies based on data-driven insights, enhancing their autonomy and self-awareness [29, 30].

GenAI has significantly transformed student engagement in content production. AI systems like ChatGPT and other advanced language models provide interactive and multimedia-driven educational resources that adapt to a student's knowledge level, interests, and cultural background. This customization includes tailored exercises, relevant examples, and narrative-driven materials that enhance the relatability and memorability of abstract ideas. Additionally, GenAI facilitates the creation of advanced virtual laboratories and immersive simulations, offering interactive learning opportunities that adapt to students' actions and learning objectives in real time. These technologies bridge theoretical knowledge and practical application by modeling intricate scenarios and enabling risk-free experimentation, improving understanding and retention of complex concepts [31, 32].

Beyond content delivery and assessment, GenAI influences the fundamental framework of educational curricula. By analyzing extensive data on learning outcomes, career paths, and industry requirements, GenAI systems assist in developing adaptable curricula that evolve with changing societal and economic needs. This ensures education remains relevant and future-oriented, equipping students for current challenges and the swiftly changing future labor market [25].

The progression of AI-enhanced Open Educational Resources (OER) in biotechnology will improve the accessibility of cutting-edge educational materials. Building upon the efforts of Rojas et al. in developing Open Educational Resources (OER) for molecular biology, we can anticipate the emergence of AI-generated and AI-curated materials that adapt to the latest scientific advancements and pedagogical standards [5].

In complex scientific fields like biology, GenAI is transforming how students and researchers examine intricate biological systems and processes. AI-based methodologies empower researchers to make high-precision predictions about the behavior of cells, tissues, and organisms. Virtual simulations allow the exploration of complex biological phenomena—such as metabolic pathways, protein–protein interactions, and gene regulatory networks—that are challenging to observe experimentally. Integrating multi-omics data and machine learning techniques enables AI models to replicate dynamic processes, advancing problem-solving and scientific discovery in biology and education [33]. Platforms like MIT's BioAutoMATED automate the creation of machine learning models for biological sequences, aiding in predicting biological processes based on sequence data, including translation efficiency and antibody–drug interactions. These tools allow students and researchers to actively explore biological systems, enhancing scientific inquiry and education in life sciences [34].

AI methodologies are also used to reconstruct biological network topologies, predict protein structures, and estimate metabolite concentrations. Deep learning systems like AlphaFold have advanced protein structure prediction based on amino acid sequences, contributing to a deeper understanding of biological systems across molecular and organismal levels [33]. While AI shows considerable potential in emulating biological systems, challenges remain in improving the quality of reconstructed networks, prediction accuracy, and model interpretability. Successful integration of AI in drug development and biological research requires careful ethical scrutiny and validation through rigorous experimental methods.

In laboratory-based learning environments, GenAI is transforming experimental frameworks. AI-driven technologies provide immediate and precise feedback on laboratory procedures and methodologies, enhancing learning. AI systems can observe and analyze student interactions with laboratory equipment, monitor experimentation progress, and furnish feedback according to established parameters. Students are promptly notified of errors or inefficiencies, allowing swift rectification and performance improvement [20, 35]. GenAI tools enable students to simulate diverse experimental settings, facilitating the exploration of novel scenarios and outcomes without limitations imposed by physical resources or safety considerations. Virtual laboratories can manipulate input parameters to produce various experimental results, helping students understand the influence of different variables. This capability enriches learning and fosters scientific investigation and analytical reasoning skills [26]. AI-powered data analysis is essential for evaluating student advancement and adjusting learning methods accordingly. These systems collect and analyze data on student performance, detecting patterns and trends not readily apparent to educators. They quantify how students tackle tasks, the duration of engagement in different experiment phases, and responses to feedback. Precise, up-to-date information enables educators to adapt teaching strategies in real time, providing customized interventions aligned with each student's learning [36].

AI can evaluate student achievement at micro and macro levels in laboratory settings. At the micro-level, the analysis includes guidance on equipment handling or procedural adherence, with computer vision algorithms analyzing video feeds to provide immediate adjustments. At the macro level, AI assesses the overall efficacy of experimental design or hypothesis testing by evaluating a student's approach's logical progression and scientific rigor [37]. AI integration in assessment also tackles grading consistency, particularly in subjective domains. By employing predefined criteria and learning from extensive datasets of previously graded work, AI systems offer more uniform and impartial assessments, ensuring all students are evaluated based on the same standards. Through its ongoing development, GenAI holds the potential to provide more efficient, engaging, and tailored learning opportunities, especially in fields like laboratory sciences, where practical experience and prompt feedback are essential.

While these technological advancements offer significant benefits, they also raise important ethical questions that must be carefully considered. As GenAI systems become more deeply integrated into educational assessment and feedback, concerns about data privacy, algorithmic bias, and the appropriate boundaries of AI intervention in student evaluation become increasingly urgent. The transformative potential of these technologies must be balanced with thoughtful consideration of their ethical implications, which we explore in the following section.

12.4 Ethical and Governance Considerations

As Generative AI (GenAI) becomes increasingly integrated into educational systems, transforming teaching and learning methodologies, it is crucial to address the intricate ethical and governance concerns that arise. The widespread implementation of GenAI in education prompts apprehensions around privacy, bias, transparency, and accountability. If these challenges are not adequately managed, the potential advantages of GenAI may be compromised. Adaptive governance frameworks must be established to ensure that the technology benefits all stakeholders somewhat [38].

Table 12.1 summarizes the principal ethical challenges faced when implementing GenAI in biotechnology education and their associated concerns and potential mitigation strategies. These ethical dimensions require careful consideration to ensure that AI integration enhances rather than undermines educational equity and effectiveness.

These ethical challenges highlight the necessity for a comprehensive governance approach that considers technical implementation and social responsibility. A multi-layered strategy involving stakeholders across educational, technological, and ethical domains is essential to navigate these complex issues effectively.

An extensive and adaptable governance framework is crucial to address these ethical dilemmas. Policymaking must involve educators, students, parents, AI developers, ethicists, and other stakeholders to ensure that diverse perspectives shape the responsible application of GenAI [54]. Ethical guidelines should address data privacy, bias reduction, informed consent, and accountability. Institutions could establish ethical review committees, like Institutional Review Boards (IRBs), to evaluate AI applications before implementation [55]. Ongoing monitoring and evaluation of AI systems are essential to maintain their integrity. Governance frameworks must adapt alongside technological advancements and societal changes [56]. Establishing systems to ensure transparency and accountability makes AI decisions understandable and open to challenge. Clearly defined responsibilities are essential for managing AI-driven outcomes and improving user confidence.

The responsible integration of GenAI into education requires coordinated efforts. Policymakers should advocate for ethical AI development, enhance digital literacy, ensure equitable technology access, and implement transparent governance structures.

A structured framework for implementing AI in educational settings becomes essential to address these complex ethical and governance challenges. Such a framework must balance the technological opportunities with ethical considerations while ensuring alignment with students' cognitive development. The Cognitive-AI Synergy Framework (CASF) was developed to meet these needs, providing educators with a systematic approach to integrating AI tools into biotechnology education. This

Table 12.1 Ethical challenges and concerns in AI-enhanced biotechnology education

Ethical challenge	Key concerns	Mitigation strategies	References
Privacy and data security	GenAI systems collect extensive student data Sensitive information about abilities and struggles Risk of unauthorized access or misuse	Adherence to GDPR and FERPA regulations Stringent cybersecurity protocols Robust data governance policies	Regan and Jesse [39] Voigt and Bussche [40]
Bias and fairness	AI may reinforce biases in training data Potential unequal treatment of students Risk of exacerbating educational disparities	Regular algorithm audits Inclusion of diverse data sources Fairness-aware machine learning Involvement of ethicists in development	Fazelpour and Lipton [41] Benjamin [42] Mehrabi et al. [43] Ferrara et al. [44]
Transparency and explainability	Lack of transparency undermines confidence "Black box" decision-making Difficulty in contesting assessments	Explainable AI (XAI) models Clear interpretability mechanisms Transparent decision pathways Informed consent protocols	Ali et al. [45] Shin and Park [46]
Balance with traditional teaching	Risk of overreliance on AI Potential reduction in human interaction Diminished development of critical thinking	AI as an assistant rather than a replacement Preserving educator's mentorship role Maintaining focus on social-emotional learning	Kurtz et al. [47]
Equitable access	Marginalization of underprivileged students Digital divide exacerbation Uneven distribution of AI benefits	Improved access to digital infrastructure Provision of affordable devices Digital literacy training Targeted support for underserved areas	Afzal et al. [48] Adeleye et al. [49]
Legal liability	Unresolved questions of accountability Ambiguity in cases of AI errors or harm Challenges in determining negligence	Development of legal frameworks specific to AI Standardized methods for assessing AI safety Clear delegation of responsibilities	Calo [50]

(continued)

Table 12.1 (continued)

Ethical challenge	Key concerns	Mitigation strategies	References
Intellectual property rights	Uncertainty regarding ownership of AI-generated content Potential for copyright conflicts Attribution challenges	More explicit legal frameworks for AI creations Policies on attribution and ownership Institutional guidelines for AI-assisted work	Taddeo and Floridi [51]
Cross-border regulation	Varying regulatory environments across countries Challenges with global platforms Data sovereignty concerns	International collaboration on standards Adoption of global ethical guidelines Regional cooperation frameworks	Taddeo and Floridi [51] Bordot [52] Morandín-Ahuerma [53]

framework, described in the following section, offers a practical roadmap for responsibly introducing AI across different educational levels while maximizing learning outcomes and maintaining ethical standards. The case study at Universidad de los Andes demonstrates how this framework can be successfully implemented in real educational contexts, providing valuable insights for institutions seeking to navigate the transformative potential of GenAI in biotechnology education.

12.5 Case Study: Implementation at Universidad de los Andes

At Universidad de los Andes, the School of Engineering is pioneering the transformative "Inteligencia Aumentada Uniandes" (Augmented Intelligence—IAU) initiative, which aims to ethically and securely integrate GenAI tools across the educational ecosystem. This initiative is designed to enhance rather than replace human intelligence. This vision aligns with the CASF, advocating for the thoughtful integration of AI to foster cognitive development across different learning stages. The IAU project's vision is forward-thinking: to create an educational environment where Augmented Intelligence seamlessly supports learning, research, and problem-solving. The mission is clear—to empower students to reach unprecedented levels of competency, equipping them with the skills and ethical understanding needed to lead in a world increasingly shaped by artificial intelligence. By leveraging GenAI tools, the project aims to revolutionize traditional education by embedding AI into curricula and teaching strategies, cultivating innovation, creativity, and critical thinking.

This initiative takes a holistic and multidisciplinary approach, benefiting the entire university community. It offers personalized learning experiences for students, enhances teaching methods for educators, and optimizes operational efficiency for administrative staff. Beyond academic improvements, the IAU project fosters institutional leadership in AI, positioning Universidad de los Andes as a trailblazer in higher education's digital transformation. As the IAU project grows, it addresses critical challenges such as preparing students for an AI-driven future, enhancing creativity, streamlining administrative tasks, and fostering ethical AI use. It transforms these challenges into opportunities for innovation in teaching, leadership, and interdisciplinary collaboration. Through close alignment with CASF, the IAU initiative ensures AI integration enhances cognitive development, preparing the university community for the future while setting an educational benchmark for institutions worldwide.

The interaction below demonstrates the use of the Cognitive-AI Synergy Framework (CASF) (https://chatgpt.com/g/g-G7I1mXhcf-casf-implementation-assistant) in facilitating the integration of Generative AI (GenAI) into biotechnology education, specifically in a course titled "Biotechnology and Biomolecular Engineering" at the undergraduate level.

Step 1: Identifying the Course and Cognitive Level

- GPT: "Could you please share the name and level of the course you are working on?"
- Educator Response: Biotechnology and Biomolecular Engineering, 4th-year undergraduate.

In this first step, the CASF GPT assistant gathers basic information about the course to determine the context of its recommendations. The assistant, in line with the CASF framework, customizes its guidance based on the level of the course—upper-level undergraduate in this case. The assumption here is that students have already acquired foundational knowledge in biotechnology but need further development in key areas such as enzyme kinetics.

Framework Implication. At the 4th-year undergraduate level, students are expected to start synthesizing knowledge and applying it to more complex scenarios. However, AI integration will vary depending on the cognitive level, starting with foundational tasks like understanding key concepts.

Step 2: Determining the Cognitive Level of the Students

- GPT: "At which cognitive level are your students currently operating? Options: Notional, Propositional, Conceptual, Formal, Categorial, Metacognitive."
- Educator Response: Notional.

The educator indicates that the students are at the notional level, which involves basic recall and understanding of foundational concepts. This cognitive level corresponds to remembering and understanding in Bloom's taxonomy.

Framework Implication. The CASF framework now adjusts its recommendations for this notional cognitive stage. Here, GenAI is primarily used to assist in basic idea generation and to provide explanatory feedback rather than problem-solving or design tasks. This ensures that the AI helps scaffold student learning without overwhelming them or bypassing critical cognitive development.

Step 3: Defining Learning Objectives

- GPT: "Could you share the main learning outcomes for the current module or topic? For example, 'Understanding the principles of enzyme kinetics.'"
- Educator Response: Understanding the principles of enzyme kinetics.

The learning outcome is clearly defined: students must comprehend enzyme kinetics principles. Given this, the CASF GPT now focuses on how AI can support students in achieving this goal, particularly at the notional level.

Framework Implication. Since enzyme kinetics involves fundamental principles that are quantitative and conceptual, the use of GenAI at this level will focus on helping students grasp the key terms, such as Vmax , Km, and the Michaelis–Menten equation. This approach ensures students can recall and understand these concepts before applying them in more advanced scenarios.

Step 4: Narrowing the Topic to Specific Concepts

- GPT: "Could you specify the particular topic or concept within enzyme kinetics you plan to focus on?"
- Educator Response: Michaelis-Menten kinetics.

By narrowing it down to Michaelis–Menten kinetics, the CASF GPT can now offer more targeted assistance, leveraging GenAI to help students understand this core topic in enzyme kinetics. This step ensures that AI-generated activities are not too broad but focused on assisting students to master specific and foundational concepts.

Framework Implication. Students are learning key terms like Vmax and Km at the notional stage and exploring how enzymes interact with substrates. The AI will be integrated to clarify these concepts through examples and analogies, reinforcing students' foundational understanding.

Step 5: Recommending AI Integration Strategies

- GPT: "Since your students are at the Notional cognitive level and the focus is on Michaelis-Menten kinetics, I'll recommend an appropriate AI integration strategy."
- CASF GPT Response: The assistant recommends two primary AI integration levels:

  1. AI-Assisted Idea Generation: The AI can help generate simplified definitions and analogies for key terms like VmaxV_{max}Vmax, KmK_mKm, and substrate concentration.
  2. AI-Assisted Feedback: After manual problem-solving, students use AI to check their work or receive feedback, reinforcing their learning.

The CASF assistant emphasizes that at this stage, AI should not be used for direct problem-solving (e.g., calculations or advanced model applications) to ensure students build their problem-solving skills.

Framework Implication. This suggestion aligns perfectly with the CASF guidelines for the notional stage—GenAI is used to generate ideas and clarify concepts, not solve problems. This ensures that students remain actively engaged in learning rather than passively relying on AI.

Step 6: Designing AI-Assisted Learning Activities

- Prompt: "Here are some activities based on AI-Assisted Idea Generation that will help your students understand Michaelis-Menten kinetics."

CASF GPT Activity Proposals

- Activity 1: Understanding Key Terms in Michaelis–Menten Kinetics: Students use GenAI to ask for simple definitions and analogies related to terms like Vmax and Km.
- Activity 2: Exploring the Michaelis–Menten Equation: Students ask AI to explain how the equation models reaction rates, reinforced through examples.
- Activity 3: Graphical Interpretation of Enzyme Kinetics: AI is used to generate explanations for enzyme kinetics graphs (reaction velocity vs. substrate concentration).

Each activity suggests specific prompts that students can use with the AI, ensuring that the learning process remains focused and directed.

Framework Implication. These activities allow for scalable AI integration, ensuring that AI helps students grasp fundamental concepts at their own pace while ensuring they critically engage with the AI's output. This aligns with CASF's principle of balancing AI use with human cognitive development.

Step 7: Implementing Activity 1—Understanding Key Terms

- Prompt: "Let's proceed with Activity 1: Understanding Key Terms in Michaelis-Menten Kinetics."

CASF GPT Detailed Activity Plan

- Introduction (5–10 min): A short lecture on the key terms (Vmax, Km, etc.).
- AI Engagement (15–20 min): Students use AI to generate definitions and analogies. Example student prompt: "What does Km represent in enzyme kinetics?"
- Peer Discussion (10–15 min): After interacting with AI, students discuss the AI-generated responses with a partner.
- Class Discussion (10–15 min): Pairs share their findings with the class, critically evaluating the clarity and accuracy of AI responses.
- Wrap-Up (5 min): The educator summarizes key takeaways, ensuring students understand core concepts.

Framework Implication. This activity exemplifies how the CASF framework guides the thoughtful integration of AI into a course. AI is used not as a crutch but as a tool to enhance student understanding, with the teacher maintaining control of critical content delivery and facilitating peer evaluation.

Step 8: Evaluation and Reflection

The class ends with a wrap-up and clarification session, where the educator reinforces the accurate scientific definitions. Students are encouraged to reflect on the AI's usefulness and how it contributed to their understanding.

Framework Implication. This final step reinforces the importance of students critically engaging with AI outputs. CASF ensures that while AI can aid understanding, it is up to the students to verify and internalize that information through discussions, class reflections, and cross-referencing with course materials.

Building on our initial exploration of integrating AI into a biotechnology course, this section delves deeper into another interaction with the CASF GPT assistant, explicitly addressing the course topic of metabolic pathways and their regulatory mechanisms. In this interaction, we focus on developing AI-enhanced activities to compare the regulatory mechanisms of glycolysis and gluconeogenesis, which are aligned with students' conceptual cognitive levels. Below is a detailed documentation of this interaction:

Step 1: Course Name and Cognitive Level

- GPT: "What is the name and level of your course?"
- Educator Response: Biotechnology and Biomolecular Engineering, 4th-year undergraduate.

The course is specified again, confirming that it's an upper-level undergraduate course. This ensures that the recommendations for AI use are tailored to advanced students familiar with foundational biotechnology concepts.

- GPT: "At which cognitive level are your students? Please choose one from the following descriptions: Notional, Propositional, Conceptual, Formal, Categorial, Metacognitive."
- Educator Response: Conceptual.

The educator identifies the students' cognitive level as Conceptual, indicating that they are ready to analyze, compare, and categorize complex concepts. This contrasts with the previous example, where students were at a more basic, notional level.

Framework Implication. At the conceptual level, AI assists students in performing higher-level cognitive tasks, such as analyzing enzyme pathways and comparing regulatory mechanisms. The assistant will now recommend more sophisticated AI integrations suited to these tasks.

Step 2: Defining Learning Outcomes

- GPT: "What are the main learning outcomes for the module or topic you teach in Biotechnology and Biomolecular Engineering?"
- Educator Response: Students should be able to analyze metabolic pathways and categorize their regulatory mechanisms.

The learning outcome focuses on students' ability to analyze and categorize metabolic pathways' regulatory mechanisms, specifically glycolysis and allosteric regulation.

Framework Implication. With this clear learning outcome, the AI assistant can now recommend activities that allow students to engage with complex biological processes. AI will be used to deepen their analytical skills rather than just to reinforce basic recall or definitions.

Step 3: Identifying the Specific Topic

- GPT: "What specific topic or concept are you planning to teach within metabolic pathways and their regulatory mechanisms?"
- Educator Response: Glycolysis and its regulation by allosteric enzymes.

By specifying the focus on glycolysis and allosteric regulation, the assistant narrows its suggestions to tailor activities around this core concept, ensuring alignment with the course's learning objectives.

Framework Implication. The specificity of the topic allows the CASF assistant to propose highly relevant AI-enhanced activities that encourage students to explore enzyme regulation in detail.

Step 4: Recommending AI Integration Strategies

- GPT: "Based on your students' conceptual cognitive level, the learning outcomes focused on analyzing metabolic pathways, and the specific topic of glycolysis and its regulation by allosteric enzymes, I suggest the following AI integration strategies:"
  - AI-Assisted Idea Generation and Feedback (Independent): Students can use AI to explore regulatory mechanisms of glycolysis and receive feedback on their understanding of allosteric regulation.
  - AI-Assisted Editing (Supervised): Students can refine their written explanations of glycolysis regulation with AI's help, focusing on clarity and scientific coherence.
  - AI Task Completion (Supervised): AI can assist in modeling how changes in enzyme regulation affect glycolysis, with students critically evaluating AI-generated models.

These suggestions focus on guiding students through a critical and analytical process rather than just providing answers. AI is positioned as a tool to explore hypotheses and refine student output.

Framework Implication. This AI integration at the conceptual level encourages students to compare AI models with their understanding, promoting active learning and critical thinking.

Step 5: Selecting AI-Enhanced Activities

- GPT: "Would you like to brainstorm new AI-enhanced activities, convert a traditional activity to include AI, or explain which AI level to use and why?"
- Educator Response: (a) Brainstorming New AI-Enhanced Activities.

The educator opts to brainstorm new activities to encourage active engagement with AI tools to facilitate student learning. The assistant proposes three detailed activities for this purpose:

- Activity 1: Analyzing Regulatory Enzyme Mutations in Glycolysis
- Activity 2: Pathway Optimization with AI
- Activity 3: Comparative Analysis of Metabolic Pathways

Framework Implication. The activity options align with the conceptual cognitive level by requiring students to analyze, compare, and categorize complex biochemical processes. AI is not a replacement for learning but a tool to enhance exploration and critical engagement.

Step 6: Choosing and Detailing Activity 3

- GPT: "Which activity would you like to explore further?"
- Educator Response: 3. Comparative Analysis of Metabolic Pathways.

The educator selects the activity that compares glycolysis with gluconeogenesis, analyzing the regulatory mechanisms of both pathways. This activity aligns well with the students' cognitive level, emphasizing categorization and analysis.

Framework Implication. The activity requires students to map out two metabolic pathways, compare their regulatory mechanisms, and evaluate the AI-generated outputs. This fosters deep analytical thinking and promotes student collaboration, enhancing their ability to evaluate AI-driven solutions.

Step 7: Detailed Plan for Comparative Analysis Activity

- GPT: "Let's dive deeper into how you can implement the Comparative Analysis of Metabolic Pathways activity using AI."

The CASF assistant outlines a detailed plan for this AI-enhanced activity:

1. Introducing the Pathways: Review glycolysis and gluconeogenesis, focusing on key enzymes and their regulatory mechanisms.
2. AI-Assisted Pathway Mapping: Students use AI to generate visual maps of both pathways. The AI will assist in mapping enzymes, substrates, and products for each step, focusing on allosteric regulation.

   a. AI Prompt Example: "Generate a comparison between glycolysis and gluconeogenesis. Map the enzymes involved in both pathways and categorize key regulatory mechanisms."

3. AI-Assisted Comparison: The AI suggests key points of comparison, focusing on differences in regulation, energy requirements, and allosteric control.

   a. AI Prompt Example: "Compare the regulation of glycolysis and gluconeogenesis, focusing on ATP's role in regulating key enzymes."

4. Critical Evaluation of AI Output: Students critically evaluate the AI-generated comparisons, refining them based on their knowledge.
5. Presentation and Discussion: Students present their findings, discussing how AI helped map the pathways, what was accurate, and how they adjusted the AI-generated outputs.

Framework Implication. This activity reinforces CASF's principle of ensuring that AI complements but does not replace critical thinking. Students must cross-check AI's output, ensuring their understanding of metabolic pathways and regulatory mechanisms deepens through active engagement.

Continuing our journey of integrating GenAI into biotechnology education, we now explore a higher-level interaction focusing on a metacognitive student cohort. In this scenario, students are tasked with designing and scaling up bioreactors for efficient fermentation processes. The interaction is designed to leverage the CASF to promote deep reflection, self-evaluation, and the refinement of complex engineering tasks through AI collaboration. Below is a detailed documentation of the interaction, illustrating the process of creating an AI-enhanced case study activity:

Step 1: Course Identification

- GPT: "Could you provide me with the course name and level?"
- Educator Response: Biotechnology and Biomolecular Engineering, 4th-year undergraduate.

Once again, the educator confirms the course as an upper-level undergraduate offering in biotechnology, preparing students for advanced engineering and scientific challenges.

Framework Implication. Given the advanced level of the course, the recommendations for AI integration will be complex, focusing on tasks that require high-level cognitive engagement, reflecting the students' readiness for sophisticated applications of biotechnology principles.

Step 2: Determining Cognitive Level

- GPT: "At which cognitive level are your students? Please choose one from the following: Notional, Propositional, Conceptual, Formal, Categorial, Metacognitive."
- Educator Response: Metacognitive.

The educator identifies the students as operating at the metacognitive level, meaning they are engaged in self-reflection, critical evaluation, and independent learning. These students can monitor their learning, assess the outcomes of their decision-making, and refine their approaches.

Framework Implication. AI tools should be used as collaborative partners at the metacognitive level. AI will assist students in generating and refining ideas and enable self-evaluation, helping students reflect on their design decisions and critically engage with AI-generated outputs.

Step 3: Defining Learning Outcomes

- GPT: "What are the main learning outcomes for the current module or topic?"
- Educator Response: Students can design bioreactors for efficient fermentation processes.

The primary outcome centers on bioreactor design, emphasizing scaling up processes for industrial use. This highly technical task requires careful consideration of engineering principles like oxygen transfer rates, heat management, and mixing dynamics.

Framework Implication. Given this complex learning outcome, AI can simulate bioreactor performance under different conditions, allowing students to explore the real-world implications of their design choices. AI tools can support the iterative design process, enabling students to assess and refine their bioreactor configurations critically.

Step 4: Identifying the Specific Topic

- GPT: "What specific topic or concept related to bioreactor design or fermentation processes are you planning to focus on?"
- Educator Response: Design considerations for scaling up bioreactors.

The focus on scaling up bioreactors adds a layer of complexity, as students must consider how laboratory-scale designs can be modified for industrial production. Oxygen transfer efficiency, mixing times, and heat distribution become critical at larger scales.

Framework Implication. Scaling up bioreactors presents multiple design challenges well-suited for AI-assisted analysis. AI can help students explore design modifications and evaluate their effects on process efficiency, fostering innovation and critical evaluation.

Step 5: AI Integration Recommendation

- GPT: "Based on your students' metacognitive level and learning objectives, I suggest integrating AI at the Full AI level."

AI Integration Strategies. The assistant recommends several ways to integrate AI into the bioreactor scaling-up process:

- AI-Assisted Idea Generation: AI helps generate multiple design scenarios for scaling up bioreactors.
- AI-Assisted Editing: AI assists in refining bioreactor designs based on simulation feedback.
- AI Task Completion: AI simulates bioreactor performance under different conditions, allowing students to evaluate oxygen transfer rates, mixing dynamics, or temperature gradients.

These approaches use AI to foster deep engagement and critical reflection, ensuring students evaluate and refine their designs based on AI outputs and engineering judgment.

Framework Implication. At the metacognitive level, AI assists with design and prompt self-evaluation. Students must critically assess AI suggestions and reflect on how those suggestions align with real-world engineering constraints.

Step 6: Selecting and Developing Activities

- GPT: "Would you like to brainstorm new AI-enhanced activities, convert a traditional activity to include AI, or explain which AI level to use and why?"
- Educator Response: (a) Brainstorming New AI-Enhanced Activities.

The educator brainstorms new activities, and the CASF assistant provides five AI-enhanced activity ideas. The educator selects an AI-guided case Study on Successful Industrial Scale-Up, which focuses on analyzing real-world bioreactor scale-up and using AI to simulate potential improvements.

Framework Implication. The chosen activity aligns well with the metacognitive level, requiring students to reflect on case studies, critically evaluate AI-generated suggestions, and propose improvements. This fosters both collaborative learning and independent analysis.

Step 7: Detailed Plan for the AI-Guided Case Study

- GPT: "Let's dive deeper into how you can implement the AI-guided case Study on Successful Industrial Scale-Up."

Steps for Implementation

1. Case Study Selection: Choose an industrial-scale bioreactor case (e.g., microbial fermentation for biofuels or biopharmaceuticals). Students will analyze how the design was scaled up and identify key process parameters.
2. Initial Student Analysis: Students review the case study, focusing on critical design and operational decisions.
3. AI-Assisted Scenario Generation: Using AI tools, students input the original design and explore alternative configurations. AI suggests changing impeller size, stirring speed, or cooling mechanisms.

    a. Example AI Prompt: "Simulate different bioreactor designs and analyze the impact of changes in oxygen transfer, mixing efficiency, and heat management during scale-up."

4. Student Evaluation of AI Outputs: Students compare the results of various AI-generated configurations, critically assessing process efficiency and operational stability implications.
5. Final Design Proposal: Based on AI outputs and their engineering knowledge, students propose a final design for the scaled-up bioreactor.
6. Reflection and Self-Evaluation: Students reflect on the AI's role in their decision-making process and write about what they learned from balancing AI recommendations with real-world constraints.

Framework Implication. The case study format allows students to engage with AI-generated outputs critically and reflect on the practical applications of their learning. This metacognitive activity strengthens their ability to monitor, regulate, and evaluate their design decisions.

Step 8: Specific AI Prompt for Case Study

- GPT: "Proceed with the specific AI prompt."

AI Prompt for Bioreactor Scale-Up

- Task: Simulate scaling up a microbial fermentation bioreactor from 10 to 5000 L. Analyze how changes in impeller size, stirring speed, and cooling systems affect oxygen transfer, heat management, and mixing efficiency.
- Request: "Simulate various design changes and provide a detailed analysis of how each design impacts key parameters. Consider trade-offs like energy consumption and mechanical stress."
- Output Format: Provide a simulation report with visual representations (e.g., charts) of the impact of each design choice on process efficiency and stability.

Example Output

- Scenario 1: Impeller size increased by 50%, improving oxygen transfer but introducing high shear forces.
- Scenario 2: Stirrer speed increased by 20%, improving mixing but causing uneven heat distribution.
- Scenario 3: Multiple smaller impellers resulted in better oxygen transfer and mixing with minimal energy increase.

Framework Implication. The AI prompt encourages students to reflect on the complex trade-offs in scaling up bioreactors. By critically evaluating AI-generated designs, students deepen their understanding of process optimization and learn to balance theoretical knowledge with practical constraints.

Integrating the Cognitive-AI Synergy Framework (CASF) into biotechnology education represents a transformative approach to teaching and learning, especially when aligned with students' cognitive levels. Across the three case studies, we witnessed how GenAI tools can be tailored to meet the specific needs of students at different stages of cognitive development—whether fostering foundational understanding, promoting critical analysis, or enhancing self-regulation and reflection in advanced learners. AI supports foundational learning for notional students by providing simplified explanations and scaffolding complex concepts. In contrast, AI becomes a collaborative partner in problem-solving, critical reflection, and design optimization for conceptual and metacognitive students. By aligning AI's role with the students' cognitive readiness, the CASF framework ensures that AI is not merely a tool for convenience but a dynamic partner in enhancing learning outcomes. This approach empowers students to engage deeply with complex biotechnological processes and cultivate critical thinking, creativity, and self-evaluation, preparing them for the evolving demands of academic inquiry and real-world applications in biotechnology.

12.6 Future Directions and Sustainability

GenAI has the potential to revolutionize biotechnology education. To fully realize this promise, educational frameworks must be adaptable, sustainable, and responsive to technological and scientific advancements. By embracing GenAI's capabilities—such as simulating complex biological systems, personalizing learning experiences, and fostering innovative research approaches—educators can enhance learning outcomes at all educational levels while addressing associated challenges.

GenAI's ability to simulate complex biological processes is one of the most promising advancements. Such simulations enable students to interact with dynamic models of cellular behaviors, genetic relationships, and biochemical pathways in unprecedented ways. For example, GenAI technologies may allow students to virtually explore CRISPR-Cas9 gene-editing methods, examining the effects of genetic alterations in simulated organisms within a safe and ethical virtual environment [57]. These immersive experiences enhance understanding of intricate concepts by allowing learners to observe and manipulate biological systems at the molecular level. Additionally, GenAI can facilitate the creation of virtual laboratories, permitting students to conduct otherwise too costly, dangerous, or ethically problematic experiments. Virtual labs democratize access to high-quality science education, providing practical experience to students regardless of location or institutional resources [58].

Advancements in GenAI also promise greater personalization in education. Adaptive AI systems can analyze learning patterns more accurately, facilitating customized learning experiences for each student. AI tutors may instantly adjust their teaching approaches, offering tailored problem sets and instructional techniques to match individual learning preferences—visual, auditory, or kinesthetic [59]. Research shows that personalized methods enhance engagement and learning outcomes by leveraging personal strengths and addressing weaknesses [60]. Moreover, adaptive AI can detect when students struggle with specific concepts, providing additional resources or alternative explanations to improve understanding, bridging learning gaps, and preventing students from falling behind [61].

Future GenAI systems will likely integrate multimodal data—text, graphics, audio, and haptic input—to provide enriched educational content, offering new ways to present complex biotechnology topics and enhancing accessibility and understanding [62]. Augmented Reality (AR) and Virtual Reality (VR) technologies, powered by GenAI, could allow students to interact with 3D models of molecules, cells, or ecosystems, improving spatial comprehension and retention [63]. Advances in natural language processing will further enable GenAI to facilitate intuitive and meaningful dialogues with students, enhancing critical thinking and problem-solving skills [64]. Conversational AI can promote inquiry-based learning by encouraging students to ask questions, explore hypotheses, and receive instant feedback [65]. While these technological developments hold significant potential, ethical considerations must be prioritized. Incorporating fairness, transparency, and accountability into GenAI's design is essential for building trust and ensuring equitable educational outcomes [66]. This includes regular algorithm evaluations to mitigate bias and strong data privacy protections for student information [67].

Institutions must adopt adaptive solutions to ensure educational frameworks evolve with technological advancements. Continuous professional development for educators is essential, enabling them to stay updated on the latest GenAI tools and methodologies [68]. Training through workshops, certifications, and AI literacy programs will help teachers integrate GenAI into their pedagogical practices [69]. Professional learning networks can facilitate the exchange of best practices and collaborative problem-solving among educators [70]. Curriculum design should remain flexible, with regular revisions and feedback mechanisms to support ongoing improvement [71]. Modular and transdisciplinary approaches can enhance curricula's adaptability to new scientific discoveries and technologies [72]. Engaging stakeholders—industry experts, alumni, and students—in curriculum development ensures educational content remains relevant and aligned with future needs [73].

Integrating academia and industry is crucial to familiarize students with advanced technologies and real-world applications. Such partnerships can provide internships, guest lectures, and joint research initiatives, bridging theoretical knowledge with practical skills [74, 75]. Digital literacy will be essential for preparing students for a future shaped by AI. Promoting lifelong learning and skill development to adapt to emerging technologies will enable students to thrive in a changing biotechnology landscape [76]. This includes technical skills, critical thinking, creativity, and ethical reasoning [77]. Adaptive learning platforms incorporating new AI models and datasets will keep educational environments updated [78]. These platforms facilitate personalized learning paths and provide data analytics to inform instructional decisions [79].

For GenAI-driven educational models to be sustainable, several issues must be addressed. Scalability is critical to ensure increased student enrollment does not compromise instructional quality. Cloud-based GenAI platforms can provide equitable access to advanced educational resources, bridging the digital divide and ensuring students from diverse backgrounds benefit equally [80]. Providing underprivileged populations with devices and internet connectivity is essential [81]. Financial viability underpins sustainability; institutions must optimize resource allocation and develop partnerships to secure funding for the ongoing development and deployment of GenAI tools [82]. Collaborations between public and private sectors, grants, and open-source projects can reduce costs and promote innovation [83].

Technical infrastructure must be robust enough to support GenAI's expanding capabilities while adhering to ethical and regulatory standards. Educational institutions must protect student data, comply with privacy laws, and ensure transparency in AI decision-making [39]. Strong cybersecurity protocols and data governance policies are essential for building trust [84]. Regular evaluations of GenAI's impact on educational outcomes are crucial for guiding future improvements and justifying continued investments. Learning analytics can help educators assess the effectiveness of GenAI tools and identify areas for enhancement [79].

Additionally, researching the pedagogical implications of GenAI will strengthen evidence-based practices [69]. Finally, the ecological impacts of AI technologies must be considered. Institutions can adopt green computing practices, such as optimizing algorithms for energy efficiency and using renewable energy in data centers, aligning GenAI applications with sustainability goals [85, 86].

GenAI offers significant opportunities to develop individualized, engaging, and effective learning experiences in biotechnology education. However, realizing this potential requires educational institutions to implement adaptable and sustainable strategies. By fostering collaboration among educators, technologists, policymakers, and industry stakeholders, the integration of GenAI can be both innovative and equitable. These efforts will ensure that GenAI-driven models are sustainable and effective, preparing students for the rapidly advancing biotechnology sector.

12.7 Conclusion

The integration of GenAI into biotechnology education represents a pivotal shift in preparing future biotechnologists. Through the CASF, aligning AI tools with students' cognitive development stages can significantly enhance learning outcomes. By providing personalized learning experiences, simulating complex biological systems, and offering real-time feedback, GenAI empowers students to engage deeply with complex biotechnological concepts, fostering critical thinking, creativity, and self-assessment. Case studies at Universidad de los Andes demonstrate the practical implementation of this approach, highlighting both opportunities and challenges. Ethical considerations and adaptive governance are crucial to ensuring responsible innovation and equitable access. The sustainability and scalability of GenAI-driven educational models will depend on ongoing collaboration among educators, technologists, policymakers, and industry stakeholders.

By embracing this balanced and forward-thinking approach, we can revolutionize biotechnology education, equipping students with the skills and ethical foundation necessary to drive scientific discovery and address global challenges in health and sustainability. The synergy between human cognition and artificial intelligence holds immense promise for the future of education and society. Our collective responsibility is to harness this potential effectively and ethically, fostering an educational environment that not only imparts knowledge but also cultivates the innovators and leaders of tomorrow.

References

1. Madden ME, Baxter M, Beauchamp H, Bouchard K, Habermas D, Huff M, Ladd B, Pearon J, Plague G (2013) Rethinking STEM education: an interdisciplinary STEAM curriculum. Procedia Comput Sci 20:541–546
2. Barnett RW, Blumner JS (1999) Writing centers and writing across the curriculum programs: building interdisciplinary partnerships. Greenwood Press
3. Tsalapatas H, Heidmann O, Houstis E (2016) A serious game for digital skill building among individuals at risk, promoting employability and social inclusion. In: Serious games, interaction and simulation. Springer, pp 91–98
4. Nietbaeva F (2023) Digital literacy in biology education through mobile learning technologies. Eurasian Sci Rev Int Peer Review Multidiscipl J 1(4)
5. Rojas AC, Aravena A, Arce A, Zapata V, Araya W, Gallardo D, Aviles J, Quero F, Nunez I, Matute T et al (2024) Open educational resources for distributed hands-on teaching in molecular biology
6. Duda HJ, Wahyuni FRE, Setyawan AE, Supiandi MI, Bustami Y (2022) Development of project based biotechnology teaching books. BIOSFER: Jurnal Pendidikan Biologi 15(2):178–191
7. Srivastava M (2023) The evolution of education: navigating 21st-century challenges. Int J Family Med Res 5(5)
8. Moralista RB, Rueda RB, Martir EM, Artajo TM, Berondo RG, Vilbar HR (2022) Students' understanding of Google Meet classroom: exploring its link with teachers' mediation. European Organization for Nuclear Research
9. Ningrum DEAF, Rofiki I, Melinda VA, Erfantinni IH, Febriani RO (2020) Development of biotechnology textbook based on bioinformatics research. Univ J Educ Res 8(11):5188–5196
10. Jumper J, Evans R, Pritzel A, Green T, Figurnov M, Ronneberger O, Tunyasuvunakool K, Bates R, Žídek A, Potapenko A et al (2021) Highly accurate protein structure prediction with AlphaFold. Nature 596(7873):583–589
11. Holstein K, Aleven V, Rummel N (2020) A conceptual framework for human–AI hybrid adaptivity in education. In: International conference on artificial intelligence in education, pp 240–254
12. Potkonjak V, Gardner M, Callaghan V, Mattila P, Guetl C, Petrović VM, Jovanović K (2016) Virtual laboratories for education in science, technology, and engineering: a review. Comput Educ 95:309–327
13. Stokes JM, Yang K, Swanson K, Jin W, Cubillos-Ruiz A, Donghia NM, MacNair CR, French S, Carfrae LA, Bloom-Ackermann Z et al (2020) A deep learning approach to antibiotic discovery. Cell 180(4):688–702
14. Zou J, Huss M, Abid A, Mohammadi P, Torkamani A, Telenti A (2018) A primer on deep learning in genomics. Nat Genet 51(1):12–18
15. Lawson CE, Martí JM, Radivojevic T, Jonnalagadda SVR, Gentz R, Hillson NJ, Peisert S, Kim J, Simmons BA, Petzold CJ et al (2021) Machine learning for metabolic engineering: a review. Metab Eng 63:34–60
16. Jobin A, Ienca M, Vayena E (2019) The global landscape of AI ethics guidelines. Nat Mach Intell 1(9):389–399
17. Makarova EA, Makarova EL, Korsakova TV (2019) The role of globalization and integration in interdisciplinary research, culture and education development. J Hist Cult Art Res 8(1):111–127

18. Stolpe K, Hallström J (2024) Artificial intelligence literacy for technology education. Comput Educ Open 6:100159
19. Karpatne A, Jia X, Kumar V (2024) Knowledge-guided machine learning: current trends and future prospects. arXiv:2403.15989
20. Holstein K, McLaren BM, Aleven V (2019) Co-designing a real-time classroom orchestration tool to support teacher–AI complementarity. J Learn Anal 6(2):27–52
21. De Zubiria Samper J (2006) Los modelos pedagógicos: Hacia una pedagogia dialogante, 2nd edn. Magisterio
22. Perkins M, Furze L, Roe J, MacVaugh J (2024) The artificial intelligence assessment scale (AIAS): A framework for ethical integration of generative AI in educational assessment. J Univ Teach Learn Pract 21(6)
23. Byrne EP (2017) Educating chemical engineers for contemporary challenges: the value of context, connection and collaboration. Educ Chem Eng 18:89–96
24. Crawford J, Cowling M, Allen K-A (2023) Leadership is needed for ethical ChatGPT: character, assessment, and learning using artificial intelligence (AI). J Univ Teach Learn Pract 20(3)
25. Lai JWM (2024) Adapting self-regulated learning in an age of generative artificial intelligence chatbots. Future Internet 16(6):218
26. Rashid SF, Duong-Trung N, Pinkwart N (2024) Generative AI in education: technical foundations, applications, and challenges. IntechOpen EBooks
27. Dong X, Fu Y, Kuo M, Sarker S, Qian L, Li X (2024) Board 262: enhancing deep knowledge tracing via diffusion models for personalized adaptive learning. In: 2024 ASEE annual conference and exposition
28. Luo H, Zhang Z, Cui L, Zhang Z, Liang Y (2024) An efficient state-aware coarse-fine-grained model for knowledge tracing. Knowl-Based Syst 302:112375
29. Seong T, Si K, Choi Y (2024) A paradigm shift and the future direction of educational assessment in the era of generative AI. Korean Soc Educ Eval 37(1):1–28
30. Zhai X, Nehm RH (2023) AI and formative assessment: the train has left the station. J Res Sci Teach 60(6):1390–1398
31. Carlon MKJ, Cross JS (2022) Knowledge tracing for adaptive learning in a metacognitive tutor. Open Educ Stud 4(1):206–224
32. Liu Q, Tong S, Liu C, Zhao H, Chen E, Ma H, Wang S (2019) Exploiting cognitive structure for adaptive learning. In: Proceedings of the 25th ACM SIGKDD international conference on knowledge discovery and data mining, pp 627–635
33. DiNuzzo M (2022) How artificial intelligence enables modeling and simulation of biological networks to accelerate drug discovery. Front Drug Discov 2
34. Valeri JA, Soenksen LR, Collins KM, Ramesh P, Cai G, Powers R, Angenent-Mari NM, Camacho DM, Wong F, Lu TK et al (2023) BioAutoMATED: an end-to-end automated machine learning tool for explanation and design of biological sequences. Cell Syst 14(6):525–542
35. Ogunleye B, Zakariyyah KI, Ajao O, Olayinka O, Sharma H (2024) A systematic review of generative AI for teaching and learning practice. Educ Sci 14(6):636
36. Lee D, Arnold M, Srivastava A, Plastow K, Strelan P, Ploeckl F, Lekkas D, Palmer E (2024) The impact of generative AI on higher education learning and teaching: a study of educators' perspectives. Comput Educ Artif Intell 6:100221
37. Pang TY, Kootsookos A, Cheng C (2024) Artificial intelligence use in feedback: a qualitative analysis. J Univ Teach Learn Pract 21(6)
38. Tang KHD (2024) Implications of artificial intelligence for teaching and learning. Acta Pedagogia Asiana 3(2):65–79
39. Regan PM, Jesse J (2018) Ethical challenges of edtech, big data and personalized learning: twenty-first century student sorting and tracking. Ethics Inf Technol 21(3):167–179
40. Voigt P, Bussche A (2017) The EU General Data Protection Regulation (GDPR): a practical guide
41. Fazelpour S, Lipton ZC (2020) Algorithmic fairness from a non-ideal perspective. In: Proceedings of the AAAI/ACM conference on AI, ethics, and society, pp 57–63
42. Benjamin R (2019) Race after technology: abolitionist tools for the new Jim code. Polity

43. Mehrabi N, Morstatter F, Saxena N, Lerman K, Galstyan A (2021) A survey on bias and fairness in machine learning. ACM Comput Surv 54(6):1–35
44. Ferrara C, Sellitto G, Ferrucci F, Palomba F, De Lucia A (2023) Fairness-aware machine learning engineering: how far are we? Empir Softw Eng 29(1):1–46
45. Ali S, Abuhmed T, El-Sappagh S, Muhammad K, Alonso-Moral JM, Confalonieri R, Guidotti R, Del Ser J, Diaz-Rodriguez N, Herrera F (2023) Explainable artificial intelligence (XAI): what we know and what is left to attain trustworthy artificial intelligence. Inf Fusion 99:101805
46. Shin D, Park YJ (2019) Role of fairness, accountability, and transparency in algorithmic affordance. Comput Hum Behav 98:277–284
47. Kurtz G, Amzalag M, Shaked N, Zaguri Y, Kohen-Vacs D, Gal E, Zailer G, Barak-Medina E (2024) Strategies for integrating generative AI into higher education: navigating challenges and leveraging opportunities. Educ Sci 14(5):503
48. Afzal A, Khan S, Daud S, Ahmad Z, Butt A (2023) Addressing the digital divide: access and use of technology in education. J Soc Sci Rev 3(2):883–895
49. Adeleye NOO, Eden NCA, Adeniyi NIS (2024) Educational technology and the digital divide: a conceptual framework for technical literacy inclusion. Int J Sci Res Arch 12(1):150–156
50. Calo R (2017) Artificial intelligence policy: a primer and roadmap. UCDL Rev 51:399
51. Taddeo M, Floridi L (2018) How AI can be a force for good. Science 361(6404):751–752
52. Bordot F (2022) Artificial intelligence, robots and unemployment: evidence from OECD countries. J Innov Econ Manag 1:117–138
53. Morandín-Ahuerma F (2023) Ten UNESCO recommendations on the ethics of artificial intelligence
54. Whittlestone J, Nyrup R, Alexandrova A, Cave S (2019) The role and limits of principles in AI ethics: towards a focus on tensions. In: Proceedings of the 2019 AAAI/ACM conference on AI, ethics, and society, pp 195–200
55. Stahl BC, Wright D (2018) Ethics and privacy in AI and big data: implementing responsible research and innovation. IEEE Secur Priv 16(3):26–33
56. Mittelstadt B (2019) Principles alone cannot guarantee ethical AI. Nat Mach Intell 1(11):501–507
57. Kubera C, Nivillac NMI, Tanner S, Le P, Begde D, Wolyniak MJ, Challa AK (2022) Workshop report: summer 2020 virtual CRISPR in the classroom. CourseSource 9
58. De Jong T, Linn MC, Zacharia ZC (2013) Physical and virtual laboratories in science and engineering education. Science 340(6130):305–308
59. Rizvi M (2023) Investigating AI-powered tutoring systems that adapt to individual student needs, providing personalized guidance and assessments. Eurasia Proc Educ Soc Sci 31:67–73
60. VanLehn K (2011) The relative effectiveness of human tutoring, intelligent tutoring systems, and other tutoring systems. Educ Psychol 46(4):197–221
61. Brusilovsky P, Millán E (2007) User models for adaptive hypermedia and adaptive educational systems. Adapt Web 3–53
62. Mercado J, Picardal JP (2023) Virtual laboratory simulations in biotechnology: a systematic review. Sci Educ Int 34(1):52–57
63. Radianti J, Majchrzak TA, Fromm J, Wohlgenannt I (2019) A systematic review of immersive virtual reality applications for higher education: design elements, lessons learned, and research agenda. Comput Educ 147:103778
64. Brown TB, Mann B, Ryder N, Subbiah M, Kaplan J, Dhariwal P, Neelakantan A, Shyam P, Sastry G, Askell A et al (2020) Language models are few-shot learners. arXiv:2005.14165
65. McGrath C, Farazouli A, Cerratto-Pargman T (2024) Generative AI chatbots in higher education: a review of an emerging research area. High Educ
66. Floridi L, Cowls J (2022) A unified framework of five principles for AI in society. In: Machine learning and the city: applications in architecture and urban design. Wiley, pp 535–545
67. Du Boulay B (2023) Artificial intelligence in education and ethics. In: Handbook of open, distance and digital education, pp 93–108
68. Trust T, Carpenter JP, Krutka DG (2017) Moving beyond silos: professional learning networks in higher education. Internet High Educ 35:1–11

69. Zawacki-Richter O, Marín VI, Bond M, Gouverneur F (2019) Systematic review of research on artificial intelligence applications in higher education—where are the educators? Int J Educ Technol High Educ 16(1):1–27
70. Vangrieken K, Meredith C, Packer T, Kyndt E (2016) Teacher communities as a context for professional development: a systematic review. Teach Teach Educ 61:47–59
71. Laurillard D (2013) Rethinking university teaching: a conversational framework for the effective use of learning technologies, 2nd edn. Routledge
72. Ruiz P (2021) Artificial intelligence in education: promises and implications for teaching and learning. CIRCLS
73. Chomunorwa S, Van Den Berg C (2023) Assessing institutional readiness for the fourth industrial revolution: using learning analytics to improve student experiences. In: Emerging technologies for developing countries. Springer, pp 17–32
74. Etzkowitz H, Leydesdorff L (2000) The dynamics of innovation: from national systems and "Mode 2" to a triple helix of university–industry–government relations. Res Policy 29(2):109–123
75. Perkmann M, Tartari V, McKelvey M, Autio E, Broström A, D'Este P, Fini R, Geuna A, Grimaldi R, Hughes A et al (2012) Academic engagement and commercialisation: a review of the literature on university–industry relations. Res Policy 42(2):423–442
76. Ng W (2012) Can we teach digital natives digital literacy? Comput Educ 59(3):1065–1078
77. Voogt J, Roblin NP (2012) A comparative analysis of international frameworks for 21st century competences: implications for national curriculum policies. J Curric Stud 44(3):299–321
78. Demartini CG, Sciascia L, Bosso A, Manuri F (2024) Artificial intelligence bringing improvements to adaptive learning in education: a case study. Sustainability 16(3):1347
79. Ifenthaler D, Yau JY-K (2020) Utilising learning analytics to support study success in higher education: a systematic review. Educ Tech Res Dev 68(4):1961–1990
80. Chiu TKF (2023) The impact of generative AI (GenAI) on practices, policies and research direction in education: a case of ChatGPT and Midjourney. Interact Learn Environ 1–17
81. Hilbert M (2016) The bad news is that the digital access divide is here to stay: domestically installed bandwidths among 172 countries for 1986–2014. Telecommun Policy 40(6):567–581
82. Labenko O (2023) Financial support for sustainable development of the society. Econ Financ Manag Top Iss Sci Pract Act 1(63):88–97
83. Kshetri N (2023) The economics of generative artificial intelligence in the academic industry. Computer 56(8):77–83
84. Jose D (2024) Data privacy and security concerns in AI-integrated educational platforms. Res Mate Chron 5(2):87–91
85. Jones N (2018) How to stop data centres from gobbling up the world's electricity. Nature 561(7722):163–166
86. Peters A, Capilla R, Coroa'a VC, Heldal R, Lago P, Leifler O, Moreira A, Fernandes JP, Penzenstadler B, Porras J, Venters CC (2024) Sustainability in computing education: a systematic literature review. ACM Trans Comput Educ 24(1):1–53

Chapter 13
Data Diversity and Sequence Length: Key Levers for Powerful Biological AI

Liam Kozma and Adrienne Hoarfrost

Abstract Biological foundation models (FMs) represent a paradigm shift in the life sciences. They enable powerful predictive capabilities on complex biological systems via transfer learning, leveraging pre-training on vast unlabeled biological datasets. Optimizing these models for diverse downstream applications requires a deep understanding of the factors influencing their performance. This chapter addresses the critical question: How does dataset biodiversity and sequence input length shape the transfer learning potential of biological foundation models? Instead of presenting novel experimental data, this work synthesizes existing literature from bioinformatics, machine learning, and genomics to evaluate the distinct and combined impacts of these two crucial parameters. We review evidence suggesting that broader dataset biodiversity (encompassing more species, environments, and functional classes) generally enhances model generalization and robustness, allowing FMs to tackle unfamiliar biological contexts more effectively. However, we also discuss the potential for diminishing returns and the need for careful dataset curation and design. Simultaneously, we explore the significance of sequence input length, highlighting how the ability to process longer contexts (from kilobases to megabases) is vital for capturing long-range biological interactions such as distal gene regulation or protein structural dependencies which encode functional information, perhaps necessitating advanced model architectures. We further discuss the fundamental tradeoff between FM performance gains due to increased biodiversity, which mostly exists in short sequence lengths, and gains due to longer sequence contexts. By providing a conceptual framework and consolidating insights from prior studies, this chapter aims to clarify these relationships and their profound implications for fields ranging from synthetic biology design to biosecurity assessment. We conclude by emphasizing the need for strategic investment and collaborative standardization in data curation and model development to harness the full potential of biological FMs

L. Kozma (✉) · A. Hoarfrost
University of Georgia, Athens, GA, USA
e-mail: liam@liamkozma.com

A. Hoarfrost
e-mail: adrienne.hoarfrost@uga.edu

C. L. Cummings et al. (eds.), *Biotechnology and AI: Technological Convergence and Information Hazards*, NATO Science for Peace and Security Series A: Chemistry and Biology, https://doi.org/10.1007/978-3-032-05246-9_13

responsibly and effectively, offering insights relevant to researchers, policymakers, and defense stakeholders navigating this rapidly evolving landscape.

Keywords Biological foundation models · Dataset diversity · Long-context sequence modeling · Transfer learning · Biosecurity

13.1 Introduction

Artificial intelligence is increasingly integral to modern life sciences, thanks in large part to foundation models (FMs)—large-scale deep learning models pre-trained on broad biological data that can be adapted to diverse tasks [1]. In bioinformatics, foundation models have demonstrated notable proficiency in learning from massive unlabeled datasets (e.g. protein or nucleotide sequences), addressing longstanding challenges of limited annotations for labeled datasets. By capturing the statistical "language" of biology, these models achieve high accuracy in representing biological entities across many downstream tasks [1]. This paradigm heralds a new era in computational biology, where a single pretrained model can be fine-tuned or prompt-guided to solve specific problems—a strategy known as transfer learning [1]. Transfer learning leverages general-purpose knowledge learned during pretraining (for example, biochemical motifs or structural patterns) and applies it to specialized tasks like predicting protein function, annotating cell types, or designing novel enzymes, often with significantly less additional labeled data than would otherwise be required. Early successes in this vein include protein language models that, after self-supervised training on large sequence databases, can be fine-tuned to predict mutational effects or even generate new proteins matching or surpassing task-specific models [1]. Such powerful reusability and versatility have cemented foundation models as a cornerstone of current bioinformatics research [1].

Despite this progress, fundamental questions remain about what factors maximize the transfer learning potential of biological foundation models. In particular, this chapter focuses on two key aspects: the biodiversity of training datasets and the length of input sequences. Dataset biodiversity refers to the breadth and variety of biological data used during a model's pretraining—for instance, does the model train on a large diversity of nucleotide or amino acid sequences from many different species, environments, or functional classes, or only a narrow subset? Intuitively, a model exposed to a wider diversity of sequences might learn more generalizable biological principles, improving its performance when transferred to new tasks or uncommon organisms. Recent studies support this notion: for example, protein language models like ESM-2 and ProGen2 utilize large sequence collections [2, 3], with some incorporating metagenomic sources to broaden diversity. In nucleotide language models as well, efforts such as the Nucleotide Transformer integrate data from >3,000 human genomes and a small number of additional diverse species to build more universally applicable DNA models [4], while the Evo suite of models train on >80,000 bacterial and archaeal genomes [5]. Notably, strong transfer learning performance has been observed even on relatively small datasets (sequences derived from 330 genomes) when data is selected uniformly from across the genomic tree of life [6].

However, it remains an open question how far increasing dataset diversity continues to yield benefits, and whether there are trade-offs (e.g. loss of specialization). Intriguingly, one large-scale analysis of single-cell transcriptomic models found that performance gains plateaued even when pretraining data reached tens of millions of cells [4], suggesting that simply adding more data (even of varied cell types) can hit diminishing returns. Clearly, understanding the optimal composition and diversity of training data is crucial for guiding the next generation of foundation models.

The second focal aspect is sequence input length—the maximum length of biological sequence context a model can effectively ingest. Biological processes often depend on long-range dependencies: in genomics, regulatory elements thousands of base pairs away can influence gene expression; in genes, distant nucleotides on the gene sequence can be close in structural space and interact with an active site or otherwise influence function; in proteins, distant residues can affect active sites. Yet most foundation models to date have practical limits on input length (for example, many protein language models were originally trained on sequence lengths of only a few hundred to ~1000 amino acids [7, 8]. If a model's context window is too short, it may truncate or ignore important long-range interactions, potentially limiting its predictive power on tasks like genome-wide function annotation or chromatin modeling. On the other hand, in focusing model attention on long-range dependencies, models may risk decreasing downstream performance on transfer learning tasks which do not require knowledge of long-range dependencies.

Researchers have begun tackling this challenge by developing architectures optimized for long sequences. Notably, transformer variants and state-space models have been introduced to extend context lengths without prohibitive computational cost. For instance, ProtHyena [9] and PTM Mamba [10] replace the standard self-attention mechanism to handle protein sequences far longer than the usual transformer limit. Similarly, genomic language models now explore context windows of tens of thousands to even millions of base pairs [5, 11]. These advances raise critical questions: Does enabling larger sequence context substantially improve downstream performance? For which tasks, and how can models avoid being overwhelmed by irrelevant parts of very long inputs? Balancing input length against model complexity is an active area of investigation, with implications for capturing the full breadth of biological information in transfer learning.

Answering how dataset biodiversity and sequence length impact transfer learning is not just an academic exercise—it is highly relevant to practical goals in synthetic biology, biosecurity, and computational biology at large. In synthetic biology, generative foundation models are emerging as design tools to create enzymes [12], genetic circuits [13], or even entire genomes [14] with desired functions. The success of generative AI models like GPT-4 and DALL·E in other domains has heightened interest in using generative AI for biological design. Foundation models could allow researchers to explore and generate biological sequences that do not naturally occur, opening up new possibilities for therapeutics and biotechnology. A model trained on highly diverse sequence data may be more likely to invent viable novel solutions (for example, a protein with a combination of traits not seen in nature), whereas a model with insufficient diversity might be stuck recombining known patterns [3]. Likewise, a model capable of handling long DNA inputs could design entire metabolic pathways or genomes in silico rather than just individual genes.

However, these opportunities come with biosecurity concerns. Powerful sequence-generating models lower the barrier to engineering pathogens or toxins, since bad actors could potentially instruct a foundation model to design harmful biological agents [15]. Indeed, the very ability of modern protein models to generate functional sequences with high success [3, 12, 16] has raised warnings that they could be misused for creating dangerous biomolecules. Recognizing how training diversity and model scope contribute to such capabilities is important for enacting safeguards (for example, curating training data to exclude known virulence factors, or watermarking model outputs to trace synthetic sequences). In the broader computational biology context, insights into these factors will guide the development of models that are both more generalizable and more secure. Ultimately, improving the foundation of these models—by training on the right mixture of data and encoding the appropriate context length—will enhance their utility in everything from disease research to environmental biology, while ensuring responsible use.

13.2 Literature Review: Key Factors and Emerging Context

13.2.1 Dataset Biodiversity: Fueling Generalizable Representations

The advent of biological foundation models (FMs) has been enabled by an explosion in available sequence data, far beyond traditional curated databases. Modern protein sequence repositories like UniProt now contain on the order of 10^8–10^9 sequences, and metagenomic initiatives contribute even larger troves of diversity [17]. For example, several metagenomic assembly efforts have produced non-redundant sets of assembled proteins from environmental samples, such as the MGnify microbiome database which provides a non-redundant set of billions of protein sequences [18], MetaClust contributing ~1.6 billion predicted protein fragments from diverse biomes [19], or environment-specific efforts such as the Ocean Microbiome Reference Gene Catalog which contributes hundreds of millions of assembled proteins [20]. By contrast, only a few years ago, the curated Swiss-Prot subset of UniProt held on the order of 10^5 sequences [21]. These metagenomically assembled proteins represent both a more diverse and unexplored sequence space (only 1–2% of these proteins align with functional annotations from Swiss-Prot). This vast, largely unexplored sequence space presents both an opportunity for FMs to learn from unprecedented biodiversity and a challenge regarding data volume.

Furthermore, these protein-based datasets themselves have limitations relative to the biodiversity present in environmental samples: they must first be assembled from metagenomic sequences, which are typically available in ca. 50–300 nucleotide fragments [22]. Mapping rates of the reads from source metagenomes to the resulting metagenomically-assembled protein databases are typically 10–30% [23]. Thus these metagenomically derived datasets still represent a small window into Earth's biodiversity. While the metagenomic reads themselves contain this biodiversity, they are available in short fragments as DNA or RNA (typical average read lengths of 100–200 base pairs), and cannot represent long-range dependencies. This limitation highlights the essentiality of identifying the factors influencing the tradeoff in biodiversity of training data with sequence length in foundation model transfer learning performance.

Overall, the literature strongly suggests that dataset biodiversity during pre-training is critical. Models exposed to broader sequence distributions encode richer, more transferable representations of biochemical properties compared to those trained on narrow datasets [1]. However, simply amassing more sequences eventually yields diminishing returns [24]. Indiscriminately training on vast, heterogeneous data can introduce noise or conflicting signals from erroneous or highly divergent sequences, potentially hindering performance. Consequently, optimal use of biodiversity requires careful consideration. Active research investigates advanced data curation strategies, such as quality filtering and balancing under-represented taxa [6, 25], alongside active learning approaches where models identify the most informative new sequences to train on, aiming to maximize signal gain and mitigate redundancy as datasets grow [26–28].

13.2.2 Sequence Length: Capturing Long-Range Biological Context

A distinguishing factor of biological sequences, especially genomes, is their length. Many crucial biological phenomena—gene regulation, splicing, chromatin organization, protein tertiary structure—depend on long-range interactions spanning kilobases to megabases. Traditional transformer architectures, with self-attention mechanisms scaling quadratically with sequence length, face severe limitations here. Most early genomic FMs were restricted to input windows of 512–4,096 tokens [5]—a tiny fraction (<0.001%) of a human chromosome. Such truncation forces models to treat biological sequences as disjoint fragments, hindering the capture of distal interactions like enhancer–promoter loops or co-evolving sites in a protein's 3D fold. While efficient attention variants (e.g., Longformer, Performer, BigBird [29–31]) offered modest length increases (e.g., into the tens of thousands of bases), they often do so at the expense of accuracy on complex tasks.

Recently, novel architectures, particularly state-space models (SSMs), have enabled dramatic scaling. HyenaDNA [5], using implicit long convolutions, achieves linear or sub-quadratic scaling, pushing context windows to 1 million base pairs while preserving single-nucleotide resolution. This unlocked capabilities like in-context learning in genomics and improved modeling of long-range regulatory syntax, allowing smaller HyenaDNA models to outperform larger transformers on some benchmarks by better leveraging context. The Mamba family of selective SSMs [32] offers further improvements, using input-dependent state dynamics to achieve linear time complexity and state-of-the-art results across modalities, including genomics. Mamba models excel on ultra-long sequences (millions of tokens) where traditional transformers fail and even other long-context models may plateau.

The Evo suite of models [5], referenced earlier, directly aligns with this advancement in handling long sequences. Like HyenaDNA and potentially leveraging similar SSM principles, Evo is explicitly designed for extended genomic contexts, enabling it to process information across vast stretches of DNA. Where Evo exemplifies the interplay between sequence length and biodiversity is in its training data: pre-trained on over 80,000 bacterial and archaeal genomes, it combines its long-context capability with significant biodiversity sourced at the genome level within these domains. This contrasts with approaches using potentially greater numbers of shorter metagenomic sequences (maximizing raw sequence diversity but losing long-range context) or models trained on fewer, less diverse genomes. Evo thus represents a strategy that prioritizes both capturing long-range genomic interactions via its architecture and learning generalizable features from substantial, albeit taxonomically focused (prokaryotic), genomic diversity.

These architectural innovations demonstrate the feasibility of modeling entire microbial genomes or megabase-scale DNA segments in a single pass, capturing hierarchical structures and genome-wide context (e.g., epigenetic states) rather than just local motifs. Still, fully exploiting long contexts remains an open challenge. Biological sequences possess latent 3D structures and looping contacts not fully captured by linear sequence models alone. Focusing solely on long-range dependencies may risk overlooking crucial short-range features important for certain tasks. Additional research should investigate novel model architectures that can bridge the biological information contained within both short- and long-range dependencies, potentially fusing both short-read metagenomic sequences and longer assembled proteins or genomes. Furthermore, processing million-token sequences poses significant computational hurdles in terms of memory and training time, potentially requiring specialized hardware or optimizations [32]. There's also a risk of new failure modes, like models overfitting to spurious long-range correlations [33, 34]. Research is therefore exploring hybrid architectures, potentially combining the strengths of sequence models with graph-based or retrieval-based methods to better capture higher-order biological structure while managing computational load [34, 35]. In essence, sequence input length is a critical design choice, requiring sophisticated model engineering to be both tractable and biologically informative.

13.2.3 Generalization, Representation Learning, and Transfer Performance

A central promise of biological FMs, enabled by training on diverse data (Sect. 13.2.1) with sufficient context (Sect. 13.2.2), is their ability to learn representations that generalize across tasks and potentially uncover fundamental biological structures. Unsupervised protein language models like ProtTrans [7] learn embeddings that correlate strongly with biophysical properties like secondary structure, achieving state-of-the-art prediction accuracy without MSAs or evolutionary profiles [7, 8]. These embeddings also prove effective for tasks like subcellular localization classification.

Similar benefits are observed in genomics. Pretrained DNA FMs (e.g., Nucleotide Transformer [4], DNABERT-2 [36], HyenaDNA [5], LookingGlass [6]) excel at diverse downstream tasks like regulatory element identification, variant effect prediction, optimal temperature prediction, and chromatin profile imputation after fine-tuning, often with minimal task-specific data [5, 37]. Benchmarks show these models substantially outperform naive baselines, confirming that pretraining captures meaningful biological signals [37]. A notable example of transfer power is ESM-2 [2]. Trained on a massive and diverse protein sequence atlas, it internalized sufficient evolutionary and structural patterns to enable ESMFold, a fine-tuned version capable of predicting 3D protein structures from single sequences at near-atomic accuracy, rivaling traditional bioinformatics methods and deep learning methods that require multiple sequence alignments [38] while performing inference orders of magnitude faster [2].

These examples highlight impressive transfer learning: pretrained models or their embeddings are adapted to new tasks, often achieving state-of-the-art results while requiring less data and compute than training specialized models from scratch [39]. However, the relationship between model scale and transfer performance is nuanced. While larger models generally have higher capacity, studies like Vieira et al. show that beyond a certain point, increasing parameter count (e.g., from 650M to 15B for ESM-2) doesn't always yield better downstream performance, especially with limited fine-tuning data [24]. Representational capacity may plateau, or other factors like task-specific data quality become bottlenecks. This suggests that for practical deployment, particularly in resource-constrained settings, moderately sized models might offer an optimal balance of performance and efficiency, or careful targeted data generation may be necessary in conjunction with model training. In summary, FMs learn generalizable patterns, but maximizing downstream benefit involves a complex interplay between data diversity, sequence length capacity, model architecture, model size, and task-specific factors.

13.2.4 Bridging Technical Advances and Practical Implications

The advancements in dataset diversity and long-range modeling described above are not merely academic; they have profound practical implications, particularly in domains relevant to NATO's interests, such as health security, bioengineering, and biodefense. Enhanced generalization means FMs pretrained on broad data can potentially recognize features of novel pathogens absent from initial training data. Increased context length allows modeling complex genetic regulation or designing entire metabolic pathways.

These capabilities drive innovation. In synthetic biology, models like ProGen [12] already accelerate the design of functional enzymes, enabling rapid prototyping for applications ranging from vaccine development to environmental remediation. However, these same powerful capabilities inherently carry dual-use risks. As Wheeler highlights, AI-driven bio-design could lower the barrier to developing harmful biological agents with "unprecedented precision and potency" [39]. A model trained on toxin databases might inadvertently aid harmful design, while long-context models understanding regulatory networks could potentially be misused to engineer more effective pathogens.

Recognizing these risks necessitates proactive governance and the integration of safety considerations into model development and deployment. This includes responsible data curation (e.g., handling or excluding sequences from known high-risk pathogens), technical safeguards like constrained decoding or output filtering to prevent generation of harmful sequences, and methods like watermarking to trace AI-generated biological designs [39]. Safety evaluation at the point of de novo gene synthesis service providers are particularly crucial. Policymakers and research agencies are increasingly engaging with these issues, seeking to balance the immense benefits of biological FMs with robust security considerations. Effectively navigating this complex landscape—optimizing model performance while mitigating risks—requires ongoing dialogue between AI researchers, biologists, ethicists, and security professionals.

13.3 Conceptual Framework and Approach

We propose a structured framework to evaluate how three key variables—dataset biodiversity, sequence input length, and model architecture—jointly influence transfer learning in biological foundation models (FMs). This framework aligns with the foundation model paradigm wherein large models are pre-trained on extensive data and then fine-tuned to diverse tasks [1]. By examining each variable and their interactions in turn, we clarify the scientific underpinnings in a policy-oriented context. The aim is to inform decision-makers about which factors most impact an

FM's ability to generalize (transfer) to new biological problems, and why a synthesized understanding of these interacting factors, drawing from existing research, can guide strategy more effectively than isolated new experiments.

13.3.1 Key Factors Influencing Transfer Learning in Biological FMs

The transfer learning potential of biological FMs is primarily shaped by the interplay of the following factors, whose detailed impacts and supporting evidence are discussed in Sect. 13.2:

- **Dataset Biodiversity**: The taxonomic and functional diversity of the data used for pre-training. As elaborated in Sect. 13.2.1, exposure to broader biological diversity generally equips models with more robust and generalizable representations.
- **Sequence Input Length**: The maximum length of sequence context the model can effectively process. The significance and tradeoffs inherent in capturing short-versus long-range biological interactions, and the architectural requirements for doing so, are detailed in Sect. 13.2.2.
- **Model Architecture and Size**: The specific neural network design (e.g., transformer vs. state-space model) and its scale (parameter count). How architecture influences computational scalability, the ability to leverage sequence length, and overall learning capacity is discussed in Sects. 13.2.2 and 13.2.3.

Figure 13.1 summarizes how these three dimensions interact. In practical terms, achieving strong transfer learning performance in a biological FM is a balancing act. An ideal model would train on a maximally diverse dataset (across many organisms and sequence types), ingest very long sequences to capture all relevant context, and employ an architecture capable of learning biologically relevant information from both short and long-range contexts efficiently from such data. In reality, trade-offs occur: one might extend sequence length but then need a suitable architecture to handle it, or use a very diverse dataset but require a larger model to avoid underfitting the variety. The diagram illustrates that each factor contributes to performance (solid arrows) and that they influence one another (dashed connections), underscoring the need for an integrated perspective. This conceptual framework will guide our analysis in the remainder of this chapter.

13.3.2 Analytical Approach: Synthesis of Existing Literature

Rather than presenting novel experimental data, this chapter synthesizes existing literature from bioinformatics, machine learning, and genomics to evaluate the distinct and combined impacts of dataset biodiversity and sequence input length,

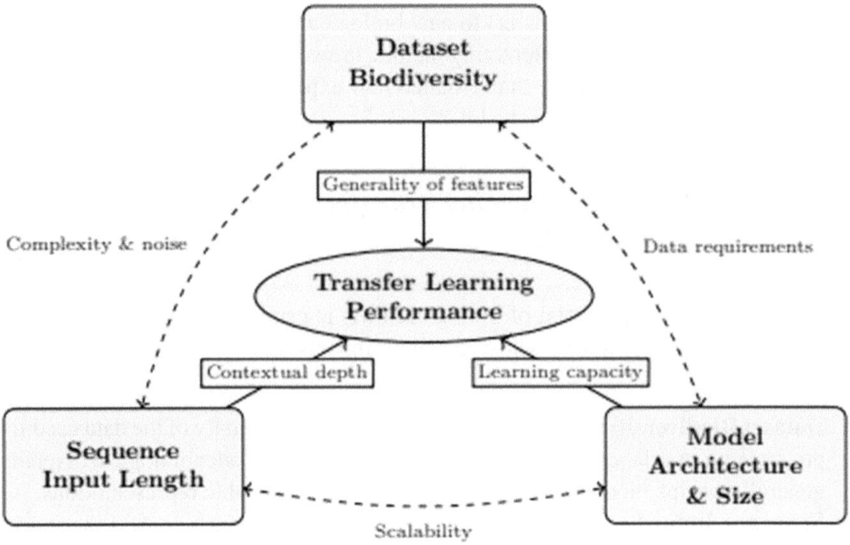

Fig. 13.1 Conceptual interplay between dataset diversity, sequence length, and model architecture in shaping biological FM transfer learning performance. Solid arrows indicate direct influences on performance (center): broader dataset biodiversity yields generalizable features and reduces bias; longer sequence length provides deeper context (e.g., long-range patterns) but requires suitable models; advanced/larger model architectures enhance learning capacity and data use. Dashed arrows show interactions, such as how well the model's design scales affects the sequence length it can effectively handle, or diverse data potentially necessitating larger models or introducing noise

guided by the conceptual framework above (Fig. 13.1). We examine how integrating insights on data diversity, sequence context, and model architecture can inform a unified understanding of transfer learning potential in biological FMs. This synthesis aims to consolidate current knowledge and highlight key relationships, providing a foundation for strategic decisions in research and policy.

13.4 Implications for Policy and Practice

The development of powerful biological foundation models (FMs), shaped significantly by the diversity of their training data and their capacity to process long sequences, carries profound implications for NATO and its member nations. Translating the scientific findings discussed previously into actionable policy requires careful consideration of strategic investments, risk mitigation, and collaborative opportunities. Based on our analysis, policymakers and practitioners should consider the following:

1. **Strategic Investment in FM Capabilities**: Understanding biological systems is fundamental to health security and defense. NATO should consider strategic investments not just in acquiring FMs, but in shaping their development, focusing on the core factors influencing their effectiveness:

    - **Prioritize Data Diversity for Robustness**: As highlighted, models trained on broader datasets (more species, environments, functional classes) are generally more adaptable to novel biological challenges (e.g., emerging pathogens or unfamiliar bioagents). Policy should encourage and potentially fund the targeted generation and curation of diverse, high-quality biological datasets relevant to NATO strategic interests. This includes data from under-represented organisms or environments pertinent to biodefense or allied health security, moving beyond simply amassing data to ensure relevance and prevent blind spots (e.g., for rare pathogens). To maximize the benefit of diversity and mitigate diminishing returns or noise from overly heterogeneous data, considcr supporting:

        – Development of guidelines for intentional inclusion of under-represented groups/contexts.
        – Establishment of cross-domain generalization benchmarks (e.g., testing models trained on human/livestock data against plant/microbial tasks) to quantify the value of diversity and guide optimal levels.
        – Exploration of adaptive training strategies (like curriculum learning or domain-balanced sampling) to effectively incorporate diverse data without performance plateaus.

    - **Support Advanced Architectures for Deeper Insight**: The ability of newer models (like HyenaDNA or Mamba) to handle much longer sequences unlocks the potential to understand complex biological regulation (e.g., whole-genome patterns, multi-gene interactions) and design larger functional units (e.g., entire pathways). R&D funding strategies should recognize that investing in novel architectures capable of processing long sequences efficiently may yield greater returns for certain strategic applications (like understanding genome-level threats or designing complex countermeasures) than simply scaling up older models. It's crucial to align input length capabilities with the biological problem at hand. Support should include:

        – Research into efficient long-context architectures (transformer variants, state-space models, recurrent networks, segmented processing, localized attention, memory augmentation) that handle kilobase to megabase sequences with lower computational cost.
        – Length-scaling experiments using consistent architectures across varying sequence lengths (e.g., 1 kb, 10 kb, 100 kb+) to systematically measure performance gains and identify plateaus for different biological tasks (e.g., gene regulation vs. protein structure).

  - Development of multi-scale modeling frameworks combining coarse-grained scans of long regions with fine-grained analysis of critical segments, balancing context with computational feasibility.

2. **Proactive Governance for Dual-Use Risks**: The same capabilities that make FMs valuable (e.g., generating novel proteins, understanding complex genetic interactions) inherently carry dual-use risks, potentially lowering the barrier for developing harmful biological agents. This necessitates proactive governance frameworks:

   - **Integrate Safety into Development**: Encourage or mandate safety considerations throughout the FM lifecycle. This could involve guidelines on responsible data curation (e.g., handling data from known high-risk pathogens), access controls for powerful generative models, and research into technical safeguards like output filtering or watermarking AI-generated sequences.
   - **Foster International Dialogue**: Given the global nature of AI development and biological threats, NATO can play a role in fostering dialogue among allies and partners on establishing norms and best practices for the safe and secure development and deployment of biological FMs.

3. **Promote Collaboration and Standardization**: Developing and training large-scale FMs is resource-intensive, often beyond the capacity of single entities. Collaboration and standardization are key to accelerating progress responsibly and efficiently, overcoming the current fragmented approach.

   - **Establish Common Standards and Guidelines**: As proposed previously, supporting the development of a standardized framework or "common playbook" among NATO allies is crucial. This should be driven by the community (e.g., via international working groups under NATO S&T or partnerships) and cover:

     - *Dataset Creation*: Minimum diversity requirements, quality control checks, provenance documentation, ethical sourcing (especially for human data).
     - *Model Development/Reporting*: Recommended evaluation metrics, transparency about limitations and training procedures, clear documentation.
     - *Shared Benchmarks*: Develop common tasks and datasets (spanning genomics, proteomics, epidemiology etc.) for objective model comparison and establishing performance baselines. Such standards enhance interoperability, facilitate burden-sharing, build trust, reduce duplication of effort, and allow for the incorporation of security/ethical guidelines relevant to policy.

   - **Explore Secure Data/Model Sharing Mechanisms**: Investigate and encourage infrastructure for collaboration, such as platforms or consortia for securely sharing relevant biological data (anonymized or public) or pre-trained model components among trusted partners. This requires well-defined legal and ethical frameworks for ownership, privacy, and security, but would create a unified ecosystem benefiting all partners, especially for shared challenges like pandemic preparedness or countering biological threats.

4. **Informed Procurement and Deployment**: When considering the use or procurement of biological FMs for specific applications (e.g., bio-surveillance analysis, medical countermeasure research), decision-makers should ask critical questions informed by this analysis:
   - Was the model trained on sufficiently diverse data relevant to the intended application?
   - Can the model handle sequence lengths appropriate for the biological phenomena being studied (e.g., local motifs vs. long-range regulation)?
   - What are the known limitations, potential biases, and associated dual-use risks of the specific FM?
   - Are appropriate security and ethical safeguards in place for its use?

In essence, harnessing the power of biological FMs for strategic advantage requires a nuanced approach that balances promoting innovation with managing risks. Understanding the technical levers of dataset diversity and sequence length provides a crucial foundation for making informed policy, guiding investments, and fostering responsible development within the Alliance.

13.5 Conclusions

The strategic potential of biological foundation models for Alliance security and technological leadership hinges critically on two controllable factors: the diversity of their training data and their architectural capacity to process long biological sequences. Our analysis demonstrates that mastering these elements is key to unlocking the transformative power of these AI systems for tasks ranging from predicting pathogen evolution to designing novel countermeasures.

Rich dataset biodiversity equips models with the generalizability needed to confront novel biological challenges, moving beyond known threats. Concurrently, the ability to process extended sequences allows for a deeper understanding of complex biological systems, enabling insights into intricate regulatory networks and the design of sophisticated synthetic constructs—capabilities vital for both biodefense and health security innovation.

However, maximizing these benefits demands more than raw scale; it requires strategic foresight and targeted investment. Simply adding more data or longer context windows yields diminishing returns without corresponding advances in model architecture. Innovative designs, like state-space models, represent crucial breakthroughs, enabling unprecedented scale but underscoring the need for continued R&D focused on efficiency, interpretability, and robustness, particularly when handling the inherent complexity and potential noise in vast biological datasets.

For NATO and its member nations, the rapid evolution of these technologies presents both immense opportunities and significant dual-use risks. Strategic engagement is therefore not optional, but essential. This necessitates proactive investment not only in cutting-edge architectures but, critically, in the collaborative curation and standardization of diverse, high-quality, and ethically sourced biological datasets relevant to Alliance priorities. Fostering international cooperation through shared guidelines, benchmarks, and responsible data-sharing mechanisms will be paramount to accelerate progress safely, mitigate risks effectively, and ensure equitable benefit across the Alliance.

Ultimately, understanding and guiding the development of biological foundation models is not merely a scientific endeavor, but a strategic imperative for maintaining technological leadership and ensuring Alliance security in the bio-digital age. By prioritizing informed investment in data diversity and long-context architectures, fostering collaborative innovation alongside rigorous oversight, and proactively shaping governance frameworks, NATO and its partners can harness these powerful tools to enhance defense, bolster public health, and advance biotechnology responsibly, ensuring these advancements serve collective security and shared values.

References

1. Guo F, Guan R, Li Y, Liu Q, Wang X, Yang C, Wang J (2025) Foundation models in bioinformatics. Natl Sci Rev nwaf028
2. Lin Z, Akin H, Rao R, Hie B, Zhu Z, Lu W, Rives A (2023) Evolutionary-scale prediction of atomic-level protein structure with a language model. Science 379(6637):1123–1130
3. Nijkamp E, Ruffolo JA, Weinstein EN, Naik N, Madani A (2023) Progen2: exploring the boundaries of protein language models. Cell Syst 14(11):968–978
4. Hoarfrost A, Aptekmann A, Farfañuk G, Bromberg Y (2022) Deep learning of a bacterial and archaeal universal language of life enables transfer learning and illuminates microbial dark matter. Nat Commun 13(1):2606
5. Dalla-Torre H, Gonzalez L, Mendoza-Revilla J, Lopez Carranza N, Grzywaczewski AH, Oteri F, Pierrot T (2024) Nucleotide transformer: building and evaluating robust foundation models for human genomics. Nat Methods 1–11
6. Elnaggar A, Heinzinger M, Dallago C, Rehawi G, Wang Y, Jones L, Rost B (2021) Prottrans: toward understanding the language of life through self-supervised learning. IEEE Trans Pattern Anal Mach Intell 44(10):7112–7127
7. Rives A, Meier J, Sercu T, Goyal S, Lin Z, Liu J, Fergus R (2021) Biological structure and function emerge from scaling unsupervised learning to 250 million protein sequences. Proc Natl Acad Sci 118(15):e2016239118
8. Zhang Y, Okumura M (2024) Prothyena: a fast and efficient foundation protein language model at single amino acid resolution. bioRxiv, 2024-01
9. Peng Z (2024) PTM-Mamba: a PTM-aware protein language model with bidirectional gated Mamba blocks. In: Proceedings of the 33rd ACM international conference on information and knowledge management, pp 5475–5478
10. Nguyen E, Poli M, Faizi M, Thomas A, Wornow M, Birch-Sykes C, Baccus S (2023) Hyenadna: long-range genomic sequence modeling at single nucleotide resolution. Adv Neural Inf Process Syst 36:43177–43201

11. Zvyagin M, Brace A, Hippe K, Deng Y, Zhang B, Bohorquez CO, Ramanathan A (2023) GenSLMs: genome-scale language models reveal SARS-CoV-2 evolutionary dynamics. Int J High Perform Comput Appl 37(6):683–705
12. Madani A, Krause B, Greene ER, Subramanian S, Mohr BP, Holton JM, Naik N (2023) Large language models generate functional protein sequences across diverse families. Nat Biotechnol 41(8):1099–1106
13. Hiscock TW (2019) Adapting machine-learning algorithms to design gene circuits. BMC Bioinform 20:1–13
14. Yelmen B, Decelle A, Ongaro L, Marnetto D, Tallec C, Montinaro F, Jay F (2021) Creating artificial human genomes using generative neural networks. PLoS Genet 17(2):e1009303
15. Chen Y, Hu Z, Wu Y, Chen R, Jin Y, Chen W, Huang H (2024) Enhancing biosecurity with watermarked protein design. bioRxiv, 2024-05
16. Ferruz N, Schmidt S, Höcker B (2022) ProtGPT2 is a deep unsupervised language model for protein design. Nat Commun 13(1):4348
17. The UniProt Consortium (2023) UniProt: the Universal Protein Knowledgebase in 2023. Nucleic Acids Res 51(D1):D523–D531
18. Richardson L, Allen B, Baldi G, Beracochea M, Bileschi ML, Burdett T, Finn RD (2023) MGnify: the microbiome sequence data analysis resource in 2023. Nucleic Acids Res 51(D1):D753–D759
19. Hou Q, Pucci F, Pan F, Xue F, Rooman M, Feng Q (2022) Using metagenomic data to boost protein structure prediction and discovery. Comput Struct Biotechnol J 20:434–442
20. Sunagawa S, Coelho LP, Chaffron S, Kultima JR, Labadie K, Salazar G, Velayoudon D (2015) Structure and function of the global ocean microbiome. Science 348(6237):1261359
21. The UniProt Consortium (2015) UniProt: a hub for protein information. Nucleic Acids Res 43(D1):D204–D212
22. CD Genomics (2024) Sequencing read length: everything you need to know. https://www.cd-genomics.com/blog/sequencing-read-length-comprehensive/
23. Nayfach S, Rodriguez-Mueller B, Garud N, Pollard KS (2016) An integrated metagenomics pipeline for strain profiling reveals novel patterns of bacterial transmission and biogeography. Genome Res 26(11):1612–1625
24. Vieira LC, Handojo ML, Wilke CO (2024) Scaling down for efficiency: medium-sized protein language models perform well at transfer learning on realistic datasets. bioRxiv, 2024-11
25. Serrano Y, Ciudad Á, Molina A (2024) Are protein language models compute optimal? arXiv: 2406.07249
26. Jain M, Bengio E, Hernandez-Garcia A, Rector-Brooks J, Dossou BF, Ekbote CA, Bengio Y (2022) Biological sequence design with GFlowNets. In: International conference on machine learning, pp 9786–9801
27. Vornholt T, Mutný M, Schmidt GW, Schellhaas C, Tachibana R, Panke S, Jeschek M (2024) Enhanced sequence-activity mapping and evolution of artificial metalloenzymes by active learning. ACS Cent Sci 10(7):1357–1370
28. Hie B, Bryson BD, Berger B (2020) Leveraging uncertainty in machine learning accelerates biological discovery and design. Cell Syst 11(5):461–477
29. Beltagy I, Peters ME, Cohan A (2020) Longformer: the long-document transformer. arXiv: 2004.05150
30. Choromanski K, Likhosherstov V, Dohan D, Song X, Gane A, Sarlos T, Weller A (2020) Rethinking attention with performers. arXiv:2009.14794
31. Zaheer M, Guruganesh G, Dubey KA, Ainslie J, Alberti C, Ontanon S, Ahmed A (2020) Big bird: transformers for longer sequences. Adv Neural Inf Process Syst 33:17283–17297
32. Gu A, Dao T (2023) Mamba: linear-time sequence modeling with selective state spaces. arXiv: 2312.00752
33. Mitros J (2024) Generalization vs. memorization in the presence of statistical biases in transformers. arXiv:2409.04654
34. Lewis P, Perez E, Piktus A, Petroni F, Karpukhin V, Goyal N, Kiela D (2020) Retrieval-augmented generation for knowledge-intensive NLP tasks. Adv Neural Inf Process Syst 33:9459–9474

35. Ying C, Cai T, Luo S, Zheng S, Ke G, He D, Liu TY (2021) Do transformers really perform badly for graph representation? Adv Neural Inf Process Syst 34:28877–28888
36. Zhou Z, Ji Y, Li W, Dutta P, Davuluri R, Liu H (2023) DNABERT-2: efficient foundation model and benchmark for multi-species genome. arXiv:2306.15006
37. Feng H, Wu L, Zhao B, Huff C, Zhang J, Wu J, Wu C (2024) Benchmarking DNA foundation models for genomic sequence classification
38. Jumper J, Evans R, Pritzel A, Green T, Figurnov M, Ronneberger O, Hassabis D (2021) Highly accurate protein structure prediction with AlphaFold. Nature 596(7873):583–589
39. Wheeler NE (2025) Responsible AI in biotechnology: balancing discovery, innovation and biosecurity risks. Front Bioeng Biotechnol 13:1537471

Index

C. L. Cummings et al. (eds.), *Biotechnology and AI: Technological Convergence and Information Hazards*, NATO Science for Peace and Security Series A: Chemistry and Biology, https://doi.org/10.1007/978-3-032-05246-9

MIX
Papier aus verantwortungsvollen Quellen
Paper from responsible sources
FSC® C105338

If you have any concerns about our products,
you can contact us on
ProductSafety@springernature.com

In case Publisher is established outside the EU,
the EU authorized representative is:
Springer Nature Customer Service Center GmbH
Europaplatz 3, 69115 Heidelberg, Germany

Printed by Libri Plureos GmbH
in Hamburg, Germany